Salmonella and Salmonellosis: Implications in Public Health

Salmonella and Salmonellosis: Implications in Public Health

Guest Editors

Rafael Jesús Astorga
Angela Galán Relaño
Antonio Valero Díaz

MDPI

Basel • Beijing • Wuhan • Barcelona • Belgrade • Novi Sad • Cluj • Manchester

Guest Editors

Rafael Jesús Astorga
Animal Health Department
University of Cordoba
Cordoba
Spain

Angela Galán Relaño
Nursing, Pharmacology and Physiotherapy Department
University of Cordoba
Cordoba
Spain

Antonio Valero Díaz
Food Science and Technology Department
University of Cordoba
Cordoba
Spain

Editorial Office
MDPI AG
Grosspeteranlage 5
4052 Basel, Switzerland

This is a reprint of the Special Issue, published open access by the journal *Animals* (ISSN 2076-2615), freely accessible at: www.mdpi.com/journal/animals/special_issues/20B075A0W1.

For citation purposes, cite each article independently as indicated on the article page online and as indicated below:

Lastname, A.A.; Lastname, B.B. Article Title. *Journal Name* **Year**, *Volume Number*, Page Range.

ISBN 978-3-7258-4072-4 (Hbk)
ISBN 978-3-7258-4071-7 (PDF)
https://doi.org/10.3390/books978-3-7258-4071-7

Contents

Preface

This issue features eleven studies with contributions from over 76 authors from six different countries. Galán-Relaño et al. [1] examine *Salmonella* Enteritidis and *S.* Typhimurium, emphasizing animals as reservoirs and the growing threat of antimicrobial resistance (AMR). They advocate for innovative control strategies—biosecurity, essential oils, bacteriophages, and vaccines—through a One Health lens, alongside predictive models to improve global control. Jordá et al. [2] explore phage-based biosanitation as an eco-friendly alternative to antibiotics for controlling persistent *Salmonella* and *Campylobacter* in poultry. Despite regulatory barriers, bacteriophages show strong potential, especially against biofilms. The third review [3] discusses *Salmonella* control in swine, highlighting the importance of pre- and post-harvest interventions. It calls for integrated One Health approaches, as pig sector efforts lag behind those in poultry. In line with the above, another research from Colombia [4] reports an 8.9% *Salmonella* prevalence in pig farms and link risk to water sources and pest control, underscoring the need for further investigation. Continuing the discussion on pre-hasvest contol, Roshen et al. [5] details the control of this pathogen in commercial layer chickens, finding that biosecurity, vaccination, and environmental management reduce colonization. These authors also reinforce the need for integrated efforts to curb AMR. Turning to the role of national surveillance programs, the sixth article [6] evaluates Spain's National *Salmonella* Control Program data from over 7,000 flocks gathered in six years, concluding that sampling timing and housing systems influence this bacterium detection. Regarding new tools of this pathogen control, Sevilla-Navarro et al. [7] study *Salmonella* Infantis in poultry, observing reduced-phage-susceptibility variants during therapy but confirming PhagoVet's effectiveness in reducing bacterial loads. In this context, the eighth study [8] assesses carvacrol, a compound in essential oils, noting its extended post-antibiotic effect against S. Typhimurium, suggesting an alternative to antimicrobials to control *Salmonella* in animals and also in with fewer side effects. The ninth article [9] examines AMR in *S.* Typhimurium, identifying integrons that boost resistance and virulence, and highlights the need for integron screening in poultry production. The tenth article [10] investigates miR-215's role in modulating immune responses during *S.* Typhimurium infection, offering insight into possible therapeutic strategies. Further, the final article [11] focuses on *S.* Heidelberg in poultry, identifying resistance patterns and genetic markers linked to virulence, stressing the need for ongoing surveillance. These studies highlight key challenges and advances in controlling *Salmonella* in poultry and pigs, emphasizing innovative, integrated strategies to combat AMR and improve food safety and public health through a One Health approach.

Rafael Jesús Astorga, Angela Galán Relaño, and Antonio Valero Díaz
Guest Editors

animals

MDPI

Review

Salmonella and Salmonellosis: An Update on Public Health Implications and Control Strategies

Ángela Galán-Relaño [1,2], Antonio Valero Díaz [2,3,*], Belén Huerta Lorenzo [1,2], Lidia Gómez-Gascón [1,2], M.ª Ángeles Mena Rodríguez [1,2], Elena Carrasco Jiménez [2,3], Fernando Pérez Rodríguez [2,3] and Rafael J. Astorga Márquez [1,2]

1 Animal Health Department, Veterinary Faculty, University of Cordoba, 14014 Cordoba, Spain; agalanr12@gmail.com (Á.G.-R.); sa2hulob@uco.es (B.H.L.); v32gogal@uco.es (L.G.-G.); mariangii53@gmail.com (M.Á.M.R.); sa1asmar@uco.es (R.J.A.M.)
2 Zoonotic and Emerging Diseases (ENZOEM), University of Cordoba, 14014 Cordoba, Spain; bt2cajie@uco.es (E.C.J.); b42perof@uco.es (F.P.R.)
3 Food Science and Technology Department, Veterinary Faculty, University of Cordoba, 14014 Cordoba, Spain
* Correspondence: bt2vadia@uco.es

Simple Summary: *Salmonella* is one of the most important zoonotic pathogen agents, causing an estimated 93.8 million cases of gastroenteritis worldwide annually, with 155,000 deaths. Efforts to reduce transmission of Salmonella by food and other routes must be implemented on a global scale. Salmonellosis control strategies are based on two fundamental aspects: (a) the reduction of prevalence levels in animals and (b) protection against infection in humans. Thus, this review will be focused on *Salmonella* and its relationship between animals and public health (one health approach). The aim is to update the status of *Salmonella* in the world, with special reference to its implications on epidemiology and public health, food chain and risk assessment, antimicrobial resistance, and control strategies. We strongly believe that this review is an opportunity to collect significant and relevant information, using an integral approach, on Animal Health, Public Health, and the relationship between the two.

Abstract: Salmonellosis is globally recognized as one of the leading causes of acute human bacterial gastroenteritis resulting from the consumption of animal-derived products, particularly those derived from the poultry and pig industry. *Salmonella* spp. is generally associated with self-limiting gastrointestinal symptoms, lasting between 2 and 7 days, which can vary from mild to severe. The bacteria can also spread in the bloodstream, causing sepsis and requiring effective antimicrobial therapy; however, sepsis rarely occurs. Salmonellosis control strategies are based on two fundamental aspects: (a) the reduction of prevalence levels in animals by means of health, biosecurity, or food strategies and (b) protection against infection in humans. At the food chain level, the prevention of salmonellosis requires a comprehensive approach at farm, manufacturing, distribution, and consumer levels. Proper handling of food, avoiding cross-contamination, and thorough cooking can reduce the risk and ensure the safety of food. Efforts to reduce transmission of *Salmonella* by food and other routes must be implemented using a One Health approach. Therefore, in this review we provide an update on *Salmonella*, one of the main zoonotic pathogens, emphasizing its relationship with animal and public health. We carry out a review on different topics about *Salmonella* and salmonellosis, with a special emphasis on epidemiology and public health, microbial behavior along the food chain, predictive microbiology principles, antimicrobial resistance, and control strategies.

Keywords: *Salmonella*; salmonellosis; animal health; public health; food chain; predictive microbiology; antimicrobial resistance; control strategies; one health

1. Introduction

Salmonella spp. is recognized as a major zoonotic foodborne pathogen of economic significance in animals and humans; it causes an estimated 90 million cases of gastroenteritis worldwide annually, with approximately 155,000 deaths [1]. Even though salmonellosis is mostly reported as a foodborne disease, it has been estimated that about 10% of the cases are due to direct contact with animals [2].

Salmonella is a genus of highly diverse bacteria that live in the digestive tract of humans and animals. They are widespread in the environment thanks to their ability to survive and adapt even under extreme conditions [3].

Among the over 2600 *Salmonella* serovars described, clinical manifestations and mortality differ depending on both serovar and host characteristics (breed, age, sex, nutrition, and/or immunity) [4].

These serovars are divided into typhoidal and non-typhoidal serovars (NTSs); all of them can cause diseases in animals and/or humans with different levels of severity. Typhoidal serovars are highly adapted to the human host, which is their exclusive reservoir. They are, thserefore, only transmittable through human-to-human contact and cause a potentially life-threatening syndrome known as typhoid (*S. typhi*) or paratyphoid fever (*S. paratyphi*). Most European cases are considered imported cases and generally involve people returning from endemic countries [4].

NTSs are known as zoonotic agents. They are spread from animals and foods to humans but also through human-to-human close contact; they are widely present in the environment and can infect animals and contaminate both water and food. Usually, zoonotic salmonellosis occurs because of a true foodborne infection of animal or plant origin or through close contact with carrier animals [4].

Regarding human infections, only a few of the NTSs are responsible for most human cases. Of these, *S. enteritidis* and *S. typhimurium* are considered the most important serovars with the greatest impact on public health, being responsible for more than 70% of human infections, as illustrated in Table 1.

Table 1. Distribution of confirmed cases of human salmonellosis acquired in the EU Member States (MSs), 2019–2021, for the six most frequent *Salmonella* serovars in 2021 [4].

Serovar	2021			2020			2019		
	Cases	MSs	%	Cases	MSs	%	Cases	MSs	%
Enteritidis	23,634	23	64.6	21,203	23	63.1	32,010	24	61.6
Typhimurium	4027	23	11.0	3702	22	11.0	6044	24	11.6
Monophasic Typhimurium 1,4,[5],12:i:-	1269	14	3.5	1530	16	4.6	2668	17	5.2
Infantis	633	23	1.7	716	21	2.1	1215	24	2.3
Derby	239	16	0.7	260	17	0.8	396	20	0.8
Coeln	315	14	0.9	201	17	0.6	270	15	0.5
Other	6462	-	17.7	6009	-	17.9	9378	-	18.0
Total	36,579	23	100	33,621	23	100	52,001	24	100

Most serovars are non-pathogenic for animals but highly pathogenic for humans. Currently, in the European Union (EU), *Salmonella enteritidis* as well as *S. typhimurium* and its monophasic variant are the main serovars responsible for human disease (Table 1). However, *S. infantis* serovar has emerged as the fourth most prevalent serovar associated with human disease [4].

The European Food Safety Authority (EFSA) has recently reported an increase in the frequency and severity of human infections caused by *S. typhimurium* and its monophasic variant, which are associated with meat products derived from swine or cattle [5–7].

A wide range of domestic and wild animals can host *Salmonella* and thereby become reservoirs: poultry, swine, cattle, wild birds, rodents, pets, and exotic animals. In fact, pets such as dogs and cats and exotic animals such as reptiles and amphibians may play a significant role in transmitting the pathogen in the environment and household by excreting *Salmonella*, which can infect other animals and humans throughout the environment [8,9]. For example, the increase in the private ownership of reptiles has led to a rise in the number of zoonotic infections [10–12]. Additionally, rodents and birds act as *Salmonella* amplifiers, playing an essential part in the dissemination of bacteria on farms [13].

The slaughter process can also constitute a source of contamination, especially when adequate hygiene conditions are not maintained. *Salmonella*, which can be present in the intestinal contents of carrier animals, may cause contamination at various stages of the slaughter process, such as in trucks, lairages, slaughter lines, and quartering [14].

The primary clinical manifestation of salmonellosis in humans is self-limiting gastroenteritis, characterized by diarrhea, abdominal pain, fever, headache, nausea, and/or vomiting, which typically resolve within 2 to 7 days. However, in certain cases, especially among children and elderly patients, the illness can progress to a severe and life-threatening condition, accompanied by systemic bacteremia [1].

In contrast, subclinical infections are common in animals, where the bacteria can easily spread between flocks without detection, and animals may become intermittent or persistent carriers. Animals can become infected through different means: (i) close contact with other infected animals; (ii) contaminated water or direct contact with feces due to farm management and/or with contaminated equipment; (iii) transmission from parents to offspring (e.g., *S. abortusovis*, *S. indiana*) [15,16]; (iv) transmission from feed and environment; (v) potential transmission by arthropods.

Regarding foodborne transmission, there is strong evidence from outbreaks in the EU of various foods acting as vehicles [4], e.g., egg and egg products (39 outbreaks), mixed food (24), bakery products (15), pig meat and associated products (14), and vegetables and juices and other products thereof (11). Other food vehicles causing foodborne outbreaks in the EU are raw milk and dairy products made with raw milk, seafood products, and processed foods (e.g., sweets and chocolate) [4].

Salmonella strain typing is a crucial component of routine laboratory investigations [17]. Phage typing and serotyping, as well as molecular methods, are essential tools for this purpose. They enable the identification and isolation of such strains from primary animal sources as well as non-animal sources (i.e., food, water, and environmental samples). The most commonly utilized methods include pulsing-field gel electrophoresis (PFGE) and multiple locus variable analysis (MLVA). New genome-based typing methods, such as whole genome sequencing (WGS), are employed to track outbreaks and determine the epidemiological origin of the infection [18–20].

Efforts to reduce the transmission of *Salmonella* through food and other routes must be implemented using a one health approach. At the food chain level, the prevention of salmonellosis requires a comprehensive approach at farm, manufacturing, distribution, and consumer levels.

This review provides an overview of the *Salmonella* and salmonellosis status. It is a useful update of concepts for health professionals involved in animal health and public health. It is a review that facilitates and improves the understanding of the epidemiology and control of the main indicator pathogen of zoonoses, *Salmonella*.

2. *Salmonella* and Its Relationship with Foodborne Outbreaks: Update in the EU

2.1. EFSA Fact Sheet

The European Food Safety Authority (EFSA) provides independent scientific support and advice by collecting and analyzing data on the prevalence of *Salmonella* in animals and foods. It does so by assessing the food safety risks posed by the bacterium for human health and advising on possible control and reduction options. EFSA's findings are used by risk managers in the EU and the member states in their decision making and support the setting of reduction targets for *Salmonella* in the food chain. EFSA supports the EU's fight against *Salmonella* using three methods (https://www.efsa.europa.eu/sites/default/files/corporate_publications/files/factsheetsalmonella.pdf, accessed on 15 August 2023) [21]: (i) annual monitoring of *Salmonella* in animals and food to measure its progress; (ii) risk assessments and recommendations; (iii) EU-wide surveys on the prevalence of *Salmonella*.

2.2. Foodborne Outbreak Dashboard in the EU

According to Directive 2003/99/EC [22], EU Member States are obliged to report information on foodborne and waterborne outbreaks. The interactive tool (https://www.efsa.europa.eu/en/microstrategy/FBO-dashboard, accessed on 15 August 2023) [23] provides the latest information on foodborne outbreaks and epidemiological information of interest in the year 2021 [4].

The dashboard shows that a total of 4088 foodborne outbreaks (FBOs) occurred in the EU in 2021. From these FBOs, there were 33,813 human cases, with 2560 hospitalizations and 33 deaths. The trends between 2016 and 2021 show a slight decrease in these parameters, with the exception of the death variable, because of the high mortality registered in 2019 and 2020 in human listeriosis cases.

The FBO-dashboard interactive tool shows all known outbreaks and cases per 100,000 people by country, the number of human cases and causative pathogen agents, and the ranking of number of outbreaks by food vehicle and place of exposure.

2.3. Salmonella Occurrence in the EU

Salmonella spp. is the second most common zoonotic pathogen, after *Campylobacter*, according to the most recent EU One Health Zoonoses Report [4], with both of them causing gastrointestinal infections in humans. The number of confirmed cases of human illness salmonellosis was 60,050, corresponding to an EU notification rate of 15.7 per 100,000 people and with a stable trend between 2017 and 2021. Among these cases, there were 11,790 hospitalizations (45.0% of outbreak-associated hospitalizations) and 71 reported deaths. Furthermore, EFSA data rank it as the leader, causing a total of 773 human cases in foodborne outbreaks (20.8% of outbreak-associated cases), with 1123 hospitalizations and 1 death. *S. enteritidis* was the predominant serovar (N = 350; 79.7% of all *Salmonella* outbreaks). The top five serovars responsible for human infections are currently *S. enteritidis*, *S. typhimurium*, *S. typhimurium* monophasic variant (mST), *S. infantis*, and *S. derby*.

Furthermore, an analysis of the most recent data released by EFSA on the distribution of serovars at the primary-sector level reveals that the majority of *Salmonella* spp. isolates originate from the production of broilers (*Gallus gallus* domesticus) (55.7%), with turkeys (*Meleagris gallopavo*) coming in second with 12.9%, pigs (*Sus scrofa* domestica) with 7.6%, and laying hens (*Gallus gallus* domesticus) with 6.0%. These data were obtained from poultry populations that fall under the purview of the *Salmonella* National Control Program (SNCP) [4]. While *S. infantis* was exclusively associated with broiler sources (95.2%), *S. enteritidis* was mainly associated with broiler flocks and meat (70.0%) and laying flocks and eggs (26.0%). On the other hand, the majority of *S. typhimurium* and mST isolates (43.2% and 65.4%, respectively) were linked to pig sources [4].

A total of 73,238 'ready-to-eat' food sampling units were collected, and they had a very low proportion of *Salmonella*-positive units (0.23%) overall. The highest proportions of positives were found for 'meat and meat products from pigs' (0.82%). For 'non-ready-to-eat' food, 466,290 sampling units were collected, and the proportion of positive samples was low (2.1%). The food categories with the highest proportions of positive units were 'meat and meat products' (2.2%), especially those from broilers (4.4%) and turkeys (3.6%) [4].

A significant increase in the estimated breeding turkey flock prevalence of *Salmonella* was noted in 2021. Flock prevalence trends for target *Salmonella* serovars have, in contrast, been stable over the last few years for all poultry populations.

3. An Update on Environmental Stresses Affecting *Salmonella* in Foods

The adaptation of *Salmonella* strains to different environmental stresses in foods has been widely reported, including known increased resistance to low pH, low water activity, and disinfectants, among others. Thus, *Salmonella* remains as an important concern in food processing environments, traditionally linked to the development of greater tolerance and cross-protection mechanisms, thus increasing the persistence along the food chain [24].

Besides the well-known effect of temperature, which is currently applied in pasteurized foods [25], recent studies have shown a growing interest in the acquired resistance mechanisms of *Salmonella* serovars against the presence of acids, low-water-activity foods, and biofilm formation on biotic or abiotic surfaces [26]. All these cumulative hurdles are applied at sub-lethal levels (especially in ready-to-eat foods), so that they promulgate an adaptative response and enable the survival of a larger fraction of *Salmonella* cells.

3.1. *Acid Resistance of Salmonella in Foods*

Acid adaptation allows *Salmonella* to withstand the challenges posed by low pH levels and potentially cause foodborne illnesses. *Salmonella* demonstrates the capacity to modify its physiological characteristics and regulate gene expression against exposure to low pH levels. This adaptation mechanism is especially interesting because of its viability in acidic foods, where conditions might inhibit the growth of other microorganisms [27]. In foodstuffs, weak organic acids like acetic, lactic, and citric acids can be present due to natural food constituents, fermentation processes, or intentional addition during food production to enhance preservation.

The optimum pH for *Salmonella* growth is generally known to be between 6.5 and 7.5, but the minimum pH value depends on many factors, such as the strain, the type of acid, or the synergistic action when combined with other stresses such as NaCl. For instance, Pye et al. [28] compared different *Salmonella* serovars in culture media supplemented with 6% NaCl, 12 mM acetic acid, or 14 mM citric acid, and they found that *S. typhimurium* showed the highest resistance to NaCl and acetic acid stress, while *S. enteritidis* showed the highest resistance level for citric acid.

The increased tolerance to a low pH following acid habituation is referred to as the Acid Tolerance Response (ATR), which has been shown to be strain dependent [29]. This pH-dependent ATR might induce a posterior acid adaptation involving bacteria growth in mildly acidic conditions [30]. These investigations also shed light on the mechanisms of acid-induced cross-protection against ethanol stress in *S. enteritidis* during the growth phase [27], which may lead to more efficient mitigation strategies.

3.2. *Survival of Salmonella in Low-Water-Activity Foods*

The high frequency of *Salmonella* in low-water-activity (a_w) foods (such as powders, flours, dried fruits, spices, oily foods, and nuts) is a cause for concern. Recent studies have extensively reported on this situation due to the growing number of salmonellosis outbreaks related to these products [31]. These matrices comprise a wide range of the so-called Low-Moisture Foods (LMFs) as being those with reduced water content, making them less favorable for the growth of most microorganisms. According to Food and Drug Administration (FDA) standards, they have an a_w at 25 °C of less than 0.85 [32]. These conditions do not allow bacterial growth; however, several studies have demonstrated the survival ability of *Salmonella* spp. in different LMFs [33,34] for months or even years, thus potentially causing adverse health effects for susceptible population groups. Microbial contamination can occur when handling and/or processing any contaminated LMF and/or from environmental contamination, suspended air particles, or inert surfaces [35]. Adaptive responses in *Salmonella* help it survive by accumulating compatible solutes, including proline, glycine, betaine, ectoine, and trehalose, leading to reduced water loss [36]. Furthermore, osmoregulation plays a vital role in maintaining the turgor pressure of the bacteria through increasing the intracellular concentration of compatible solutes [37,38].

Industrial interventions to effectively control *Salmonella* in LMFs have been mainly oriented toward thermal processing. Yet, there is a lack of knowledge on the main factors involved in the thermal resistance of *Salmonella* in LMFs. Liu et al. [39] presented an overview on the factors affecting the microbial safety of LMFs together with the latest developments in analytical methods for the detection of pathogens in dried food commodities. Microbial resistance of *Salmonella* in LMFs can differ according to the type of strain, physiological conditions of the pathogen, food composition (e.g., sugar or fat content), a_w, and heating temperature [40].

Importantly, the use of cocktails may ensure that novel processes can remove the most resistant strains, as reported in previous studies [41,42]. Some of the latest studies deal with the relationship between moisture content, as a better indicator than temperature, and a_w in the thermal inactivation of *Salmonella* in LMFs [43], the effect of food structure combined with emerging technologies [44], and the design of novel test cells to better estimate a_w in the thermal resistance of *Salmonella* strains [45]. Despite these recent developments, food industries might still be faced with the randomness and variety of environmental factors associated with *Salmonella* contamination in LMFs, combined with its persistence for a long-term storage period and the difficulties of current methods and sampling strategies in its detection.

3.3. *The Biofilm Formation of Salmonella in Food-Processing Environments*

The different survival mechanisms of *Salmonella*, such as the formation of biofilms, are hypothesized as possible factors for the onset of foodborne diseases. There is clear evidence of the formation of biofilms by *Salmonella* in foods and in different materials present in food processing environments [46–48]. *Salmonella* produces a biofilm matrix that is mainly composed of fimbriae (curli) and cellulose [49]. The ability of *Salmonella* to adhere and form biofilms is influenced by multiple factors, such as the composition of the growth medium, the developmental stage of the cells, the characteristics of the inert material, the contact time, the presence of organic substances, and environmental conditions like temperature and pH [50].

Control strategies against *Salmonella* and microbial biofilms overall have been traditionally based on the use of chemical disinfectants, widely applied in the meat industry [51]. However, their effectiveness may differ depending on the type of surface. Other drawbacks such as the increased bacterial resistance to sub-lethal concentrations of disinfectants and the presence of chemical residues preclude their use as a valid antibiofilm strategy. Antimicrobial resistance and toxicity issues have been associated with the use of antibiotics or nanoparticles. Other control strategies still require the application mode and targeted dose to be optimized, as the use of enzymes and quorum-sensing inhibitors are of dubious efficacy against relevant biofilms [52]. Among the physical treatments, pulsed light and UV-C radiation could inactivate the formation of *Salmonella* biofilms [53]. Gao et al. used a combined pulsed light treatment with sodium hypochlorite at moderate levels (100 ppm for 30 min), and it was found to be effective in deactivating a six-cocktail strain of *Salmonella* spp. However, the induction of sub-lethal cells caused by pulsed light deserves further investigation. UV-C radiation alone has overall limited efficacy in reducing *Salmonella* biofilm cells, but its combined use with organic acids with chemical sanitizers seems to be a promising strategy in industrial facilities [54].

Recent developments in biofilm eradication are based on biocontrol strategies such as the use of bacteriophages. Ashrafudoulla et al. [55] evaluated specific lytic bacteriophages against *S. thompson* biofilms on eggshells, which showed better efficacy when using bacteriophage cocktails. In contrast, temperate *Salmonella* bacteriophages can confer greater virulence and resistance to adverse factors, as shown by *S. typhimurium* biofilms. Therefore, the expression of virulence genes and metabolic pathways of *Salmonella* induced by the presence of bacteriophages deserves to be studied further [56]. Another comprehensive study by Asma et al. [57] on natural strategies for biofilm control highlighted the use of plant-based and bee products as antibiofilm molecules. Interestingly, the development of plant-derived nanoparticles (NPs) has arisen as a promising strategy against various bacterial biofilms, including the use of liposomes, cyclodextrins, or hydrogels.

Finally, dual-species biofilms using lactic acid bacteria and/or bacteriocins have been extensively explored as a strategy for the competitive exclusion of *Salmonella* during processing. Research trends are oriented toward the study of extracellular polymeric substances (EPS) to better understand the mechanisms of *Salmonella* biofilm inhibition by LAB and to further explore combinations of LAB biofilms with other LAB metabolites (hydrogen peroxide or bacteriocins) in industrial environments [58].

3.4. Predictive Microbiology Models for Estimation of the Microbial Behavior of Salmonella *in Foods*

Since *Salmonella* can be present in several food commodities, microbial behavior along the food chain has been extensively studied over recent decades. Predictive microbiology is a field that involves using mathematical models and statistical tools to predict the behavior of microorganisms in various environments [59]. Predictive models aim to describe the effect of a certain process (e.g., disinfection, heat treatment, storage, etc.) modulated by a range of environmental factors (e.g., pH, temperature, a_w, etc.) on the microbial population of interest. Predictions can be quantified through different parameters describing microbial growth, survival, or inactivation, such as maximum growth rate, lag phase, inactivation rate, etc. [60]. As such, applications of predictive microbiology may be oriented to different areas, including food innovation, process control, risk management, reduction of food wastage, design of experiments, and training. There is a wide range of predictive models for describing *Salmonella* behavior in various food categories. Growth, survival, or inactivation ability has been extensively explored in eggs and egg products [61–63], meat products [64–66], melons [67–69], low-moisture foods [40,70–73], and leafy vegetables [74,75], among others. Furthermore, an extensive review of existing growth/no growth models of *Salmonella* was presented by Carrasco et al. [76] as well as other cross-contamination models [77,78].

While the effect of the most representative environmental factors, such as temperature, pH, and water activity, on *Salmonella* behavior has been properly characterized by the use of dedicated models, research efforts are focused on the effect of emerging preservation technologies or novel antimicrobial agents, as shown in some recent papers. Shahdadi et al. [79] conducted a systematic review and modelling of the role of bacteriophages against *Salmonella* in meat products, while Austrich-Comas et al. [80] evaluated a combined strategy using starter cultures, storage, and high-pressure processing in dry fermented chicken sausages. The use of radio frequency as an inactivation technology against *Salmonella* in treated eggs was successfully modelled by Bermúdez-Aguirre and Niemira [81]. Other models for *Salmonella* using pulsed ohmic heating, UV-radiation, ultrasound, and microwave technologies were reviewed by Alvarenga et al. [82]. It is clear that predictive models can aid in decision making to establish standards for processing by using emerging technologies. Future work should be oriented toward incorporating specific parameters that accurately quantify the effectiveness of emerging technologies in food preservation.

With the advent of dedicated predictive microbiology software, the integration of computational elements is crucial to providing an applicability dimension to predictive models. Machine Learning (ML) algorithms enable computers to learn from and make predictions or decisions based on data. ML models can be trained to identify *Salmonella* genes relevant to disease outcome, thus facilitating the integration of genomic data in microbial risk assessment [83]. Other applications of ML techniques are related to pathogen source attribution [84] and gene-based risk assessments [85]. The inclusion of meteorological factors in ML algorithms on *Salmonella*'s infectivity and outbreak scale was recently reported by Karanth et al., 2023 [86]. The integration of these data with well-defined metadata offers the opportunity for ML models to forecast future trends in antibiotic resistance, determine the sources of pathogens, aid in the investigation of foodborne outbreaks, and enhance risk assessment protocols.

The routine and successful use of mathematical models by the food industry as well as governmental or educational agencies will depend on the development of appropriate and useful applications (software tools) with easy management. There has been an effort to harmonize data formats and model annotations to increase transparency and fit-for-purpose use of predictive microbiology models in a real system. Recent software developments were reviewed by Possas et al. [87]. The authors highlighted the novel fitting shiny apps and improved algorithms that provide better data visualization and graphical representation.

For industrial and health authorities, the use of web interfaces is becoming crucial for a more effective interpretation of predictive models. MicroHibro software (www.microhibro.com, accessed on 15 August 2023) [88] was developed by the University of Córdoba and includes a range of freely available applications related to predictive modelling (safety and shelf life), sampling plans, and risk assessment tools [89]. Currently, MicroHibro software is being updated to include quality or shelf-life models and more advanced risk assessment features.

To illustrate the different predictive model applications for estimating *Salmonella* behavior in the animal-derived foodstuffs supply chain, we show two examples using validated models of pork meat and egg yolk.

MicroHibro contains 27 primary and secondary predictive models of *Salmonella*, including raw vegetable products (tomato, lettuce, avocado, apple, strawberry, cantaloupe), beverages (soya milk), and animal foods (fresh salmon, pork meat, and egg byproducts). Additional models can be included using information from published studies or experimental works. Nevertheless, data curation and inclusion of new models are performed by experts from the University of Cordoba upon request.

The model of Pin et al. [90] was developed for ground pork using data from the literature on different *Salmonella* serovars. The model can describe the microbial fate in the pork supply chain considering that, according to the product formulation, *Salmonella* could tentatively grow or survive during storage. For this case study, growth will be assumed as a function of different pH, a_w, and temperature conditions. It is well known that before industrial application, predictive models must be validated in the food of interest [59]. Thus, to apply this model to a real processing condition, validation (i.e., comparing predictions of growth responses from the model to actual measures of growth or survival published in the scientific literature) should be performed using observed data for various levels of the environmental factors included within the model domain (Table 2).

Table 2. Dataset used for the model validation of [90] against temperature, pH, and a_w conditions and their respective observed maximum growth rates (μ_{max}, log CFU/h).

T (°C)	pH	a_w	μ_{max}
5	5.2	0.970	0.005
8	5.7	0.970	0.010
10	5.7	0.976	0.030
15	5.9	0.980	0.080
20	6.0	0.990	0.200

In MicroHibro, the user can define validation conditions to assess the closeness of the observed and predicted values. This can be done visually using the equivalence line graph, where predictions are equal to observations, and by comparing the effect of temperature on the μ_{max}. In such a way, the validation indices (e.g., bias and accuracy factors) reported by Ross et al. [91] are facilitated in MicroHibro. In Figure 1, the model predictions are relatively close to microbial observations; thus, the model can be used effectively for assessing *Salmonella* growth in the pork supply chain.

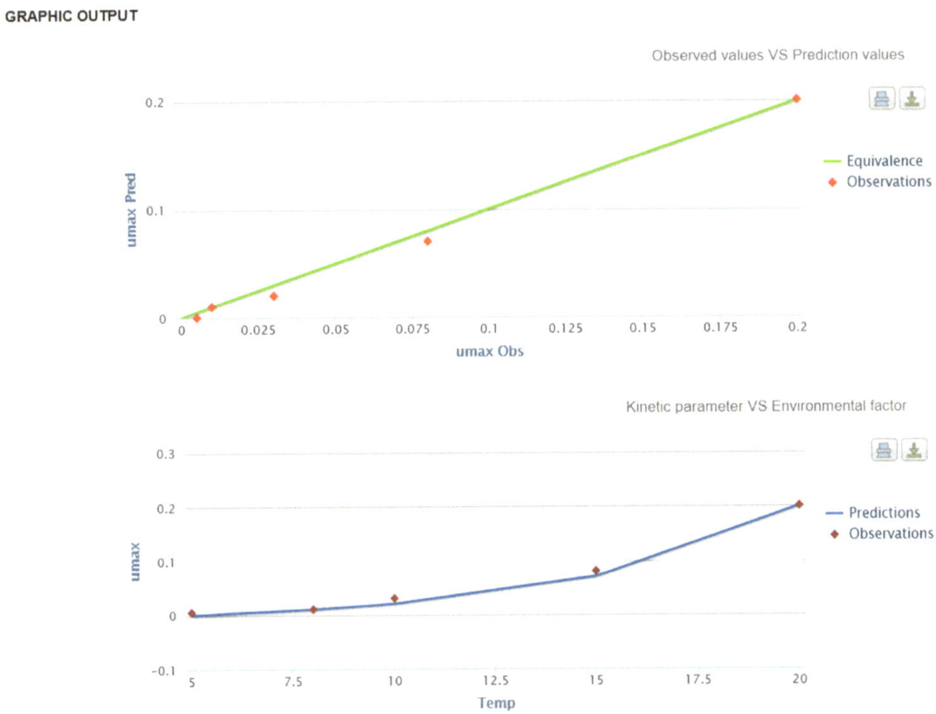

Figure 1. Validation of the model [90] for estimating the growth of *Salmonella* spp. in the pork supply chain using MicroHibro. Top: comparison between observed and predicted maximum growth rates (*y*-axis) (μ_{max}, log CFU/h); bottom: evolution of maximum growth rates (μ_{max}, log CFU/h) against storage temperature(*x*-axis).

Other modelling applications can be used for comparing static and dynamic temperature conditions. As an example, a secondary model for *S. enteritidis* in egg yolk using a dynamic (non-isothermal) profile has been developed [92]. The model can predict the effect of temperature on *S. enteritidis* growth (10–43 °C). Dynamic models can predict the effect of temperature changes over time. Figure 2 shows the comparison between *S. enteritidis* growth in short-term storage (4 h between 15 and 20 °C) and static temperature storage (10 °C). Figure 2 shows that *S. enteritidis* can grow in more than 0.5 log units at dynamic temperatures, thus increasing the probability of a foodborne outbreak through the ingestion of contaminated egg yolk samples (assuming that there are not additional treatments for *Salmonella* inactivation before consumption).

Through these examples, the use of expert computational systems, such as MicroHibro software 3.0, is a powerful tool for supporting food safety and quality activities by Health Authorities and the food industry. This represents a breakthrough in the assessment and management of food safety based on scientific evidence.

Figure 2. Predictions of *S. enteritidis* growth in egg yolk at dynamic temperatures obtained using the model of Gumudavelli [92] in MicroHibro. Top: representation of the dynamic temperature profile; middle: growth of *S. enteritidis* at dynamic temperatures; bottom: comparison of *S. enteritidis* growth in egg yolk under dynamic conditions and at a static temperature of 10 °C.

4. Antimicrobial Resistance

4.1. Salmonella and Antimicrobial Resistance: Preface

Several factors of bacterial chromosomes or plasmids may be the cause of *Salmonella* antibiotic resistance [93,94]. These genetic determinants might be in charge of expressing the intrinsic resistance mechanisms linked to the synthesis of beta-lactam antibiotics, alterations in the composition of antimicrobials caused by bacterial enzymes, differences in the permeability of bacteria, the existence of efflux pumps, or changes in target receptors.

The expression of acquired resistance mechanisms, which arise from point mutations in chromosomal genes (e.g., monophasic strains of *S. typhimurium*) or the acquisition of mobile elements like plasmids, transposons, or genomic islands can also result in antimicrobial resistance (AMR) [94]. Resistance transmission can happen vertically between different bacteria or horizontally within the same species or genus. Moreover, it can also occur indirectly through environmental factors [93]. Antimicrobial substances of varying classes, doses, and exposure frequencies are often administered to the gut microbiota of both people and animals for the purposes of treatment, prophylaxis, or metaphylaxis. Additionally, the environment or animal feed sources may contribute to this exposure [95].

A variety of reasons, including improper use of antibiotics in human and veterinary medicine, unhygienic environments and practices in healthcare settings, and pathogens that are resistant to treatment spreading via the food chain, can lead to the development of resistance. Antimicrobials become less effective over time and eventually worthless as a result of this [96]. The primary selective pressure resulting from antibiotic overuse and abuse is still thought to be responsible for the appearance, selection, and spread of microorganisms resistant to antibiotics [97]. Gut bacteria can develop resistance to certain antibiotic substances, which they can then vertically transfer to *Salmonellae* sharing the same ecological niche.

Humans and animals are both impacted by the significant health issue of antibiotic-resistant strains spreading. AMR is still regarded as a zoonosis and poses a major threat to public health, despite efforts in recent decades to decrease the use of antibiotics [98]. Regarding this, a significant problem is the occurrence of Multiple Drug Resistance (MDR) in bacteria, such as *Salmonella* spp., that cause foodborne illnesses that are common around the world. In fact, multidrug resistance (MDR) complicates the use of antibiotics to treat infections and increases the cost of healthcare, lengthens hospital stays, and increases mortality [98].

One of the main causes of global concern for health authorities has been the increase in cases of gastroenteritis and sepsis linked to *Salmonella* strains that are becoming more resistant or even multi-resistant to conventional antimicrobials (e.g., beta-lactams, aminoglycosides, and quinolones, among others). These strains include mainly the Typhimurium serovar and its monophasic variants (mST) [99–101] as well as *S. infantis* and *S. kentucky* [102]. Consequentially, resistant infections are on the rise, causing therapeutic failures and longer hospital stays and thus heavily affecting public health and the economy. With over 90,000 salmonellosis cases reported every year in the EU, the EFSA has estimated that the overall economic burden of human salmonellosis could be as high as EUR 3 billion per year (https://www.efsa.europa.eu/sites/default/files/corporate_publications/files/factsheetsalmonella.pdf, accessed on 28 June 2023).

When comparing the presence of *Salmonella* spp. MDR in strains isolated from food and bacteria obtained from animals, relevant research has shown a statistically significant difference. These findings tend to suggest that *Salmonella* strains isolated from food are the main source of MDR [98], which is of great importance when taking into account the fundamental role that this pathogen plays in the food industry and the resistance that has been demonstrated to exist to conventional disinfectants [103]. In light of this, it is crucial to identify the actual animals and people that are the sources of MDR strains in order to reduce their prevalence and enhance public health protection [98].

4.2. Key Findings

The annual collection of data on Antimicrobial Resistance (AMR) pertaining to zoonotic and indicator bacteria from humans, animals, and food sources is a collaborative effort undertaken by Member States (MSs) and reporting countries. The resultant datasets are jointly analyzed by the European Food Safety Authority (EFSA) and the European Centre for Disease Prevention and Control (ECDC) and are published in an annual EU Summary

Report. The most recent report provides an overview of the key findings of the period from the harmonized AMR monitoring conducted between 2020 and 2021, with a specific focus on *Salmonella* spp. in humans and food-producing animals, including broilers, laying hens and turkeys, fattening pigs, and cattle under 1 year of age along with the associated meat products [104].

Among the reporting countries, the number of *Salmonella* spp. in isolates from human cases varied considerably. Of the 26 reporting countries, including those within the EU and EAA countries, six countries reported very few (<100) human isolates, while three countries reported more than 1000 isolates.

In 2021, overall resistance to ampicillin, sulfonamides, and tetracyclines was noticeably high in *Salmonella* spp. isolates from humans. Similar resistance patterns were observed in isolates from food-producing animals and poultry carcasses, except in the case of laying hens, where resistance levels to these antibiotics were comparatively lower.

Over the period 2013–2021, declining trends in resistance to ampicillin and tetracyclines in isolates from humans was observed in 13 and 11 countries, respectively, coinciding with a decrease in the prevalence of *S. typhimurium*, a serotype commonly associated with pigs and calves. From data reported in 2021, the resistance to fluoroquinolones, specifically ciprofloxacin, was moderate in *Salmonella* isolates from fattening pigs (10.1%) and cattle under 1 year of age (calves) (12.7%). In contrast, in 2020, resistance to ciprofloxacin was noticeably high in isolates recovered from broilers (57.5%), fattening turkeys (65.0%), broiler carcasses (69.3%), and turkey carcasses (46.9%). In 2021, *Salmonella* isolates from humans displayed an average rate of 14.9%, with the lowest levels observed in *S. typhimurium* (7.6%) and *S. typhimurium* monophasic variant (8.9%) and high to extremely high levels in *S. infantis* (33.9%) and *S. kentucky* (78.1%).

It is noteworthy that approximately 95% of isolated *S. infantis* serovars identified in the EU were traced back to broilers and their derived products [102]. Recent research has demonstrated a strong association between the *S. infantis* serovar and elevated antimicrobial and multidrug resistance, resistance to disinfectants, increased tolerance to environmental mercury, heightened virulence, and an enhanced ability to form biofilms and attach to host cells [102].

In contrast, *S. kentucky* isolates from human cases demonstrated consistently high to extremely high resistance levels for ampicillin (62%), ciprofloxacin (77%), tetracycline (57%), sulfamethoxazole (51%), and gentamicin (27.9%). However, resistance to cefotaxime/ceftazidime (6%) and chloramphenicol (12.6%) was observed at low to moderate levels. Similarly, extremely high resistance to ciprofloxacin was reported in *S. kentucky* isolates from broilers (78.0%), laying hens (91.9%), fattening turkeys (96.6%), broiler carcasses (100%), and turkey carcasses (93.3%). In the case of *S. enteritidis*, the most prevalent serovar identified in human cases, resistance to quinolones (ciprofloxacin and nalidixic acid) was 22.6% and 24.8%, respectively [4].

Resistance to third-generation cephalosporins remained notably low in isolates from humans in 2021 (1.1% to ceftazidime and 1.1% to cefotaxime, on average) and was seldom detected in isolates from animal and carcass origins in 2020–2021, except for calves (2.6% to cefotaxime and 1.3% to ceftazidime) and broiler flocks (2.1% to cefotaxime and 2.0% to ceftazidime). Conversely, combined resistance to fluoroquinolones and cephalosporins was very low in isolates from both humans and animals but exhibited higher prevalence in certain *Salmonella* serovars (e.g., *S. kentucky* and *S. infantis*) [4].

Among isolates from human cases, *S. enteritidis* displayed the highest levels of resistance to ciprofloxacin and colistin (22.6% and 17.6%, respectively) when compared to other serovars (14.9% and 5.1%, respectively). Colistin resistance was similarly pronounced in certain *Salmonella* serovars (2020–2021) derived from food-producing animals, particularly laying hens (55.4%) and broilers (53.1%). In these cases, *S. enteritidis* was the predominant serovar [4].

Multidrug resistance was high (22.6%) among *Salmonella* spp. reported in human cases in the EU, ranging from low levels among *S. enteritidis* (1.9%) to very high among *S.*

kentucky (54.8%) and extremely high for monophasic *S. typhimurium* 1,4,[5],12:i:- (78.4%). Similarly, MDR was observed at moderate to very high levels in *Salmonella* spp. recovered from carcasses of food-producing animals such as turkeys and broilers (19.1% and 51.2%, respectively) and high levels for all food-producing animals checked, including fattening broilers (41.8%), fattening pigs (39.1%), fattening turkeys (38.2%), and calves (30.4%); the exception was laying hens, which had low-level MDR (6.3%) [4].

The contributions of selected multi-resistant serovars to overall MDR levels in *Salmonella* isolated from animals in 2020–2021 were as follows: broiler carcasses and *S. infantis* (79.4%), broilers and *S. infantis* (73.3%), fattening pigs and Monophasic ST (53.1%), calves and Monophasic ST (41.7%), laying hens and *S. infantis* (34.4%), turkey carcasses and Monophasic ST (24.6%), and fattening turkeys and *S. infantis* (22.1%) [4].

Finally, reporting at the EU level showed that the overall proportion of presumptive ESBL—or AmpC—producers ranged from very low to low among *Salmonella* isolates recovered from all food-producing animal populations and broiler carcasses. It was also very low in isolates from human cases, although higher resistance was observed in specific *Salmonella* serovars (*S. typhimurium* and its monophasic variant and *S. infantis*). No carbapenemase-producing *Salmonella* spp. were isolated from human cases in 2021, nor in animal isolates from 2020 to 2021 [4].

5. Control Strategies in Animal Health

Several environmental and management factors have been associated with high levels of *Salmonella* spp. in the animal population. Based on these risk factors, different prevention and control methods related to hygiene and management, health and biosafety, animal welfare, and feeding strategies have been proposed [105,106].

5.1. Feeding Strategies

In the case of the poultry and pig industry, the main reservoirs of *Salmonella*, feeding strategies aimed at optimizing intestinal functions may have an impact on the colonization of *Salmonella* in the digestive tract. Among them, we must highlight the acidification of feed by means of organic acids, the use of probiotics, prebiotics, or phytobiotics, and the new lines of research on the incorporation of essential oils (EOs) extracted from plants [107,108]. Most of these products are used in animal health as feed additives, and their approval as therapeutics requires proven scientific studies that demonstrate their antimicrobial efficacy, effect on animal production, and safety for public and environmental health.

The efficacy of EOs obtained mainly from oregano, cinnamon, thyme, and citrus fruits have been evaluated against *Salmonella* serovars [109–111]. As an example, the effect of EOs against *Salmonella* serovars isolated from human outbreaks and river water has recently been investigated [108]. This research showed that oregano best inhibited the growth of clinical and environmental Saintpaul, Oranienburg, and Infantis serovars, followed by thyme and grapefruit EOs. The antimicrobial property of the oregano EO, higher than even antibiotic ampicillin, may be attributed to the terpenoids thymol and carvacrol. Therefore, this study concludes that the use of oregano and thyme EOs in conjunction with other oils or bactericidal agents may enhance their effectiveness against infections caused by atypical *Salmonella*. Furthermore, other studies have provided new data on the susceptivity distribution of *Salmonella enterica* strains involved in animal and public health to EOs and a first estimation of the MIC_{90} and MBC_{90} (understood as the Minimum Inhibitory and Bactericidal Concentrations, respectively, able to inhibit or kill 90% of the bacterial population) [109]. The results supported the bactericidal potential of EOs of oregano, common thyme, and red thyme against this bacterium and significant differences between the susceptibility of Typhimurium and Enteritidis serovars. The presence of *S. typhimurium* strains with possible multiple essential oil resistance was also demonstrated.

In addition, different authors have continued their research on assessing the combined effect of these natural substances with traditional antimicrobials (AMBs) as an effective option to reduce bacterial resistance and administration doses [112,113]. In this sense, the synergistic effect between EOs with the main AMBs used against *Salmonella* (enrofloxacin, ceftiofur, and trimethoprim-sulfamethoxazole) has been reported, highlighting the higher percentage of total synergies of trimethoprim-sulfamethoxazole with four EOs (cinnamon, clove, oregano, and red thyme), the most effective combination being enrofloxacin and cinnamon EO [107,113]. These results support the need to expand these trials to more clinical strains and to investigate the mechanisms of action of these synergies.

Based on the above, we believe that the research supports the potential use of EOs (especially oregano, thyme, and cinnamon), alone or in combination, with traditional AMBs, as an effective alternative for the control of *Salmonella* infections of animal or other origin and as a strategy to reduce the development of new bacterial resistance. In addition, we conclude that it is necessary to continue the in vitro studies of susceptibility distribution, the mechanisms that determine the synergy, the in vivo toxicity, and the development of possible resistance mechanisms.

5.2. Non-Feeding Strategies

In addition to the feeding-based approach, non-feeding alternatives focus on the use of bacteriophages, vaccines, and the application of biosecurity measures. These strategies are common in poultry and pigs to minimize *Salmonella* prevalence in farms [105,106].

5.2.1. Bacteriophages or Phages

Bacteriophages or phages are viruses that infect and replicate in bacteria until they lyse. They have a capsule and genetic material like eukaryotic viruses. They are natural bactericides and probably one of the most widely distributed microorganisms in the biosphere. Despite their potential usefulness in the treatment of infections, the study of their feasibility has been relegated to the use of antibiotics. In the current context, with the reduction and/or withdrawal of antibiotics from the medical-veterinary scene, alternatives such as phages or bacteriophages may be useful for the treatment and control of bacterial infections such as *Salmonella* [3].

When it comes to prophylaxis, animal therapy, and reducing the number of bacteria in animal-based food products, bacteriophages are thought to be a valuable alternative to antibiotics [114]. Their host-specificity makes them natural, non-toxic, and feasible for therapeutic application, allowing them to attack only the targeted bacteria while safeguarding the rest of the microbiota. Since the immune system can tolerate phages well, they also have the advantage of preventing host allergies [105]. Moreover, they are able to combat resistance to antimicrobial bacteria [115]. *Salmonella* and other foodborne infections have been successfully treated in a number of experiments involving germ-free chickens raised in battery cages [116].

Phage-based methods of controlling *Salmonella* have been tested in poultry [117–120] and, to a considerably lesser extent, in pigs [121]. In fact, the environment found in chicken farms may be a valuable source of *Salmonella* phages. It has been found that broiler chicken farms in Spain have more diversified *Salmonella* bacteriophages than layer ones based on the most common serovars [118]. However, more research is required to understand the epidemiology of phages in relation to other serovars.

Furthermore, some researchers have recently investigated the use of microencapsulated bacteriophages incorporated into feed for *Salmonella* control in poultry [119,120]. In a first study, in vitro and in vivo gastrointestinal survival of non-encapsulated and microencapsulated *Salmonella* bacteriophages and its implications for bacteriophage therapy in poultry were reported. Significant differences were observed in the results between the phage delivery of in vitro studies compared with in vivo studies [119].

A second study showed that adding the L100 encapsulated phage as a feed additive to the starting diet during rearing could significantly reduce the incidence of flock contamination with *S. enteritidis*. At the conclusion of the rearing period, this pathogen had been fully eradicated from the environment, and there was a decrease in *Salmonella* colonization and excretion. Nevertheless, higher phage doses, better delivery protocols, and/or the combination of different approaches might be required [120].

Finally, other studies have assessed the effect of bacteriophages against *Salmonella* Infantis and *Salmonella enteritidis* on farm surfaces, evaluating bacteriophage application as a complementary tool for cleaning and disinfection procedures [117].

5.2.2. Vaccines

Strategies based on vaccination for the control of *Salmonella* spp. have proven to be a very effective tool for controlling salmonellosis in species such as poultry. For that reason, the manufacturing of vaccines for the poultry industry is based on strains of *S. enteritidis* and *S. typhimurium* [122].

On the contrary, in swine there are currently no effective commercial vaccines. The main problem with medical prophylaxis against *Salmonella* in swine is that there is no cross-immunity between the different serovars (e.g., Typhimurium, Rissen, Derby, Anatum, Bredeney, etc.); therefore, it would be necessary to use specific vaccines (autologous or autovaccines) against the serotype involved in the infection/disease on the farm or to design vaccine candidates that included the predominant serotypes in the geographical area and/or farms involved [3].

The different types of vaccines available on the market are live-attenuated, inactivated, and subunit vaccines [123]. The protection conferred by live vaccines is theoretically greater since they promote a cellular-based response, which *a priori* is ideal for facultative intracellular pathogens such as *Salmonella*. In addition, if they are administered orally, they will manage to produce immunoglobulin-A in the intestine, the main component of the immune system in the control of digestive pathogens. However, these vaccines have certain disadvantages, such as the need to withdraw any antibiotic treatment during oral administration of the vaccine, their cost, and the potential risk of reversion to virulence and biosafety [3]. In fact, secondary mutations in live vaccines can cause reversion to virulence, which affects the overall health of flocks and thus contaminates the environment [124]. To conclude, it is important to note that a vaccine should be safe, give protection against various serovars, and stimulate the host's immunity system.

Finally, the application of vaccines can have negative effects, such as the development of antibodies (because of vaccination) that interfere with or mask the antibodies developed by the infection. This becomes a problem in countries with a control program based on serological analysis, since the techniques used do not allow for the differentiation of vaccine antibodies from those produced by the infection. There are alternatives to this, such as ELISA techniques that make it possible to differentiate vaccine antibodies from antibodies produced by natural infection (DIVA strategy, Differentiating Infected from Vaccinated); however, these entail additional costs in the surveillance and control of *Salmonella* [3].

5.2.3. Biosecurity

Biosecurity is the most effective and inexpensive disease control measure, aimed at managing the risks posed by diseases to the economy, environment, and human health [125]. The application of biosecurity measures to reduce the levels of prevalence of infections/diseases, with special attention to those that pose a risk to public health (e.g., salmonellosis), should be one of the main objectives of health authorities [126].

The application of strict hygiene and biosecurity measures not only improves the situation of farms with respect to specific pathogens but also improves the overall health of farms. In the specific case of *Salmonella* spp., in intensively reared white pigs and in intensive poultry farming, numerous biosafety protocols and practical guides have been described both at the farm and slaughterhouse level. The most critical points are related to cleaning and disinfection protocols [3].

Salmonella Cleaning and Disinfection Protocols

The objective of sanitation is to clean and disinfect equipment and materials that enter or remain on farms, including the personal hygiene of farm staff. Following the sanitation program guidelines helps to exclude the presence of pathogens on the farms before they can be spread [126]. As an example, the efficacy of disinfectant misting in the lairage of a pig abattoir to reduce *Salmonella* in pigs prior to slaughter has been reported [127]. This comprises the following: (1) application of high-pressure water to remove organic matter; (2) use of detergent with rinse (e.g., sodium hydroxide or hypochlorite); (3) use of disinfectant without rinsing (e.g., chlorocresol or quaternary ammonium); (4) drying for at least 24–48 h; and (5) fumigation based on cypermethrin.

Other Aspects Related to Biosecurity

In this sense, it is necessary to highlight the correct control of rodents on the farm as a basic topic within the *Salmonella* control program. Rodents can carry not only *Salmonella* but also a host of microorganisms: *Campylobacter* spp., *Lawsonia intracellularis*, *Leptospira* spp., *Brucella* spp., *Triquinella spiralis*, and porcine reproductive and respiratory syndrome (PRRS) virus [126].

The control of rodents must be based on good knowledge of their ethology, to locate the refuge points, breeding nests, and passage areas on the farm and to effectively use baits with authorized rodenticide products. All the actions carried out in the rat extermination program must be registered. Likewise, evaluation and verification of the program should be carried out periodically to make modifications if a reduction in the effectiveness of the strategy followed or product used is detected. Finally, it is necessary to indicate that rodenticides must be replaced periodically to avoid tolerances [126].

Wild birds can also act as authentic amplifying reservoirs of different *Salmonella* serotypes. The implication of different *Salmonella* serovars transmitted through birds (e.g., pigeons, turtledoves) acting as the main vectors in disease outbreaks in farms has been reported [16]. Therefore, in livestock farms it is necessary to use anti-bird mesh on windows and access points as well as other preventive biosecurity measures such as closed warehouse doors, permanently closed silo lids, closed feed, and raw material stores, to prevent the access of birds [3].

6. Conclusions

Efforts to reduce the transmission of *Salmonella* through food and other routes must be implemented using a One Health approach. The control of salmonellosis is based on two fundamental aspects: the reduction of prevalence levels in animals and the protection of humans from infection.

At the food chain level, the prevention of salmonellosis requires a comprehensive approach at farm, manufacturing, distribution, and consumer levels. Food operators and health authorities play a crucial role in preventing *Salmonella* transmission to consumers by ensuring safe food handling, monitoring and enforcing hygiene standards, and swiftly responding to foodborne outbreaks. Their collaboration safeguards public health and reduces the risk of foodborne illness, underscoring the importance of their roles in safeguarding food safety.

A significant concern is the rise of *Salmonella* MDR strains, which are responsible for foodborne illnesses that are common throughout the world. In fact, multidrug resistance (MDR) complicates the use of antibiotics to treat infections, raises healthcare expenses, lengthens hospital stays, and increases mortality.

Finally, several environmental and management factors have been associated with high levels of *Salmonella* spp. in the animal population. Based on these risk factors, different prevention and control methods related to hygiene and management, health and biosafety, animal welfare, and feeding strategies have been proposed.

Author Contributions: Conceptualization, R.J.A.M.; methodology, R.J.A.M., Á.G.-R., B.H.L., L.G.-G., M.Á.M.R., E.C.J., F.P.R. and A.V.D.; investigation, R.J.A.M., Á.G.-R., B.H.L., L.G.-G., M.Á.M.R., E.C.J., F.P.R. and A.V.D.; writing original—draft preparation, R.J.A.M., Á.G.-R. and A.V.D.; writing—review and editing R.J.A.M., Á.G.-R. and A.V.D.; supervision, R.J.A.M. All authors have read and agreed to the published version of the manuscript.

Funding: This research received no external funding.

Institutional Review Board Statement: Not applicable.

Informed Consent Statement: Not applicable.

Data Availability Statement: Not applicable.

Acknowledgments: The authors would like to acknowledge the help of Barry John Fagan for the English translation.

Conflicts of Interest: The authors declare that they have no known competing financial interest or personal relationships that could have appeared to influence the work reported in this review.

References

1. World Health Organization (WHO). *Critically Important Antimicrobials for Human Medicine, 6th Revision*; World Health Organization: Geneva, Switzerland, 2018.
2. Majowicz, S.E.; Musto, J.; Scallan, E.; Angulo, F.J.; Kirk, M.; O'Brien, S.J.; Jones, T.F.; Fazil, A.; Hoekstra, R.M. The global burden of nontyphoidal *Salmonella gastroenteritis*. *Clin. Infect. Dis.* **2010**, *50*, 882–889. [CrossRef]
3. Astorga, R.J. *Sanidad Animal y Salud Pública: El Paradigma de Salmonella*; Editorial Amazing Books: Zaragoza, Spain, 2020; pp. 1–282, ISBN 978-84-17403-70-6.
4. EFSA (European Food Safety Authority); ECDC (European Centre for Disease Prevention and Control). The European Union One Health 2021 Zoonoses Report. *EFSA J.* **2022**, *20*, 7666. [CrossRef]
5. Gosling, R.J.; Mueller-Doblies, D.; Martelli, F.; Nunez-García, J.; Kell, N.; Rabie, A.; Wales, A.D.; Davies, R.H. Observations on the distribution and persistence of monophasic *Salmonella Typhimurium* on infected pig and cattle farms. *Vet. Microbiol.* **2018**, *227*, 90–96. [CrossRef]
6. Meneguzzi, M.; Pissetti, C.; Rebelatto, R.; Trachsel, J.; Kuchiishi, S.S.; Reis, A.T.; Guedes, R.M.C.; Leão, J.A.; Reichen, C.; Kich, J.D. Re-Emergence of Salmonellosis in Hog Farms: Outbreak and Bacteriological Characterization. *Microorganisms* **2021**, *9*, 947. [CrossRef]
7. Bonifait, L.; Thépault, A.; Baugé, L.; Rouxel, S.; Le Gall, F.; Chemaly, M. Occurrence of *Salmonella* in the Cattle Production in France. *Microorganisms* **2021**, *9*, 872. [CrossRef]
8. Wei, L.; Yang, C.; Shao, W.; Sun, T.; Wang, J.; Zhou, Z.; Chen, C.; Zhu, A.; Pan, Z. Prevalence, and drug resistance of *Salmonella* in dogs and cats in Xuzhou, China. *J. Vet. Res.* **2020**, *64*, 263. [CrossRef]
9. Rosario, I.; Calcines, M.I.; Rodríguez-Ponce, E.; Déniz, S.; Real, F.; Vega, S.; Marín, C.; Padilla, D.; Martín, J.L.; Acosta-Hernández, B. *Salmonella enterica* subsp. *enterica* serotypes isolated for the first time in feral cats: The impact on public health. *Comp. Immunol. Microbiol. Infect. Dis.* **2022**, *84*, 101792. [CrossRef]
10. Zajac, M.; Skarzynska, M.; Lalak, A.; Kwit, R.; Smialowska-Weglinska, A.; Pasim, P.; Szulowski, K.; Wasyl, D. *Salmonella* in Captive Reptiles and Their Environment—Can We Tame the Dragon? *Microorganisms* **2021**, *9*, 1012. [CrossRef]
11. Marin, C.; Martín-Maldonado, B.; Cerdà-Cuéllar, M.; Sevilla-Navarro, S.; Lorenzo-Rebenaque, L.; Montoro-Dasi, L.; Manzanares, A.; Ayats, T.; Mencía-Gutiérrez, A.; Jordá, J.; et al. Antimicrobial Resistant *Salmonella* in Chelonians: Assessing Its Potential Risk in Zoological Institutions in Spain. *Vet. Sci.* **2022**, *9*, 264. [CrossRef]
12. Santana-Hernández, K.M.; Rodríguez-Ponce, E.; Rosario Medina, I.; Acosta-Hernández, B.; Priestnall, S.L.; Vega, S.; Marin, C.; Cerdà-Cuéllar, M.; Marco-Fuertes, A.; Ayats, T.; et al. One Health Approach: Invasive California Kingsnake (*Lampropeltis californiae*) as an Important Source of Antimicrobial Drug-Resistant *Salmonella* Clones on Gran Canary Island. *Animals* **2023**, *13*, 1790. [CrossRef]
13. Graziani, C.; Losasso, C.; Luzzi, I.; Ricci, A.; Scavia, G.; Pasquali, P. Salmonella. In *Foodborne Diseases*, 3rd ed.; Elsevier Inc.: Amsterdam, The Netherlands, 2017; pp. 133–169.
14. Hernández, M.; Gómez-Laguna, J.; Luque, I.; Herrera-León, S.; Maldonado, A.; Reguillo, L.; Astorga, R.J. *Salmonella* prevalence and characterization in a free-range pig processing plant: Tracking in trucks, lairage, slaughter line and quartering. *Int. J. Food Microbiol.* **2013**, *162*, 48–54. [CrossRef] [PubMed]
15. Valdezate, S.; Astorga, R.; Herrera-León, S.; Usera, M.A.; Huerta, B.; Echeita, A. Epidemiological tracing of *Salmonella enterica* serotype Abortusovis from Spanish ovine flocks by PFGE fingerprinting. *Epidemiol. Infect.* **2007**, *135*, 695–702. [CrossRef] [PubMed]
16. Luque, I.; Echeita, A.; León, J.; Herrera-León, S.; Tarradas, C.; González-Sanz, R.; Huerta, B.; Astorga, R.J. *Salmonella Indiana* as a cause of abortion in ewes: Genetic diversity and resistance patterns. I. *Vet. Microbiol.* **2009**, *134*, 396–399. [CrossRef] [PubMed]

17. Chui, L.; Ferrato, C.; Li, V.; Christianson, S. Comparison of Molecular and In Silico *Salmonella* Serotyping for *Salmonella* Surveillance. *Microorganisms* **2021**, *9*, 955. [CrossRef] [PubMed]
18. Didelot, X.; Fraser, C.; Gardy, J.; Colijn, C. Genomic Infectious Disease Epidemiology in Partially Sampled and Ongoing Outbreaks. *Mol. Biol. Evol.* **2017**, *34*, 997–1007. [CrossRef] [PubMed]
19. Arrieta-Gisasoa, A.; Atxaerandio-Landa, A.; Garrido, V.; Grilló, M.J.; Martínez-Ballesteros, I.; Laorden, L.; Garaizar, J.; Bikandi, J. Genotyping Study of *Salmonella* 4,[5],12:i:- Monophasic Variant of *Serovar Typhimurium* and Characterization of the Second-Phase Flagellar Deletion by Whole Genome Sequencing. *Microorganisms* **2020**, *8*, 2049. [CrossRef] [PubMed]
20. Wan Makhtar, W.R.; Bharudin, I.; Samsulrizal, N.H.; Yusof, N.Y. Whole Genome Sequencing Analysis of *Salmonella enterica* Serovar Typhi: History and Current Approaches. *Microorganisms* **2021**, *9*, 2155. [CrossRef]
21. EFSA Fact Sheet. Available online: https://www.efsa.europa.eu/sites/default/files/corporate_publications/files/factsheetsalmonella.pdf (accessed on 15 August 2023).
22. EC (European Commission). *Regulation (EC) N° 2160/2003 of the European Parliament and of the Council of 17 November 2003 on the Control of Salmonella and Other Specified Food-Borne Zoonotic Agents*; EC (European Commission): Brussels, Belgium, 2003.
23. EFSA. FBO Dashboard. Available online: https://www.efsa.europa.eu/en/microstrategy/FBO-dashboard (accessed on 15 August 2023).
24. Álvarez-Ordóñez, A.; Broussolle, V.; Colin, P.; Nguyen-The, C.; Prieto, M. The adaptive response of bacterial food-borne pathogens in the environment, host, and food: Implications for food safety. *Int. J. Food Microbiol.* **2015**, *213*, 99–109. [CrossRef]
25. Silva, F.V.M.; Gibbs, P.A. Thermal pasteurization requirements for the inactivation of *Salmonella* in foods. *Food Res. Int.* **2012**, *45*, 695–699. [CrossRef]
26. Dimakopoulou-Papazoglou, D.; Lianou, A.; Koutsoumanis, K.P. Modelling biofilm formation of *Salmonella enterica* ser. Newport as a function of pH and water activity. *Food Microbiol.* **2016**, *53*, 76–81. [CrossRef]
27. He, S.; Ye, B.; Zhang, Z.; Cui, Y.; Wang, S.; Shi, X. Cross-protective effect of acid adaptation on ethanol tolerance in *Salmonella* Enteritidis. *Food Sci. Hum. Wellness* **2023**, *12*, 1402–1407. [CrossRef]
28. Pye, H.V.; Thilliez, G.; Acton, L.; Kolenda, R.; Al-Khanaq, H.A.; Grove, S.; Kingsley, R.A. Strain and serovar variants of *Salmonella enterica* exhibit diverse tolerance to food chain-related stress. *Food Microbiol.* **2023**, *112*, 104237. [CrossRef] [PubMed]
29. Lianou, A.; Nychas, G.J.E.; Koutsoumanis, K.P. Variability in the adaptive acid tolerance response phenotype of *Salmonella enterica* strains. *Food Microbiol.* **2017**, *62*, 99–105. [CrossRef] [PubMed]
30. Álvarez-Ordóñez, A.; Fernández, A.; Bernardo, A.; López, M. Acid tolerance in *Salmonella typhimurium* induced by culturing in the presence of organic acids at different growth temperatures. *Food Microbiol.* **2010**, *27*, 44–49. [CrossRef] [PubMed]
31. Acuff, J.C.; Dickson, J.S.; Farber, J.M.; Grasso-Kelley, E.M.; Hedberg, C.; Lee, A.; Zhu, M. Practice and Progress: Updates on Outbreaks, Advances in Research, and Processing Technologies for Low-moisture Food Safety. *J. Food Prot.* **2023**, *86*, 100018. [CrossRef]
32. Beuchat, L.R.; Komitopoulou, E.; Beckers, H.; Betts, R.P.; Bourdichon, F.; Fanning, S.; Joosten, H.M.; Kuile, B.H.T. Low-water activity foods: Increased concern as vehicles of foodborne pathogens. *J. Food Prot.* **2013**, *76*, 150–172. [CrossRef]
33. Hiramatsu, R.; Matsumoto, M.; Sakae, K.; Miyazaki, Y. Ability of Shiga toxin-producing *Escherichia coli* and *Salmonella* spp. to survive in a desiccation model system and in dry foods. *Appl. Environ. Microbiol.* **2005**, *71*, 6657–6663. [CrossRef]
34. Podolak, R.; Enache, E.; Stone, W.; Black, D.G.; Elliot, P.H. Sources and risk factors for contamination, survival, persistence, and heat resistance of *Salmonella* in low-moisture foods. *J. Food Prot.* **2010**, *73*, 1919–1936. [CrossRef]
35. Liu, S.; Keller, S.E.; Anderson, N.M. Transfer of *Salmonella* from Inert Food Contact Surfaces to Wheat Flour, Cornmeal, and NaCl. *J. Food Prot.* **2022**, *85*, 231–237. [CrossRef]
36. Burgess, C.M.; Gianotti, A.; Gruzdev, N.; Holah, J.; Knochel, S.; Lehner, A.; Marges, E.; Esser, S.S.; Sela, S.; Tresse, O. The response of foodborne pathogens to osmotic and desiccation stresses in the food chain. *Int. J. Food Microbiol.* **2016**, *221*, 37–53. [CrossRef]
37. Finn, S.; Condell, O.; McClure, P.; Amezquita, A.; Fanning, S. Mechanisms of survival, responses, and sources of *Salmonella* in low-moisture environments. *Front. Microbiol.* **2013**, *4*, 331. [CrossRef] [PubMed]
38. Bremer, E.; Krämer, R. Responses of microorganisms to osmotic stress. *Annu. Rev. Microbiol.* **2019**, *73*, 313–334. [CrossRef] [PubMed]
39. Liu, S.; Roopesh, M.S.; Tang, J.; Wu, Q.; Qin, W. Recent development in low-moisture foods: Microbial safety and thermal process. *Food Res. Int.* **2022**, *155*, 111072. [CrossRef] [PubMed]
40. Santillana-Farakos, S.M.; Schaffner, D.W.; Frank, J.F. Predicting survival of *Salmonella* in low-water activity foods: An analysis of literature data. *J. Food Prot.* **2014**, *77*, 1448–1461. [CrossRef] [PubMed]
41. Hsu, H.; Sheen, S.; Sites, J.; Huang, L.; Wu, J.S.B. Effect of high-pressure treatment on the survival of Shiga toxin-producing *Escherichia coli* in strawberry puree. *Food Microbiol.* **2014**, *40*, 25–30. [CrossRef]
42. Gautam, B.; Govindan, B.N.; Ganzle, M.; Roopesh, M.S. Influence of water activity on the heat resistance of *Salmonella enterica* in selected low-moisture foods. *Int. J. Food Microbiol.* **2020**, *334*, 108813. [CrossRef]
43. Garcés-Vega, F.J.; Ryser, E.T.; Marks, B.P. Relationships of Water Activity and Moisture Content to the Thermal Inactivation Kinetics of *Salmonella* in Low-Moisture Foods. *J. Food Prot.* **2019**, *82*, 963–970. [CrossRef]
44. Steinbrunner, P.J.; Limcharoenchat, P.; Suehr, Q.J.; Ryser, E.T.; Marks, B.P.; Jeong, S. Effect of Food Structure, Water Activity, and Long-Term Storage on X-Ray Irradiation for Inactivating *Salmonella* Enteritidis PT30 in Low-Moisture Foods. *J. Food Prot.* **2019**, *82*, 1405–1411. [CrossRef]

45. Tadapaneni, R.K.; Syamaladevi, R.M.; Villa-Rojas, R.; Tang, J. Design of a novel test cell to study the influence of water activity on the thermal resistance of *Salmonella* in low-moisture foods. *J. Food Eng.* **2017**, *208*, 48–56. [CrossRef]
46. Fu, Y.; Deering, A.J.; Bhunia, A.K.; Yao, Y. Pathogen biofilm formation on cantaloupe rface and its impact on the antibacterial effect of lauryl arginate ethyl. *Food Microbiol.* **2017**, *64*, 139–144. [CrossRef]
47. Lee, K.H.; Lee, J.Y.; Roy, P.K.; Mizan, M.F.R.; Hossain, M.I.; Park, S.H.; Ha, S.D. Viability of *Salmonella* typhimurium biofilms on major food-contact surfaces and eggshell treated during 35 days with and without water storage at room temperature. *Poult. Sci.* **2020**, *99*, 4558–4565. [CrossRef] [PubMed]
48. Tadielo, L.E.; Bellé, T.H.; Dos Santos, E.A.R.; Schmiedt, J.A.; Cerqueira-Cézar, C.K.; Nero, L.A.; Yamatogi, R.S.; Pereira, J.G.; Bersot, L.S. Pure and mixed biofilms formation of *Listeria monocytogenes* and *Salmonella Typhimurium* on polypropylene surfaces. *LWT* **2022**, *162*, 113469. [CrossRef]
49. Solomon, E.B.; Niemira, B.A.; Sapers, G.M.; Annous, B.A. Biofilm Formation, Cellulose Production, and Curli Biosynthesis by *Salmonella* Originating from Produce, Animal, and Clinical Sources. *J. Food Prot.* **2005**, *68*, 906–912. [CrossRef] [PubMed]
50. Speranza, B.; Corbo, M.R.; Sinigaglia, M. Effects of nutritional and environmental conditions on *Salmonella* sp. biofilm formation. *J. Food Sci.* **2011**, *76*, 12–16. [CrossRef] [PubMed]
51. Merino, L.; Procura, F.; Trejo, F.M.; Bueno, D.J.; Golowcyc, M.A. Biofilm formation by *Salmonella* sp. in the poultry industry: Detection, control, and eradication strategies. *Food Res. Int.* **2019**, *119*, 530–540. [CrossRef] [PubMed]
52. Brackman, G.; Coenye, T. Quorum sensing inhibitors as anti-biofilm agents. *Curr. Pharm. Des.* **2015**, *21*, 5–11. [CrossRef] [PubMed]
53. Gao, F.; Lyu, C.; Ning, Z.; Zhao, S.; Shao, L.; Xu, X.; Wang, H. Inactivation of *Salmonella* biofilms formed on stainless steel surfaces by pulsed light. *Food Control* **2023**, *153*, 109955. [CrossRef]
54. Byun, K.H.; Na, K.W.; Ashrafudoulla, M.; Choi, M.W.; Ha Han, S.; Kang, I.; Park, S.H.; Ha, S.D. Combination treatment of peroxyacetic acid or lactic acid with UV-C to control *Salmonella* Enteritidis biofilms on food contact surface and chicken skin. *Food Microbiol.* **2022**, *102*, 103906. [CrossRef]
55. Ashrafudoulla, M.; Kim, H.J.; Her, E.; Shaila, S.; Park, S.D.; Ha, S.D. Characterization of *Salmonella* Thompson-specific bacteriophages, and their preventive application against *Salmonella* Thompson biofilm on eggshell as a promising antimicrobial agent in the food industry. *Food Control* **2023**, *154*, 110008. [CrossRef]
56. Wang, S.; Mirmiran, S.D.; Li, X.; Li, X.; Zhang, F.; Duan, X.; Gao, D.; Chen, Y.; Chen, H.; Qian, P. Temperate phage influence virulence and biofilm-forming of *Salmonella Typhimurium* and enhance the ability to contaminate food product. *Int. J. Food Microbiol.* **2023**, *398*, 110223. [CrossRef]
57. Asma, S.T.; Imre, K.; Morar, A.; Imre, M.; Acaroz, U.; Ali Shad, S.R.; Hussain, S.Z.; Arslan-Acaroz, D.; Istanbullugil, F.R.; Madani, K.; et al. Natural strategies as potential weapons against bacterial biofilms. *Life* **2022**, *12*, 1618. [CrossRef] [PubMed]
58. Mgomi, F.C.; Yang, Y.; Cheng, G.; Yang, Z. Lactic acid bacteria biofilms and their antimicrobial potential against pathogenic microorganisms. *Biofilm* **2023**, *5*, 110118. [CrossRef] [PubMed]
59. Pérez-Rodríguez, F.; Valero, A. Predictive Microbiology in Foods. In *Predictive Microbiology in Foods. SpringerBriefs in Food, Health, and Nutrition*; Springer: New York, NY, USA, 2015; Volume 5, 128p. [CrossRef]
60. Lobacz, A.; Zulewska, J.; Kowalik, J. Chapter 3—Predictive microbiology and risk analysis. In *Dairy Foods. Processing, Quality and Analytical Techniques*; Elsevier Inc.: Amsterdam, The Netherlands, 2022; pp. 47–68. [CrossRef]
61. McQuestin, O.J.; Musgrove, M.T.; Tamplin, M.L. Kinetics of growth and inactivation of *Salmonella* enterica serotype Typhimurium DT104 in pasteurised liquid egg products. *Food Microbiol.* **2010**, *27*, 396–402. [CrossRef] [PubMed]
62. Pasquali, F.; Klein, G.; Reich, F.; Manfreda, G.; Valero, A. Modelling survival behaviour of *Salmonella enterica* ser. Enteritidis, Typhimurium and Tennessee on table eggs during storage at different temperatures. *Food Control* **2016**, *59*, 314–319. [CrossRef]
63. Bermúdez-Aguirre, D.; Niemira, B.A. Effect of nisin, EDTA, and abuse temperature on the growth of Salmonella Typhimurium in liquid whole egg during refrigerated storage. *Food Res. Int.* **2023**, *174*, 113568. [CrossRef] [PubMed]
64. Oscar, T.P. Neural network model for survival and growth of Salmonella enterica serotype 8, 20:–:z6 in ground chicken thigh meat during cold storage: Extrapolation to other serotypes. *J. Food Prot.* **2015**, *78*, 1819–1827. [CrossRef] [PubMed]
65. Coroller, L.; Jeuge, S.; Couvert, O.; Christieans, S.; Ellouze, M. Extending the gamma concept to non-thermal inactivation: A dynamic model to predict the fate of *Salmonella* during the dried sausages process. *Food Microbiol.* **2015**, *45*, 266–275. [CrossRef]
66. Milkievicz, T.; Badia, V.; Souza, V.B.; Longhi, D.A.; Galvao, A.C.; Robazza, W.S. Modeling *Salmonella* spp. inactivation in chicken meat subjected to isothermal and non-isothermal temperature profiles. *Int. J. Food Microbiol.* **2021**, *344*, 109110. [CrossRef]
67. Li, D.; Friedrich, L.M.; Danyluk, M.D.; Harris, L.J.; Schaffner, D.W. Development and validation of a mathematical model for growth of pathogens in cut melons. *J. Food Prot.* **2013**, *76*, 953–958. [CrossRef]
68. Ma, C.; Li, J.; Zhang, Q. Behavior of *Salmonella* spp. on fresh-cut tropical fruits. *Food Microbiol.* **2016**, *54*, 133–141. [CrossRef]
69. Saha, J.; Topalcengiz, Z.; Sharma, V.; Friedrich, L.M.; Danyluk, M.D. Fate and growth kinetics of Salmonella and Listeria monocytogenes on mangoes during storage. *J. Food Prot.* **2023**, *86*, 100151. [CrossRef]
70. Santillana-Farakos, S.M.; Frank, J.F.; Schaffner, D.W. Modeling the influence of temperature, water activity and water mobility on the persistence of *Salmonella* in low-moisture foods. *Int. J. Food Microbiol.* **2013**, *166*, 280–293. [CrossRef]
71. Santillana-Farakos, S.M.; Pouillot, R.; Keller, S.E. Salmonella survival kinetics on pecans, hazelnuts, and pine nuts at various water activities and temperatures. *J. Food Prot.* **2017**, *80*, 879–885. [CrossRef] [PubMed]
72. Igo, M.J.; Schaffner, D.W. Models for factors influencing pathogen survival in low water activity foods from literature data are highly significant but show large unexplained variance. *Food Microbiol.* **2021**, *98*, 103783. [CrossRef] [PubMed]

73. Dhaliwal, H.K.; Gänzle, M.; Roopesh, M.S. Influence of drying conditions, food composition, and water activity on the thermal resistance of *Salmonella enterica*. *Food Res. Int.* **2021**, *147*, 110548. [CrossRef] [PubMed]
74. Mishra, A.; Guo, M.; Buchanan, R.L.; Schaffner, D.W.; Pradhan, A.K. Development of growth and survival models for *Salmonella* and *Listeria monocytogenes* during non-isothermal time-temperature profiles in leafy greens. *Food Control* **2017**, *71*, 32–41. [CrossRef]
75. Possas, A.; Posada-Izquierdo, G.D.; Tarlak, F.; Jiménez-Jiménez, F.; Pérez-Rodríguez, F. Inactivation of *Salmonella Typhimurium* in fresh-cut lettuce during chlorine washing: Assessing the impacts of free chlorine concentrations and exposure times. *LWT* **2023**, *184*, 115069. [CrossRef]
76. Carrasco, E.; Del Rosal, S.; Racero, J.C.; García-Gimeno, R.M. A review on growth/no growth *Salmonella* models. *Food Res. Int.* **2012**, *47*, 90–99. [CrossRef]
77. 2012b: Carrasco, E.; Morales-Rueda, A.; García-Gimeno, R.M. Cross-contamination and recontamination by *Salmonella* in foods: A review. *Food Res. Int.* **2012**, *45*, 545–556. [CrossRef]
78. Possas, A.; Carrasco, E.; García-Gimeno, R.M.; Valero, A. Models of microbial cross-contamination dynamics. *Curr. Opin. Food Sci.* **2017**, *14*, 43–49. [CrossRef]
79. Shahdadi, M.; Safarirad, M.; Berizi, E.; Mazloomi, S.M.; Hosseinzadeh, S.; Zare, M.; Derakhshan, Z.; Rajabi, S. A systematic review and modeling of the effect of bacteriophages on *Salmonella* spp. Reduction in chicken meat. *Heliyon* **2023**, *9*, e14870. [CrossRef] [PubMed]
80. Austrich-Comas, A.; Jofré, A.; Gou, P.; Bover-Cid, S. Assessing the impact of different technological strategies on the fate of *Salmonella* in chicken dry-fermented sausages by means of challenge testing and predictive models. *Microorganisms* **2023**, *11*, 432. [CrossRef] [PubMed]
81. Bermúdez-Aguirre, D.; Niemira, B.A. Microbial inactivation models of *Salmonella Typhimurium* in radio frequency treated eggs. *Food Control* **2023**, *148*, 109634. [CrossRef]
82. Alvarenga, V.O.; Brito, L.M.; Lacerda, I.C.A. Application of mathematical models to validate emerging processing technologies in food. *Curr. Opin. Food Sci.* **2022**, *48*, 100928. [CrossRef]
83. Karanth, S.; Tanui, C.K.; Meng, J.; Pradhan, A. Exploring the predictive capability of advanced machine learning in identifying severe disease phenotype in *Salmonella enterica*. *Food Res.* **2022**, *151*, 110817. [CrossRef] [PubMed]
84. Munck, N.; Njage, P.M.K.; Leekitcharoenphon, P.; Litrup, E.; Hald, T. Application of Whole-Genome Sequences and Machine Learning in Source Attribution of *Salmonella Typhimurium*. *Risk Anal.* **2020**, *40*, 1693–1705. [CrossRef]
85. Fritsch, L.; Guillier, L.; Augustin, J.C. Next generation quantitative microbiological risk assessment: Refinement of the cold smoked salmon-related listeriosis risk model by integrating genomic data. *Microb. Risk Anal.* **2018**, *10*, 20–27. [CrossRef]
86. Karanth, S.; Patel, J.; Shirmohammadi, A.; Pradhan, A.K. Machine learning to predict foodborne salmonellosis outbreaks based on genome characteristics and meteorological trends. *Curr. Res. Food Sci.* **2023**, *6*, 100525. [CrossRef]
87. Possas, A.; Valero, A.; Pérez-Rodríguez, F. New software solutions for microbiological food safety assessment and management. *Curr. Opin. Food Sci.* **2022**, *44*, 100814. [CrossRef]
88. MicroHibro 3.0. Available online: www.microhibro.com (accessed on 30 August 2023).
89. Cubero González, S.; Possas, A.; Carrasco, E.; Valero, A.; Bolivar, A.; Posada-Izquierdo, G.D.; García-Gimeno, R.M.; Zurera, G.; Pérez-Rodríguez, F. 'Micro-Hibro': A software tool for predicitive microbiology and microbial risk assessment in foods. *Int. J. Food Microbiol.* **2019**, *290*, 226–236. [CrossRef]
90. Pin, C.; Avendaño-Pérez, G.; Cosciani-Cunico, E.; Gómez, N.; Gounadakic, A.; Nychas, G.J.; Skandamis, P.; Barker, G. Modelling *Salmonella* concentration throughout the pork supply chain by considering growth and survival in fluctuating conditions of temperature, pH, and a(w). *Int. J. Food Microbiol.* **2011**, *145*, S96–S102. [CrossRef] [PubMed]
91. Ross, T. Indices for performance evaluation of predictive models in food microbiology. *J. Appl. Microbiol.* **1996**, *81*, 501–508. [CrossRef]
92. Gumudavelli, V.; Subbiah, J.; Thippareddi, H.; Velugoti, P.R.; Froning, G. Dynamic predictive model for growth of *Salmonella enteritidis* in egg yolk. *J. Food Sci.* **2007**, *72*, M254–M262. [CrossRef] [PubMed]
93. Foley, S.L.; Lynne, A.M. Food animal-associated *Salmonella* challenges: Pathogenicity and antimicrobial resistance. *J. Anim. Sci.* **2008**, *86*, 173–187. [CrossRef] [PubMed]
94. Switt, A.I.M.; Soyer, Y.; Warnick, L.D.; Wiedmann, M. Emergence, distribution and molecular and phenotypic characteristics of *Salmonella enterica* serotype 4,5,12:i:-. *Foodborne Pathog. Dis.* **2009**, *6*, 407–415. [CrossRef]
95. Martins da Costa, P.; Oliveira, M.; Bica, A.; Vaz-Pires, P.; Bernardo, F. Antimicrobial resistance in *Enterococcus* spp. and *Escherichia coli* isolated from poultry feed and feed ingredients. *Vet. Microbiol.* **2007**, *120*, 122–131. [CrossRef]
96. European Centre for Disease Prevention and Control (ECDC). *Assessing the Health Burden of Infections with Antibiotic-Resistant Bacteria in the EU/EEA, 2016–2020*; ECDC: Stockholm, Sweden, 2022.
97. Sayah, R.S.; Kaneene, J.B.; Johnson, Y.; Miller, R.A. Patterns of antimicrobial resistance observed in *Escherichia coli* isolates obtained from domestic- and wild-animal fecal samples, human septage, and surface water. *Appl. Environ. Microbiol.* **2005**, *71*, 1394–1404. [CrossRef]
98. Gargano, V.; Gambino, D.; Migliore, S.; Vitale, M.; Sciortino, S.; Costa, A.; Vicari, D. Can Human Handling Increase the Presence of Multidrug Resistance (MDR) in *Salmonella* spp. Isolated from Food Sources? *Microorganisms* **2021**, *9*, 2018. [CrossRef]

99. Astorga Márquez, R.J.; Echeita Salaberria, A.; Maldonado García, A.; Valdezate-Jiménez, S.; Carbonero Martínez, A.; Aladueña García, A.; Arenas Casas, A. Surveillance and Antimicrobial Resistance of *Salmonella* Strains Isolated from Slaughtered Pigs in Spain. *J. Food Protec.* **2007**, *70*, 1502–1506. [CrossRef]
100. Gómez-Laguna, J.; Hernández, M.; Creus, E.; Echeita, A.; Otal, J.; Herrera-Léon, S.; Astorga, R. Prevalence and antimicrobial susceptibility of Salmonella infections in free-range pigs. *Vet. J.* **2011**, *190*, 176–178. [CrossRef]
101. Galán-Relaño, A.; Sánchez-Carvajal, J.M.; Gómez-Gascón, L.; Vera Salmoral, E.; Huerta Lorenzo, B.; Cardoso-Toset, F.; Gómez-Laguna, J.; Astorga, R.J. Phenotypic and genotypic antibiotic resitance patterns in *Salmonella Typhimurium* and its monophasic variant form pigs in southern Spain. *Res. Vet. Sci.* **2022**, *152*, 596–603. [CrossRef]
102. Montoro-Dasi, L.; Lorenzo-Rebenaque, L.; Marco-Fuertes, A.; Vega, S.; Marín, C. Holistic Strategies to Control *Salmonella* Infantis: An Emerging Challenge in the European Broiler Sector. *Microorganisms* **2023**, *11*, 1765. [CrossRef] [PubMed]
103. Moretro, T.; Heir, E.; Nesse, L.L.; Vestby, L.K.; Langsrud, S. Control of *Salmonella* in food related environments by chemical disinfection. *Food Res. Int.* **2012**, *45*, 532–544. [CrossRef]
104. The European Food Safety Authority (EFSA); European Centre for Disease Prevention and Control (ECDC). The European Union Summary Report on Antimicrobial Resistance in zoonotic and indicator bacteria from humans, animals, and food in 2020/2021. *EFSA J.* **2023**, *21*, 7867. [CrossRef]
105. Ruvalcaba-Gómez, J.M.; Villagrán, Z.; Valdez-Alarcón, J.J.; Martínez-Núñez, M.; Gomez-Godínez, L.J.; Ruesga-Gutiérrez, E.; Anaya-Esparza, L.M.; Arteaga-Garibay, R.I.; Villarruel-López, A. Non-Antibiotics Strategies to Control *Salmonella* Infection in Poultry. *Animals* **2022**, *12*, 102. [CrossRef] [PubMed]
106. Raut, R.; Maharjan, P.; Foudladkhah, A.C. Practical preventive considerations for reducing the public health burden of poultry-related salmonellosis. *Int. J. Environ. Res. Public Health* **2023**, *20*, 6654. [CrossRef] [PubMed]
107. Solarte, A.L.; Astorga, R.J.; de Aguiar, F.C.; Tarradas, C.; Luque, I.; Gómez-Gascón, L.; Huerta, B. Reduced Susceptibility of *Salmonella Typhimurium* Strains to Oregano Essential Oil and Enrofloxacin: An In Vitro Assay. *Foodborne Pathog. Dis.* **2020**, *17*, 29–34. [CrossRef]
108. Rochín-Medina, J.J.; Mendoza-López, I.A.; Campo, N.C.-D.; Bastidas-Bastidas, P.J.; Ramírez, K. Activity of plant essential oils against clinically and environmentally isolated *Salmonella enterica* serotypes: In vitro assays and molecular docking. *Lett. Appl. Microbiol.* **2023**, *3*, ovad045. [CrossRef]
109. Solarte, A.L.; Astorga, R.J.; de Aguiar, F.C.; De Frutos, C.; Barrero-Domínguez, B.; Huerta, B. Susceptibility Distribution to Essential Oils of *Salmonella enterica* Strains Involved in Animal and Public Health and Comparison of the Typhimurium and Enteritidis Serotypes. *J. Med. Food.* **2018**, *21*, 946–950. [CrossRef]
110. Čabarkapa, I.; Čolović, R.; Đuragić, O.; Popović, S.; Kokić, B.; Milanov, D.; Pezo, L. Anti-biofilm activities of essential oils rich in carvacrol and thymol against *Salmonella* Enteritidis. *Biofouling* **2019**, *35*, 361–375. [CrossRef]
111. Al-Nabulsi, A.A.; Osaili, T.M.; Olaimat, A.N.; Almasri, W.E.; Ayyash, M.; Al-Holy, M.A.; Jaradat, Z.W.; Obaid, R.S.; Holley, R.A. Inactivation of *Salmonella* spp. in tahini using plant essential oil extracts. *Food Microbiol.* **2020**, *86*, 103338. [CrossRef] [PubMed]
112. Da Silva, J.P.L.; de Souza, E.F.; Della Modesta, R.C.; Gomes, I.A.; Freitas-Silva, O.; de Melo Franco, B.D.G. Antibacterial activity of nisin, oregano essential oil, EDTA, and their combination against *Salmonella* Enteritidis for application in mayonnaise. *Vigil. Sanit. Debate Soc. Cienc. Tecnol.* **2016**, *4*, 83–91. [CrossRef]
113. Solarte, A.L.; Astorga, R.J.; Aguiar, F.; Galán-Relaño, Á.; Maldonado, A.; Huerta, B. Combination of antimicrobials and essential oils as an alternative for the control of *Salmonella enterica* multiresistant strains related to foodborne disease. *Foodborne Pathog. Dis.* **2017**, *14*, 558–563. [CrossRef] [PubMed]
114. Abedon, S.T.; Loc-Carrillo, C. Pros, and cons of phage therapy. *Bacteriophage* **2011**, *1*, 111–114.
115. Nilsson, A.S. Phage therapy—Constraints and possibilities. *Upsala J. Med. Sci.* **2014**, *119*, 192–198. [CrossRef] [PubMed]
116. Wong, C.L.; Sieo, C.C.; Tan, W.S.; Abdullah, N.; Hair-Bejo, M.; Abu, J.; Ho, Y.W. Evaluation of a lytic bacteriophage, Φ st1, for biocontrol of Salmonella enterica serovar Typhimurium in chickens. *Int. J. Food Microbiol.* **2014**, *172*, 92–101. [CrossRef]
117. Sevilla-Navarro, S.; Catalá-Gregori, P.; García, C.; Cortés, V.; Marin, C. *Salmonella* Infantis and *Salmonella* Enteritidis specific bacteriophages isolated form poultry faeces as a complementary tool for cleaning and disinfection against *Salmonella*. *Comp. Immunol. Microbiol. Infect. Dis.* **2019**, *68*, 101405. [CrossRef]
118. Sevilla-Navarro, S.; Catalá-Gregori, P.; Marin, C. *Salmonella* Bacteriophage Diversity According to Most Prevalent *Salmonella* Serovars in Layer and Broiler Poultry Farms from Eastern Spain. *Animals* **2020**, *10*, 1456. [CrossRef]
119. Lorenzo-Rebenaque, L.; Malik, D.J.; Catalá-Gregori, P.; Marin, C.; Sevilla-Navarro, S. In Vitro and In Vivo Gastrointestinal Survival of Non-Encapsulated and Microencapsulated *Salmonella* Bacteriophages: Implications for Bacteriophage Therapy in Poultry. *Pharmaceuticals* **2021**, *6*, 434. [CrossRef]
120. Lorenzo-Rebenaque, L.; Malik, D.J.; Catalá-Gregori, P.; Torres-Boncompte, J.; Marin, C.; Sevilla-Navarro, S. Microencapsulated bacteriophages incorporated in feed for *Salmonella* control in broilers. *Vet. Microbiol.* **2022**, *274*, 109579. [CrossRef]
121. Sáez, A.C.; Zhang, J.; Rostagno, M.H.; Ebner, P.D. Direct feeding of microencapsulated bacteriophages to reduce *Salmonella* colonization in pigs. *Foodborne Pathog. Dis.* **2011**, *8*, 1269–1274. [CrossRef] [PubMed]
122. Renu, S.; Han, Y.; Dhakal, S.; Lakshmanappa, Y.S.; Ghimire, S.; Feliciano-Ruiz, N.; Senapati, S.; Narasimhan, B.; Selvaraj, R.; Renukaradhya, G.J. Chitosan-adjuvanted *Salmonella* subunit nanoparticle vaccine for poultry delivered through drinking water and feed. *Carbohydr. Polym.* **2020**, *243*, 116434. [CrossRef] [PubMed]
123. Rabie, N.S.; Amin Girh, Z.M.S. Bacterial vaccines in poultry. *Bull. Natl. Res. Cent.* **2020**, *44*, 15. [CrossRef] [PubMed]

124. Jia, S.; McWhorter, A.R.; Andrews, D.M.; Underwood, G.J.; Chousalkar, K.K. Challenges in Vaccinating Layer Hens against *Salmonella* Typhimurium. *Vaccines* **2020**, *8*, 696. [CrossRef]
125. Conan, A.; Goutard, F.L.; Sorn, S.; Vong, S. Biosecurity measures for backyard poultry in developing countries: A systematic review. *BMC Vet. Res.* **2012**, *8*, 240. [CrossRef]
126. Astorga, R.J.; Vega, S.; Argüello, H. *Bioseguridad en Producción Porcina*; Editorial SERVET; Grupo Asís Biomedia SL: Zaragoza, Spain, 2021; pp. 1–82. ISBN 978-84-18498-98-5.
127. Walia, K.; Argüello, H.; Lynch, H.; Grant, J.; Leonard, F.C.; Lawlor, P.G.; Gardiner, G.E.; Duffy, G. The efficacy of different cleaning and disinfection proce- dures to reduce *Salmonella* and *Enterobacteriaceae* in the lairage environment of a pig abattoir. *Int. J. Food Microbiol.* **2017**, *4*, 64–71. [CrossRef]

animals

MDPI

Review

Phage-Based Biosanitation Strategies for Minimizing Persistent *Salmonella* and *Campylobacter* Bacteria in Poultry

Jaume Jordá [1], Laura Lorenzo-Rebenaque [2], Laura Montoro-Dasi [1], Ana Marco-Fuertes [1], Santiago Vega [1] and Clara Marin [1,*]

[1] Departamento de Producción y Sanidad Animal, Salud Pública Veterinaria y Ciencia y Tecnología de los Alimentos, Facultad de Veterinaria, Instituto de Ciencias Biomédicas, Universidad Cardenal Herrera-CEU, CEU Universities, Calle Santiago Ramón y Cajal 20, 46115 Alfara del Patriarca, Spain; jaume.jorda@uchceu.es (J.J.); laura.montoro@uchceu.es (L.M.-D.); ana.marcofuertes@uchceu.es (A.M.-F.); svega@uchceu.es (S.V.)

[2] Institute of Animal Science and Technology, Universitat Politècnica de València, 46022 Valencia, Spain; laulore@upv.es

* Correspondence: clara.marin@uchceu.es

Simple Summary: Implementing strategies to reduce harmful bacteria in food animal production plays a vital role in ensuring safer food for consumers. The food industry, especially in poultry and swine farming, faces significant challenges, including antibiotic-resistant and disinfection-resistant zoonotic bacteria. In this context, bacteriophages, which are viruses that attack bacteria, have emerged as a promising tool to control these bacteria throughout the food production process, from the animals and farms to the end product. Bacteriophages offer several advantages as a biocontrol agent, including their precision in targeting specific bacteria, ability to replicate, adaptability, low risk of toxicity and ease of isolation. Developing them as a biocontrol tool is particularly appealing because it aligns with the use of environmentally friendly "green" technology to combat harmful bacteria. This review provides a comprehensive overview of how phage-based strategies can help reduce persistent *Salmonella* and *Campylobacter* bacteria in poultry, contributing to safer food production.

Abstract: Control strategies to minimize pathogenic bacteria in food animal production are one of the key components in ensuring safer food for consumers. The most significant challenges confronting the food industry, particularly in the major poultry and swine sectors, are antibiotic resistance and resistance to cleaning and disinfection in zoonotic bacteria. In this context, bacteriophages have emerged as a promising tool for zoonotic bacteria control in the food industry, from animals and farm facilities to the final product. Phages are viruses that infect bacteria, with several advantages as a biocontrol agent such as high specificity, self-replication, self-limitation, continuous adaptation, low inherent toxicity and easy isolation. Their development as a biocontrol agent is of particular interest, as it would allow the application of a promising and even necessary "green" technology to combat pathogenic bacteria in the environment. However, bacteriophage applications have limitations, including selecting appropriate phages, legal restrictions, purification, dosage determination and bacterial resistance. Overcoming these limitations is crucial to enhance phage therapy's effectiveness against zoonotic bacteria in poultry. Thus, this review aims to provide a comprehensive view of the phage-biosanitation strategies for minimizing persistent *Salmonella* and *Campylobacter* bacteria in poultry.

Keywords: bacteriophages; *Salmonella*; *Campylobacter*; persistent clones; poultry

1. Introduction

Salmonella and *Campylobacter* constitute the most common zoonotic pathogens involved in human outbreaks of foodborne disease in the developed world, followed by other important pathogens such as *Escherichia coli*, *Listeria* and *Vibrio cholerae* [1,2]. Even though many recent outbreaks have been linked to pork meat, fresh products and fruit, human clinical cases and outbreaks are still often associated with and attributed to poultry meat and eggs [3,4].

Since the mid-20th century, *Salmonella* has been acknowledged as a zoonotic pathogen with a mass impact on public health worldwide [5,6]. In Europe, a total of 773 salmonellosis foodborne outbreaks were reported across 27 European countries in 2021, resulting in 6755 cases, 1123 hospitalizations and 1 death [1]. Moreover, it is assumed that only 1 in 7 cases is diagnosed [7–9], since up to 80% of salmonellosis cases are not associated with a known outbreak; they are indeed considered sporadic illnesses, and the majority go undiagnosed [10]. Although different serotypes have been associated with salmonellosis, recent studies have reported *Salmonella enterica* subsp. *enterica* serovar Infantis as being widely spread through chicken meat [1,11] and *Salmonella enterica* subsp. *enterica* serovar Enteritidis through eggs [1,12]. Epidemiological surveillance suggests that the majority of outbreaks are associated with antibiotic-resistant bacteria, leading to increased severity of systemic disease, treatment failures and a surge in hospitalizations [10].

Public health implications of *Campylobacter* infections have evolved over more than a century, and currently, they are one of the major causes of foodborne illness worldwide [13,14]. In Europe, a total of 249 campylobacteriosis foodborne outbreaks were reported across 27 European countries in 2021, resulting in 1051 cases, 134 hospitalizations and 6 deaths [1]. Poultry, especially chicken meat, has been linked to 50 to 80% of human cases [15]. The role of poultry in the epidemiology of human disease was proven in Belgium in 1999, during the dioxin crisis when high levels of dioxins in chicken feed produced a reduction in human poultry meat consumption, resulting in a 40% reduction in human cases of campylobacteriosis [16]. Since then, poultry has been considered the natural reservoir of *Campylobacter jejuni* [17].

Both bacteria can colonize the poultry gut, especially the caeca, at high levels without exhibiting symptoms or a loss in production parameters, potentially leading to contamination of poultry meat during slaughter (in the case of *Salmonella* and *Campylobacter*) and the eggs (in the case of *Salmonella*), subsequently leading to human infections [4,17–19]. To control this situation, *Salmonella* National Control Programmes (SNCP) is aimed at on-farm *Salmonella* control (Directive 2003/2160/EC), and the process hygiene criterion for *Campylobacter* focused on slaughterhouse-level *Campylobacter* control (Regulation 2017/1495/EC) in each European Member State (MS). In this farm-to-table conceptualization, biosecurity measures, vaccination, alternative antibiotic products, cleaning and disinfection are the key tools to control these bacteria from the field [15,20]. Although strict measures have been proposed and regulations have improved, both bacteria continue to persist in the poultry industry [21]. This situation is closely related to the emergence of stress-tolerant and biofilm-forming *Salmonella* and *Campylobacter* which enhance their capacity to survive rigorous cleaning and disinfection protocols, facilitating their persistence in the facilities' environment, even for more than one year [21,22]. In this context, cleaning and disinfection play a crucial role in eliminating bacteria. However, new sanitation techniques must be continually developed in order to control the emerging resistant pathogens and their impact on a global scale of any food safety program [23].

Currently, there is growing societal awareness and concern for environmental conservation, with an increasing intolerance towards the use of products that may have a detrimental impact on the environment and the surrounding ecosystem. In this regard, bacteriophages (or phages) represent an emerging up-and-coming "green" technology that could improve food safety with the potential to act against persistent *Salmonella* and *Campylobacter* [24]. The application of phages as a biocontrol tool emerges as a promising approach, not only as an animal treatment alternative to antibiotics, which has been undergoing an extensive body of work [25–32], but also as a complementary cleaning and disinfection tool for persistent strains, where they have been less studied.

2. General Considerations about Phages

Bacteriophages are ubiquitous in nature. They are a group of viruses whose life cycle is strictly associated with the bacterial cell. They are often called bacterial parasites, as they lack the cellular structure and enzyme systems needed to survive on their own. These incomplete organisms can only replicate inside the bacteria host [33,34]. Bacteriophages are specifically associated with a particular bacterial strain and exhibit strong bactericidal activity against Gram-positive and Gram-negative bacteria. Some phages show a specific affinity for individual types of bacteria, while others have a wide range of activity. Their specificity and range of activity are determined by the presence of receptors located on the bacteria's surface, among which we can distinguish fragments of lipopolysaccharide (LPS), fimbriae and other surface proteins [35,36]. In this sense, monovalent phages target a single bacterial species, while polyvalent phages can attack two or more bacterial species. Studies indicate that phages that attack Gram-positive bacteria typically exhibit reduced efficacy against Gram-negative bacteria. Phages encode endolysins, hydrolases engaged in the lysis of the bacterial cell wall during the lytic cycle [37]. Endolysins can also destroy the peptidoglycan layer externally, making them effective antimicrobial agents, especially against Gram-positive bacteria. Nevertheless, the lysis of Gram-negative bacteria poses greater challenges due to their outer membrane, which impedes the access of endolysins to peptidoglycan. Lysis can be achieved by using permeabilizing agents [38–41].

Phages can be divided into lytic phages and temperate phages, which have different life cycles, as shown in Figure 1. Lytic activity is typical of virulent phages, while lysogenic activity involves the integration of the bacteriophage genetic material with the bacterial chromosome and its replication as part of the bacteria DNA, leading to the appearance of a prophage [42]. The lytic cycle begins after the integration of the genetic material in the cell [43]. Once inside the cell, the phage's genetic material replicates and assembles into new viral particles. To do this, during the eclipse period, the host's biosynthetic machinery controls the synthesis of coating proteins, such as lysis proteins that cause lysis of the host cell [44]. The genetic material is packaged into the capsid, and the tail is attached to the head to form mature virions. When the number of viral particles in the bacterial cell reaches a critical level, the lysis process is triggered, which involves the rupture of the cell membrane of the bacterium and the release of the new viral particles into the environment, releasing new phages into the medium to begin another cycle of infection [43].

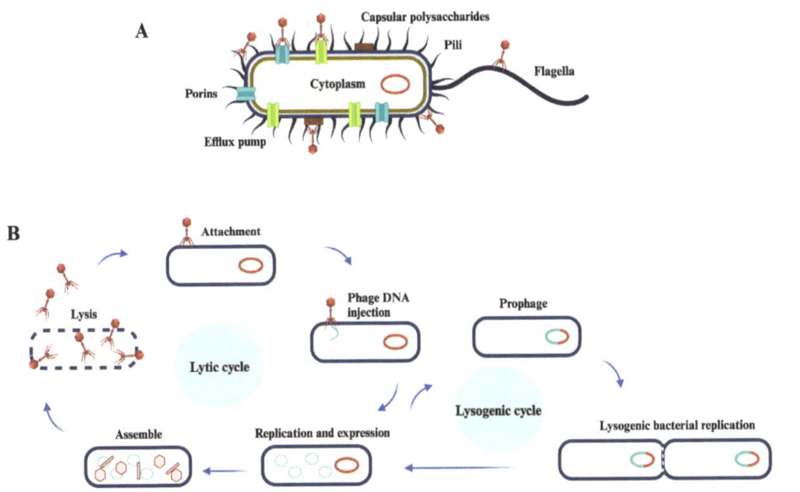

Figure 1. Mechanism of phage infection in bacteria. (**A**) Receptors for phage adsorption on bacteria. The color scheme distinguishes different components: blue represents porins, green represents efflux pumps, dark blue represents flagella and pili, and brown represents capsular polysaccharide.

(**B**) Description of the phases of a phage lytic cycle and lysogenic cycle. The life cycle initiates with the attachment of the bacteriophage to receptors on the host bacteria's cell membrane, followed by the injection of its genetic material. Subsequent stages include genome replication within the host cell, assembly of new phage progeny and their release. In contrast, the lysogenic cycle entails phage attachment to the host bacteria's cell membrane receptors, genome injection, integration into the bacterial genome and replication alongside the host cell. Adapted from Huang et al. [45].

Conversely, lysogenic or temperate phages have the ability to exploit the host and deposit their genetic material within the infected bacterium for multiple generations [46–48]. In the lysogenic cycle, phage DNA replicates along with bacterial DNA, thus establishing a stable relationship. Usually in this cycle, the viral DNA is integrated into the bacterial chromosomal DNA of the host. In later stages during this life cycle, the change from lysogen to lytic form occurs, leading to activation of the lytic life cycle [49,50]. Lysogenic bacteriophages are typically not utilized directly as treatments because they have the ability to transfer genetic material from one bacterial cell to another, a phenomenon known as transduction. They can also transmit genes that enhance the virulence of the host, a process referred to as lysogenic conversion. Due to their replication cycle, they do not eliminate all the bacteria they infect, and a cell that carries a prophage within its genetic material becomes resistant to infection by it, a process known as immunity to superinfection. In contrast to lysogenic phages, lytic or virulent phages can multiply exponentially in bacterial culture and swiftly eradicate bacteria, irrespective of their resistance to antibiotics.

3. Current Utilities and Legislations in the Use of Phage in Poultry

It is well known that the world has an urgent need for access to new effective treatments for bacterial infections in order to replace the miracle drugs or "antibiotics" of the last century [23]. Phage therapy has been considered a promising tool in eliminating bacterial infections in poultry. Research is ongoing to reduce on-farm pathogen occurrence in broilers for *Salmonella* [25,26,29,51–53], *Campylobacter* [54–58], *Escherichia coli* [53,59–61] or *Clostridium perfringens* [62–64], and in laying hens for *Salmonella* [65–67]. Other on-farm interventions included the control of bacteria in drinking water, shavings and on plastic surfaces [68]. Moreover, studies on post-harvest interventions using bacteriophages were conducted to control pathogen contamination in raw chicken meat against *Salmonella* [69,70] and *Clostridium perfringens* [71,72] in raw turkey meat [73] and eggs against *Salmonella* [74,75].

Nevertheless, one of the major limiting factors for the widespread use of phages is the regulatory framework of bacteriophage products [76]. Since 2011, phages have been categorized as drugs in the United States and as medicinal products in the European Union [77–79]. Thus, the Food and Drug Administration (FDA) and the European Medicines Agency (EMA) are responsible for their marketing and manufacturing authorization in the United States and European Union, respectively [79]. In the United States, the FDA has granted bacteriophage-based products GRAS (Generally Recognized As Safe) approval, and countries such as Switzerland, Israel, Canada, Australia, New Zealand, China and Brazil have approved bacteriophage-based products for application in foodstuffs [45,76,80]. However, there is currently no established regulatory pathway to register a phage-based product. There is a lack of consensus on how it should be regulated, regardless of whether they are intended for use as a feed additive, in pre-harvest intervention or in post-harvest application [76]. The lack of legal consensus and clarity slows the development of commercially available bacteriophage products in Europe. Although countries like Georgia, Russia and Poland have been using phages to treat infections since their discovery, there are no regulatory guidelines that can be readily adopted [78,81]. In fact, in Poland, phage therapy is categorized as an "experimental treatment" according to the *Polish Law Gazette*, 2011, item 1634 and article 37 of the Declaration of Helsinki [78]. Nevertheless, on a positive note, based on Regulation 2019/6, in 2022, the EMA published the conceptual paper on the quality, safety and efficacy of phage products as veterinary medicines in Europe. The document set out the starting point for developing the guidelines for the approval of phage-based products, and the key challenges were outlined. Among its main conclusions was the establishment of a task force for novel therapies. In 2023, the first concept paper on the quality, safety and efficacy of bacteriophages as veterinary medicines (EMA/CVMP/NTWP/32862/2022) was published, which laid down the applicable requirements for phage-based veterinary products. Moreover, in the European Union, the EFSA deemed the application of phages against *Listeria* as safe [82]. In this sense, EFSA positively evaluated the safety and efficacy of Listex™ P100 for food safety applications on different ready-to-eat (RTE) food products [80,82]. For poultry, Żbikowska [39] reviewed the different commercially available phage products, both for pre-harvest steps, such as Bafasal® (Proteon Pharmaceutical, Piotrkowska, Poland), Biotector® S (CJBio, Seoul, Korea), SalmoFREE® (Nofima, Tromsø, Noruega), Ecolicide PX™ (Intralitix, Columbia, MD, USA), ListShieldTM Listex™ P100 (PhageGuard) (Intralytix, Columbia, MD, USA), and for the post-harvest step, such as SalmoFresh™ (Intralytix, Baltimore, MD, USA), SalmoPro® (Phagelux, Shanghai, China), Salmonelex™ (PhageGuard) (Micreos, Wageningen, The Netherland), PhageGuard S™ (Micreos, Wageningen, The Netherland), BacWash™ (Elanco, Greenfield, MA, USA) and EcoShield™ (Intralytix, Baltimore, MD, USA).

However, there is an important factor in *Salmonella* epidemiology: the environmental persistence of the bacteria. Because of its ubiquitous nature, *Salmonella* may cycle through a flock of broilers or layers into the livestock facilities environment and back into another flock of broilers or layers [83]. In this regard, as pointed out by the EFSA and ECDC, it was observed that one out of every two *Salmonella* isolates from broilers belonged to the Infantis serovar [84], constituting up to 90% of all *Salmonella* isolates in broilers at slaughter [85]. Numerous researchers are studying the rise of this serovar in an effort to tackle the virulence factors that make it so persistent and how to effectively address it. Notably, *S.* Infantis has been found to present the pESI-like mega-plasmid (the plasmid of emerging *S.* Infantis) [11, 86]. The pESI-like plasmid has been linked with superior biofilm formation, adhesion and invasion into avian and mammalian host cells. This plasmid increased *S.* Infantis fitness and antibiotic resistance, acquisition and transmission of AMR and resistance to quaternary ammonium compounds and heavy metals under various environmental conditions [11,86–90]. Hence, the elimination of *S.* Infantis from farms or slaughterhouses remains a current challenge, even with thorough cleaning and disinfection [86,91]. *Salmonella* persistence highlights the importance of developing novel biocontrol agents, such as bacteriophages, to combat these continuously evolving organisms. Bacteriophage-based products can serve as biosanitizers in hatcheries, farms, transport crates, poultry processing plants and food contact surfaces [39]. For example, a phage-based surface disinfectant against *Salmonella* has been marketed by the US company OmniLytics Inc. (Sandy, UT, USA) [39]. Researchers are highlighting the role of bacteriophages as a complement to the cleaning and disinfection processes across all facilities involved in the production of food, ranging from incineration and disinfection of equipment to the sanitation of the food products themselves.

4. Current Characteristics of Cleaning and Disinfection Used in Poultry Production and the Main Challenges: Biofilms

Good internal and external biosecurity measures are key components to prevent the introduction or reduce the infection pressure of *Salmonella* and *Campylobacter* on the farm. Hygiene measures (cleaning and disinfection) play a significant role in the epidemiology of both bacteria on two levels. Persistent environmental contamination is a major factor in the reinfection of poultry flocks [23]. Firstly, inadequate equipment maintenance, surfaces and unhygienic factory design can create niches where bacteria could find nutrients, water and protection from cleaning, allowing their survival and growth. Secondly, surface sanitization in poultry facilities and food industries can contribute to the emergence of antimicrobial resistance indirectly, mainly due to the existence of cross-resistance between biocides and antibiotics [92,93].

Current best practices for eliminating the growth of resistant bacteria require on-site treatment, including good hygiene practices and different antimicrobial agents. It is widely acknowledged that the cleaning process alone is responsible for the removal of around 90% of bacteria, and subsequent disinfection eliminates an additional 6–7% [94,95]. The presence of organic material in the environment enhances the survival of microorganisms, so its comprehensive removal is an essential step in the cleaning and disinfection process, due to its influence on the effectiveness of disinfectants. These practices must be applied at all stages of the food production process to guarantee safety and quality. When commonly used disinfectants are applied correctly, they could inhibit the colonization of introduced bacteria [96]. However, failures in disinfectant dosing and/or application to wet surfaces can lead to inadequate equipment disinfection and bacteria exposure to subinhibitory chemical levels [96–98]. Although desiccation processes have been shown to enhance the effectiveness of disinfection procedures [96,99], managing them can be complex when continuous or even daily production runs are required [96]. Notably, bacteria could develop resistance, when an increase in concentration or time becomes necessary in order to exert the same reduction, or tolerance, when the bacteria's susceptibility changes due to its exposure to subinhibitory levels [96].

Disinfectants are chemicals used to remove pathogenic microorganisms and other infectious agents from surfaces and equipment. Disinfectants can have different modes of action, but in general, they act by damaging the cell membrane of microorganisms or disrupting their metabolic processes [39]. Among the various commercially used disinfectants is chlorine. Although the disinfectant properties of chlorine have been recognized since the latter half of the nineteenth century, it was not until 1988 that chlorine-containing compounds were also found to possess oxidative properties. Various chlorine-containing compounds are employed as biocides, often chosen for their combination of affordability, ease of use and high efficacy. Upon addition to water, chlorine reacts with the hydrogen and oxygen in water molecules, leading to the formation of hydrochloric acid (HCl) and hypochlorous acid (HOCl). HOCl further undergoes dissociation to generate hypochlorite (OCl^-) and hydrogen (H^+) ions. Both HOCl and OCl^- constitute the "free chlorine" in a solution and are the primary compounds responsible for the antimicrobial action of chlorine supplementation [100–104]. Slow-release chlorine dioxide (SRCD) is frequently applied in poultry processing to reduce *Salmonella* levels in carcasses. The necessity for utilizing SRCD as an antimicrobial stems from the reality that, despite meticulous endeavors to minimize the prevalence of live birds carrying *Salmonella*, the mechanical actions of plucking machines and unintentional damage to entrails during the evisceration stage inevitably contribute to the dissemination of *Salmonella*. In addition, the contamination levels of carcasses that end up in the supermarket are strongly correlated with the amount of cross-contamination that occurs during processing. While these chemicals have shown good efficacy, they are not allowed in the EU due to potential hazards associated with their chemical action [105–107].

Organic acids are a widely chosen antimicrobial option due to their high efficacy, affordability and ease of application, leading to their extensive utilization for various purposes. Organic acids inhibit bacterial growth by reducing the pH. Fundamentally, organic acids hinder the bacterial cell by inducing an accumulation of anions in the bacterial cytoplasm, adversely impacting the proton motive force (PMF) and, consequently, the cell's capacity to sustain an optimal pH. As a result, this perturbation modifies the internal environment of the cell, hindering DNA synthesis, normal enzyme activity and cell reproduction. Multiple studies have been conducted regarding which organic acids are the most efficient in their action, resulting in significance [108–115]. A solution of lactic acid (1%) resulted in a 66% reduction in CFU/g, whereas acetic acid (1%) and citric acid (1%) demonstrated reductions of 55% and 51%, respectively. Fumaric acid is a technically effective option, but it affects the sensory properties of chicken breast more. Other options tested include succinic acid, also with good results. These studies emphasize the importance of employing organic acids in combination with other methods, such as refrigeration and freezing, to ensure a satisfactory level of reduction [108–115]. Furthermore, the antimicrobial effectiveness of organic acids is significantly influenced by factors such as contact time, temperature, acid concentration, or its combination with other agents. This complexity poses challenges, particularly in the context of the high prevalence of *Salmonella* resistance and the ongoing necessity to guarantee effective pathogen elimination through specific application methods [116,117]. While organic acids often exhibit relatively high efficacy, there is a potential risk associated with incorporating them at certain levels and temperatures, as this may impact the sensory properties of the meat [116,118–120].

Regardless of the many preventative measures in place, bacteria tend to live attached to surfaces and to create a complex structure known as biofilm [121–123]. In this sense, bacteria living under the biofilm state present a different phenotype from their planktonic counterparts [124,125] and demonstrate a high tolerance to antimicrobials, desiccation and heat [124,126]. Biofilms are complex three-dimensional structures that protect microbial communities from biotic and abiotic factors produced by the microorganisms themselves. The extracellular matrix is composed of a mixture of various polymeric compounds, such as polysaccharides, proteins, nucleic acids and lipids [127]. As in ancient Chinese warfare, where soldiers often formed an organized formation in a fortified circle to protect themselves and attack their opponents, bacteria build a protective system that is hard to achieve for individual bacteria [128]. It serves the purpose of maintaining bacteria in close proximity to one another while also establishing channels to distribute water, nutrients, oxygen, enzymes and cell debris [127]. They have inherent resistance to antimicrobial agents and are challenging to remove or eradicate with the application of disinfectants [129]. Their multilayered structure acts as a barrier and limits the diffusion of the antimicrobial agent [129]. Moreover, it includes microorganisms at various stages of metabolic dormancy in cells with different susceptibility, which favors the acquisition of new genetic traits of resistance [129]. Biofilm formation is especially relevant in *Salmonella* and *Campylobacter* epidemiology, thus also in poultry production. Approximately half of the *Salmonella* strains isolated in poultry farms present the ability to generate biofilms in poultry farm processing areas and on contact surfaces [130,131]. *Salmonella* biofilm production is mainly related to the presence of sgD, adrA and gcpA genes, which are responsible for the production of cellulose and curli fimbriae [130], while *Campylobacter* could develop monoculture and mixed-culture biofilms. Indeed, oxidative stress has been linked not only with the transformation to viable but not cultivable forms but also with biofilm formation to favor *Campylobacter* survival [132]. These bacteria are resistant to most traditional antimicrobial strategies, making it necessary to design and implement novel biocontrol agents.

5. Biofilms and Bacteriophages

Effective intervention on the farm will likely require a multifaceted approach, and in the light of bacterial resistance to cleaning and disinfection protocols, it is unlikely that there will be any single "silver bullet" approach to eliminate *Campylobacter* and *Salmonella* in poultry processing. Subsequently, a prophylactic phage-based biocontrol/disinfectant agent will ensure the safety level throughout the farm-to-fork process (Figure 2).

Phages are described as nontoxic, natural agents, able to withstand extreme physiological conditions while also maintaining a narrow spectrum of antimicrobial activity [24]. Phages exert their antibacterial activity through the action of depolymerase and lysin enzymes. Depolymerases are responsible for breaking down capsular polysaccharides, while lysins target the peptidoglycan in bacterial cells [128,133,134]. Despite biofilms providing resistance to phages due to the impermeability of the biofilm matrix, phages exhibit a different mode of action on bacteria within biofilms compared to antibiotics or biocides [135]. Phages have the ability to destroy bacterial hosts, thereby preventing biofilm formation [135,136]. They can also infiltrate existing biofilms, eliminating the biofilm structure with or without destroying the resident bacteria [135,136]. The removal of biofilms using phages in nature can be classified as intra- to extracellular degradation of the bacterial structure, extra- to intracellular degradation of the bacterial structure and chemical dispersion of the biofilm matrix [135,137]. Phage-based treatment acts through three corresponding modes: basic phage therapy, phage-derived lysins and phage-derived depolymerases [133,134,138].

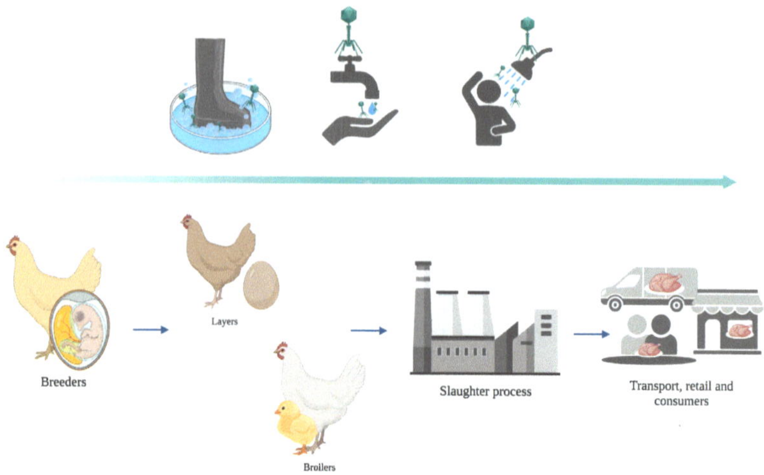

Figure 2. Schematic diagram illustrating phage biocontrol measures "from farm to fork".

As mentioned above, phages can carry and express depolymerizing enzymes [135]. Depolymerases perform various activities, such as degrading EPS or cleaving structural polysaccharides like LPS or the PG glycan chain [139]. In this sense, to target biofilms, phages infiltrate biofilm using depolymerases, which are enzymes specialized in hydrolyzing the polysaccharides and polysaccharide derivatives found in the biofilm's outer membrane [24]. Usually, this enzyme is encoded at the tail structure of the phage, a structure involved in the phage infection, thus giving phage a notable advantage over other antimicrobial agents [24].

The intra- to extracellular degradation of the host cell is characteristic of lytic phages [43,128], when phages induce the release of progeny phages from the bacteria host at the final stage cycle [43,128]. Throughout bacteriophage replication, the growing population of infectious progeny phages in the biofilm eliminates the bacteria responsible for the production of extracellular polymeric substance (EPS), which composes the biofilm matrix [135]. This process gradually removes the biofilm and reduces its capacity for regeneration. In fact, phages could coevolve with the bacterial biofilms, and consequently, their ability to infect adherent bacterial populations is to be expected [135].

Another interesting feature is that phages can induce the expression of these enzymes in their host bacteria [139]. Phages could also infect persistent cells and remain within them until reactivation occurs, which then destroys the cells [135].

6. The Role of on-Farm Use of Bacteriophages as Disinfectants

Bacteriophages as Disinfectants in Poultry Pre-Harvest Stages

The phage anti-biofilm activity in an experimental poultry model has been explored. At pre-slaughter stages, phages could serve as biosanitizers in hatcheries, farms and transport crates [39]. During these phases, contamination and the dissemination of *Salmonella* are of particular relevance. Due to its epidemiology and biofilm formation, it is not only transmitted vertically but also persists in the environment, spreading horizontally and among successive flocks. Indeed, *Salmonella* persists in the litter, dried feces and feed within empty poultry houses for up to two years [140]. Furthermore, following cleaning and disinfection procedures, its survival is notably significant in drinker and feeder lines, the anteroom (including electrical panels, floors, and surfaces), farmers' boots and fomites in the lysogenic form, the bacterium [141]. Furthermore, in caged layer houses, an additional challenge arises due to the difficulty of cleaning cage tiers [95]. However, the exterior of persistently infected houses remained substantially contaminated [141], and this fact is not usually the focus of control studies or legislation. Indeed, the surrounding farm environment sustains *Salmonella* circulation with wildlife vectors, including rodents [95,142] and insects, including flies and mealworms [143], highlighting its significant role in transmitting *Salmonella* between successive flocks. Hence, it is necessary to focus on the elimination of persistent environmental *Salmonella* from facilities, materials and the surrounding farm environment [11].

For example, Korzeniowski [144] studied the phage's ability to eradicate *Salmonella* from poultry drinkers (the main point in pathogen horizontal transmission within a poultry flock) and on a stainless-steel surface. They showed that the biofilm formed by *S.* Enteritidis was eradicated from poultry drinkers and was reduced in the range of 60–97% on a stainless-steel surface. Moreover, *Salmonella* phages isolated from a chicken farm and slaughter plant demonstrated a reduction in developing and mature biofilms in 96-well microplates [145]. Similarly, a phage isolated from poultry fecal samples reduced the concentration of *S.* Enteritidis excreted by chicks through feces and in their surrounding environment and from metal surfaces [146]. Gong [147] evaluated the application of bacteriophages to reduce *Salmonella* contamination on workers' boots in both the laboratory and the rendering-processing environments. They combined and compared phage application (cocktail), disinfection with sodium hypochlorite (400 ppm) and 30 s brush scrubbing to assess the efficacy of phages for reducing *Salmonella* on workers' boots [147]. Under laboratory conditions, the phage treatment of *Salmonella* biofilms on boot soles resulted in a reduction of 91.5% CFU/boot when applied alone, and 97.0% and 99.2% CFU/boot when combined with hypochlorite or brushing, respectively. In a rendering-processing plant, the phage treatment of *Salmonella* biofilms on workers' boots resulted in a reduction of 82.2% CFU/boot when applied alone, and 92.9% and 93.2% CFU/boot when combined with hypochlorite or brushing, respectively. Although the prevalence of *Campylobacter* in poultry flocks has been associated with several factors, including inadequate disinfection between chick placements and the presence of rodents and insects [148], studies on phages as disinfectants against *Campylobacter* have primarily focused only on post-harvest stages.

7. Bacteriophages as Disinfectants in Poultry Post-Harvest Stages

At post-harvest stages, phages could serve as biosanitizers in poultry processing plants and on food contact surfaces. During these phases, contamination of *Salmonella* and *Campylobacter* are relevant. For both bacteria, just a few infected chickens at the slaughterhouse can contaminate the entire processing line [23]. Throughout the slaughter process, carcasses may become contaminated by bacteria present in the intestinal content of the animals, either from the same flock or in previously slaughtered flocks [21,149]. Cross-contamination of broiler meat can also take place in the consumer´s home. This is a particularly critical risk stage for *Campylobacter*. When the meat is stored under refrigeration conditions, it allows bacteria to infect other meats stored in the same packaging [21]. In this sense, *Campylobacter* can survive for up to 18 days under standard refrigeration temperatures of 4 °C without showing any decrease in bacterial counts [21,150–152]. On the other hand, the risk of *Salmonella* contamination in eggs extends to the post-collection, storage, transport and food handling stages [153]. Consequently, post-harvest practices are focused on the use of physical and chemical approaches, which may not succeed in reducing pathogen loads for several reasons, such as application-dependent and bacterial-resistance-related factors [76].

Phage treatment can be used to inactivate *Salmonella* and *Campylobacter* attached to food contact surfaces or grown as biofilms. In this sense, de Ornellas [154] highlighted the relation to the contact surface and the contaminant pathogen, as both are capable of interfering in the bacteria's persistence in industrial environments through biofilm production and the ability of phages isolated from hospital wastewater and poultry wastewater to reduce

Salmonella biofilm producer. The effectiveness of different phages in removing *Campylobacter, Listeria* and *E. coli* O157:H7 from the surfaces of stainless-steel, polypropylene and ceramic materials has been evaluated [127,155–157]. Its use is promising, although very challenging due to the diversity of bacteria found in different environments [127,155–157]. Siringan [158] first studied the effect of bacteriophage treatments of *Campylobacter* biofilms on glass as a matrix and demonstrated that these bacteriophages can reduce the numbers of viable bacteria and disperse the matrix. After that, Siringan [158] highlighted that in phage treatment, despite an equilibrium between host and phage, bacteria can act as expendable vehicles for the delivery of phages to new host bacteria within pre-colonized chickens. Hence, the application of phages is not expected to replace the use of disinfectants, but under particular circumstances, it could act as a complement [129,159].

8. Main Limitations to the Use of Bacteriophages

Despite its numerous advantages, the use of phage therapy is substantially limited due to different causes. A summary of all the information gathered in this review is presented in the following table (Table 1).

Table 1. Summary of main limitations to the use of bacteriophages.

Limitation	Description
Phage of Choice	Individual phages are insufficient for broad-spectrum infections; complex identification is needed. Elimination by the reticuloendothelial system reduces half-life, limiting efficacy.
Lysogenic Forms	Lysogenic phages confer poor results due to acquired immunity. Transduction of bacterial genome and potential transmission of harmful genes are concerns. Using multiple phages is often more effective.
Legal Limitations	Global regulatory variation in phage utilization. Permitted as a processing aid in certain countries, limited in the EU.
Purification and Stabilization	Phage characterization is essential for toxicity removal. Purification by ultracentrifugation or chromatography. Stability is crucial but varies among phages.
Dosage	High phage concentrations are needed for bacterial removal; lower doses are ineffective. Timing and delivery are critical, with potential for induced antibodies.
Terms of Use	Bacteriophage persistence varies with type, application conditions, and environmental factors. Refrigeration enhances persistence.
Resistance Mechanisms	Increased phage application may lead to bacterial resistance mechanisms. Coevolution cycles involve various resistance strategies.
Effectiveness	Efficacy is a major limitation; initial reduction is observed, but complete eradication is challenging. Factors influencing effectiveness include food matrix, pH, temperature, MOI, phagoresistance, and combination with other measures.

8.1. Limitation Due to the Phage of Choice

Individual bacteriophages cannot be used to fight broad-spectrum infections. In many cases, complex identification and characterization of the etiological agent is necessary. Another adverse phenomenon in phage therapy is that phages can be eliminated by the reticuloendothelial system, reducing their half-life in the body and limiting the effectiveness of the treatment [34,160,161].

8.2. Lysogenous Forms

In the lysogenic state, the bacterium gains immunity against superinfection by phages of the same type, which is an unfavorable outcome for the goals of phage therapy. Consequently, lysogenic phages are not employed in therapeutic applications [162]. An additional drawback is its capacity to transduce segments of the bacterial genome post-infection, potentially leading to the dissemination of detrimental or virulent genes within the bacterial population. The analysis of phage genomes might be a time-consuming technique, posing challenges in urgent infection treatments. Due to the heightened specificity of phage infection to a particular bacterial strain, employing multiple phages often proves more effective in managing an infection [48,163,164].

8.3. Legal Limitations

The regulatory approval for phage utilization lacks global standardization, exhibiting variations across different countries. In the United States, Canada, Switzerland, New Zealand, Australia and Israel, the application of phages as a processing aid is permitted. However, within the European Union, this practice is exclusively allowed in the Netherlands and is not listed in the qualified presumption of safety (QPS) list [120,165–168].

8.4. Need for Purification and Stabilization

Only complete characterization and screening of phages can remove those that encode toxic proteins or proteins that allow the behavior of temperate (integrative) phages. To achieve an adequate level of purification for animal model studies, ultracentrifugation using a CsCl gradient is employed, followed by the removal of endotoxins. Phage purification can also be accomplished through chromatography methods [34,169,170]. When chromatography is used for purification, endotoxin levels decrease by 10 to 30 times compared to the traditional method, albeit often resulting in a lower final phage titer. The stability of phage preparations is crucial for effective delivery over time. However, due to the unique sensitivity of each phage to chemical and environmental factors, a universal preparation strategy is not currently feasible. Typically, phages are resuspended in simple aqueous solutions. Nevertheless, prolonged storage of phage solutions may lead to a gradual loss of phage activity, necessitating the addition of stabilizers. An alternative approach involves freeze-drying phage solutions, converting them into a stable powder with a high degree of stability [48,171].

8.5. Dosage

One of the main obstacles to the removal of bacteria from poultry is that a significant number of phages are needed to adsorb individual host cells. The application of phages at lower doses did not provide statistically significant protection. In addition, preventive treatment in phage therapy did not prevent colonization. In many cases, the efficacy of phage therapy should be maximized by using a high titer of bacteriophages to reduce colonization [56,60,172]. An additional hurdle in the use of phage therapy is that colonization of the chicken caecum by *S.* Enteritidis and Typhimurium is inhibited for only 24 to 48 h after phage treatment, so it seems necessary to determine the optimal timing and delivery of bacteriophages in a real-life poultry industry setting [34,173]. The presented data illustrate that only elevated phage concentrations prove effective in achieving a notable reduction in both mortality rates and foodborne pathogens. Conversely, administering high doses over extended periods can lead to the development of neutralizing antibodies [174,175]. It is imperative that each phage preparation designed for application in poultry veterinary medicine, poultry production and the poultry industry be, above all, safe and efficacious. Critical considerations include the dosage, route of administration (including the preparation of standardized formulations), timing of phage-based product administration and the concurrent use of other preparations (e.g., competitive exclusion) or vaccines [39].

8.6. Terms of Use

The endurance of bacteriophages on or within food is subject to variability among different bacteriophages and is influenced by application conditions (e.g., dosage) and environmental factors (e.g., temperature). Refrigeration temperatures have the potential to enhance the longevity of bacteriophages on the surfaces of meat products [39,176].

8.7. Resistance Mechanisms

A potential concern of phage resistance arises from increased or prolonged application of phages in the food industry. In the environment, bacteria and bacteriophages exist in a cycle of coevolution, in which hosts insensitive to phages survive or prevent phage predation by transmitting corresponding resistance mechanisms [177–179]. Prevention of phage infection is not the only bacterial response used. Other phage resistance mechanisms focus not on preventing phage entry but on the bacterial survival of the host once infected by phages. Bacteria are able to use different pathways of action to achieve this evolutionary survival [158,180–189]. Occasionally, when phage resistance develops, bacteria increase sensitivity to antibiotics. These results demonstrate the potential to reverse existing antibiotic resistance and potentially alleviate some of the public health problems associated with antibiotic treatment [46,190].

8.8. Effectiveness

The primary constraint in the application of phages is their efficacy. Numerous studies indicate an initial reduction in bacteria, yet no further reduction thereafter, underscoring that phages can diminish bacteria but may fall short of complete eradication. This limitation could stem from the incapacity of phages to reach and infiltrate bacteria after progeny, emphasizing the crucial role of adequate moisture to facilitate phage dispersion [120,191,192].

Among the factors that can influence the bactericidal effectiveness of bacteriophages in food are the following:

- Food matrix: In solid foods, it will depend on the ability of the food to adsorb the phage suspension and that it is not diluted [193]. One solution that would avoid this problem would be phage immobilization. In plant compounds, substances such as organic acids and tannins would inactivate bacteriophages [193–195].
- pH: Given their nature, they do not seem useful when the food has a pH not between 5 and 8, extending this range to between 4 and 10 when the temperature is low [196,197].
- Temperature: Phages have a lower effect at ambient temperatures (20 °C) than at refrigeration temperatures (4 °C) [175,198].
- Multiplicity of infection (MOI): The greater the MOI, the greater the bacterial lysis, which usually requires high concentrations of phages [199,200].
- Phagoresistance: Bacteria have been described that exhibit resistance to lytic bacteriophages due to mutations [36,201].
- Combination with other control measures: Although phages are a good alternative for biocontrol, they do not usually achieve a complete elimination of the pathogen, so it is always recommended that they be applied together with other measures [202].

9. Key Insights into Bacteriophage Application in Poultry Farming

9.1. Pathogen Threat

Salmonella and *Campylobacter* are significant zoonotic pathogens that continue to be a major public health threat, primarily through the consumption of poultry meat and eggs. These bacteria have been responsible for numerous foodborne outbreaks. The growing concern is their antibiotic resistance, which results in more severe illnesses, treatment failures and increased hospitalizations.

9.2. Emerging Phage Technology

Bacteriophages, often referred to as phages, are emerging as a promising "green" technology to enhance food safety and combat persistent *Salmonella* and *Campylobacter* infections. They offer potential solutions to reduce these pathogens in poultry, both in on-farm and post-harvest applications. However, the regulatory framework for bacteriophage products is still evolving, posing a challenge to their widespread use in the food industry.

9.3. Persistent Bacteria and Phage Solutions

The persistence of zoonotic bacteria in the environment remains a significant challenge despite rigorous cleaning and disinfection efforts. The presence of certain plasmids can enhance their resistance and virulence. Bacteriophages are considered potential biocontrol agents to combat the persistence of *Salmonella* at various stages of poultry production, from hatcheries to farms and processing plants. These phage-based products can complement existing tools for maintaining food safety in the poultry industry.

9.4. Biofilm Challenge

The formation of biofilms by these zoonotic bacteria is a major obstacle in poultry production. Biofilms are complex three-dimensional structures that shield bacterial communities, making them highly resistant to antimicrobials and creating a barrier that hinders the effectiveness of disinfectants. Biofilm formation is particularly relevant in the epidemiology of *Salmonella* and *Campylobacter* in poultry production, emphasizing the need for innovative biocontrol agents to combat these persistent pathogens.

9.5. Multifaceted Phage Approach

Addressing the challenge of *Campylobacter* and *Salmonella* in poultry processing will likely require a multifaceted strategy, as there is no single "silver bullet" solution. Given bacterial resistance to traditional cleaning and disinfection methods, a preventive phage-based biocontrol/disinfectant approach is proposed to ensure safety from farm to fork. Phages, which are natural and nontoxic agents, are effective tools against these bacteria. They can withstand extreme conditions and have a narrow antimicrobial spectrum. Phages act through depolymerase and lysin enzymes, breaking down bacterial structures. Importantly, they have distinct modes of action on biofilms compared to antibiotics. They can prevent biofilm formation, infiltrate existing biofilms and disrupt their structure. The phage-based treatment includes three modes: basic phage therapy, phage-derived lysins and phage-derived depolymerases. These approaches offer the potential to target and eliminate biofilms, providing a valuable tool in combating *Salmonella* and *Campylobacter* in poultry processing. Moreover, phages can coevolve with bacterial biofilms and infect persistent cells, offering an adaptable solution to the problem.

9.6. Effective Phage-Based Treatment

Phage-based treatments have proven effective in reducing and eradicating *Salmonella* and *Campylobacter* from various surfaces, including those in direct contact with food and biofilms. These treatments show promise in preventing cross-contamination during poultry processing, which is especially critical for *Campylobacter*. Additionally, their effectiveness in addressing the persistence of these pathogens in the poultry environment, including on farms and equipment, has been highlighted. Phages offer a promising approach to eliminating persistent environmental contamination.

9.7. Main Phage Limitations

The use of bacteriophages for combating bacterial infections comes with several constraints that need to be addressed. These limitations include the selection of appropriate phages, the presence of lysogenic forms, legal restrictions, the need for phage purification and stabilization, determining the correct dosage, terms of use, the development of bacterial resistance mechanisms and the overall effectiveness of phage therapy. Recognizing and overcoming these limitations is crucial to enhance the effectiveness of phage therapy in the fight against bacterial infections.

10. Conclusions

In summary, *Salmonella* and *Campylobacter* continue to pose significant challenges to food safety in the developed world, with antibiotic-resistant strains contributing to persistent issues. Bacteriophages (phages) emerge as a promising "green" technology, demonstrating potential applications in mitigating on-farm pathogen occurrence, particularly in the context of poultry production. While regulatory hurdles exist, recent developments suggest progress in establishing guidelines for phage-based veterinary products. Phage-based treatments exhibit efficacy in eradicating *Salmonella* biofilms, providing a complementary approach to traditional cleaning and disinfection methods, thereby addressing the complex challenges associated with persistent bacterial contamination in poultry environments.

Effective biosecurity measures, on-site treatments and innovative biocontrol agents are essential components in preventing *Salmonella* and *Campylobacter* in poultry farms. The unique modes of action of phages, particularly their ability to target and eliminate biofilms, make them a promising solution. Post-harvest, phages show potential as biosanitizers, addressing contamination risks on food contact surfaces and during meat storage. Despite facing limitations such as regulatory uncertainties and bacterial resistance concerns, addressing these constraints is crucial for the successful integration of phage therapy into poultry farming practices and broader food safety initiatives.

Author Contributions: Conceptualization, J.J., C.M. and L.L.-R.; data curation, J.J.; writing—original draft preparation, J.J. and C.M.; writing—review and editing, J.J., C.M., L.L.-R., A.M.-F., S.V. and L.M.-D.; visualization, J.J., C.M., L.L.-R., A.M.-F., S.V. and L.M.-D. All authors have read and agreed to the published version of the manuscript.

Funding: This research was funded by the University CEU-UCH (INDI 23/39).

Institutional Review Board Statement: Not applicable.

Informed Consent Statement: Not applicable.

Data Availability Statement: Not applicable.

Acknowledgments: The authors wish to thank the "Improvement of Production System-related Food Safety and End Products" research group (Veterinary Faculty, University CEU-Cardenal Herrera) for their technical support.

Conflicts of Interest: The authors declare no conflict of interest.

References

1. EFSA; ECDC. The European Union Summary Report on Antimicrobial Resistance in zoonotic and indicator bacteria from humans, animals and food in 2019–2020. *EFSA J.* **2022**, *20*, e07209. [CrossRef]
2. World Health Organization. Food Safety. 2023. Available online: https://www.who.int/news-room/fact-sheets/detail/food-safety (accessed on 1 August 2023).
3. Sosnowski, M.; Osek, J. Microbiological safety of food of animal origin from organic farms. *J. Vet. Res.* **2021**, *65*, 87–92. [CrossRef]
4. Stewart, J.; Pavic, A. Advances in enteropathogen control throughout the meat chicken production chain. *Compr. Rev. Food Sci. Food Saf.* **2023**, *22*, 2346–2407. [CrossRef] [PubMed]
5. Venkitanarayanan, K.; Thakur, S.; Ricke, S.C. Salmonella in Poultry Meat Production. In *Food Safety in Poultry Meat Production*; Springer International Publishing AG: Cham, Switzerland, 2019; pp. 1–24.
6. Liu, B.; Zhang, X.; Ding, X.; Bin, P.; Zhu, G. The vertical transmission of *Salmonella* Enteritidis in a One-Health context. *One Health* **2023**, *16*, 100469. [CrossRef] [PubMed]
7. Majowicz, S.E.; Musto, J.; Scallan, E.; Angulo, F.J.; Kirk, M.; O'Brien, S.J.; Jones, T.F.; Fazil, A.; Hoekstra, R.M. The Global Burden of Nontyphoidal Salmonella Gastroenteritis. *Clin. Infect. Dis.* **2010**, *50*, 882–889. [CrossRef]
8. Andino, A.; Hanning, I. Salmonella enterica: Survival, Colonization, and Virulence Differences among Serovars. *Sci. World J.* **2015**, *2015*, 520179. [CrossRef] [PubMed]
9. Gut, A.M.; Vasiljevic, T.; Yeager, T.; Donkor, O.N. Salmonella infection–prevention and treatment by antibiotics and probiotic yeasts: A review. *Microbiology* **2018**, *164*, 1327–1344. [CrossRef]
10. Teklemariam, A.D.; Al-Hindi, R.R.; Albiheyri, R.S.; Alharbi, M.G.; Alghamdi, M.A.; Filimban, A.A.R.; Al Mutiri, A.S.; Al-Alyani, A.M.; Alseghayer, M.S.; Almaneea, A.M.; et al. Human Salmonellosis: A Continuous Global Threat in the Farm-to-Fork Food Safety Continuum. *Foods* **2023**, *12*, 1756. [CrossRef]

11. Montoro-Dasi, L.; Lorenzo-Rebenaque, L.; Marco-Fuertes, A.; Vega, S.; Marin, C. Holistic Strategies to Control Salmonella Infantis: An Emerging Challenge in the European Broiler Sector. *Microorganisms* **2023**, *11*, 1765. [CrossRef]
12. Gast, R.K.; Dittoe, D.K.; Ricke, S.C. Salmonella in eggs and egg-laying chickens: Pathways to effective control. *Crit. Rev. Microbiol.* **2022**, 1–25, *ahead-of-print*. [CrossRef]
13. Altekruse, S.; Koehler, J.; Hickman-Brenner, F.; Tauxe, R.V.; Ferris, K. A comparison of *Salmonella enteritidis* phage types from egg-associated outbreaks and implicated laying flocks. *Epidemiol. Infect.* **1993**, *110*, 17–22. [CrossRef] [PubMed]
14. World Health Organization. Campylobacter. 2020. Available online: https://www.who.int/news-room/fact-sheets/detail/campylobacter (accessed on 1 August 2023).
15. Taha-Abdelaziz, K.; Singh, M.; Sharif, S.; Sharma, S.; Kulkarni, R.R.; Alizadeh, M.; Yitbarek, A.; Helmy, Y.A. Intervention Strategies to Control Campylobacter at Different Stages of the Food Chain. *Microorganisms* **2023**, *11*, 113. [CrossRef] [PubMed]
16. Moore, J.E.; Corcoran, D.; Dooley, J.S.G.; Fanning, S.; Lucey, B.; Matsuda, M.; McDowell, D.A.; Mégraud, F.; Millar, B.C.; O'Mahony, R.; et al. Campylobacter. *Vet. Res.* **2005**, *36*, 351–382. [CrossRef]
17. Hermans, D.; Pasmans, F.; Messens, W.; Martel, A.; Van Immerseel, F.; Rasschaert, G.; Heyndrickx, M.; Van Deun, K.; Haesebrouck, F. Poultry as a Host for the Zoonotic Pathogen *Campylobacter jejuni*. *Vector Borne Zoonotic Dis.* **2012**, *12*, 89–98. [CrossRef]
18. Lee, M.D.; Newell, D.G. Campylobacter in Poultry: Filling an Ecological Niche. *Avian Dis.* **2006**, *50*, 1–9. [CrossRef] [PubMed]
19. Wagenaar, J.A.; French, N.P.; Havelaar, A.H. Preventing Campylobacter at the Source: Why Is It So Difficult? *Clin. Infect. Dis.* **2013**, *57*, 1600–1606. [CrossRef] [PubMed]
20. Marin, C.; Balasch, S.; Vega, S.; Lainez, M. Sources of Salmonella contamination during broiler production in Eastern Spain. *Prev. Vet. Med.* **2011**, *98*, 39–45. [CrossRef]
21. Thames, H.T.; Theradiyil Sukumaran, A. A Review of Salmonella and Campylobacter in Broiler Meat: Emerging Challenges and Food Safety Measures. *Foods* **2020**, *9*, 776. [CrossRef]
22. Davies, R.H.; Wray, C. Observations on Disinfection Regimens Used on *Salmonella enteritidis* Infected Poultry Units. *Poult. Sci.* **1995**, *74*, 638–647. [CrossRef]
23. Jassim, S.A.A.; Limoges, R.G. *Bacteriophages: Practical Applications for Nature's Biocontrol*; Springer International Publishing: Cham, Switzerland, 2017.
24. Gildea, L.; Ayariga, J.A.; Robertson, B.K. Bacteriophages as Biocontrol Agents in Livestock Food Production. *Microorganisms* **2022**, *10*, 2126. [CrossRef]
25. Andreatti Filho, R.L.; Higgins, J.P.; Higgins, S.E.; Gaona, G.; Wolfenden, A.D.; Tellez, G.; Hargis, B.M. Ability of Bacteriophages Isolated from Different Sources to Reduce *Salmonella enterica* Serovar Enteritidis In Vitro and In Vivo. *Poult. Sci.* **2007**, *86*, 1904–1909. [CrossRef]
26. Borie, C.; Sánchez, M.L.; Navarro, C.; Ramírez, S.; Morales, M.A.; Retamales, J.; Robeson, J. Aerosol Spray Treatment with Bacteriophages and Competitive Exclusion Reduces *Salmonella* Enteritidis Infection in Chickens. *Avian Dis.* **2009**, *53*, 250–254. [CrossRef]
27. Kim, J.H.; Kim, J.W.; Shin, H.S.; Kim, M.C.; Lee, J.H.; Kim, G.-B.; Kil, D.Y. Effect of dietary supplementation of bacteriophage on performance, egg quality and caecal bacterial populations in laying hens. *Br. Poult. Sci.* **2015**, *56*, 132–136. [CrossRef]
28. Grabowski, Ł.; Węgrzyn, G.; Węgrzyn, A.; Podlacha, M. Highly different effects of phage therapy and antibiotic therapy on immunological responses of chickens infected with Salmonella enterica serovar Typhimurium. *Front. Immunol.* **2022**, *13*, 956833. [CrossRef]
29. Lorenzo-Rebenaque, L.; Malik, D.J.; Catalá-Gregori, P.; Torres-Boncompte, J.; Marin, C.; Sevilla-Navarro, S. Microencapsulated bacteriophages incorporated in feed for Salmonella control in broilers. *Vet. Microbiol.* **2022**, *274*, 109579. [CrossRef] [PubMed]
30. Peh, E.; Szott, V.; Reichelt, B.; Friese, A.; Rösler, U.; Plötz, M.; Kittler, S. Bacteriophage cocktail application for Campylobacter mitigation—From in vitro to in vivo. *BMC Microbiol.* **2023**, *23*, 209. [CrossRef] [PubMed]
31. Sarrami, Z.; Sedghi, M.; Mohammadi, I.; Bedford, M.; Miranzadeh, H.; Ghasemi, R. Effects of bacteriophage on *Salmonella* Enteritidis infection in broilers. *Sci. Rep.* **2023**, *13*, 12198. [CrossRef]
32. Thanki, A.M.; Hooton, S.; Whenham, N.; Salter, M.G.; Bedford, M.R.; O'Neill, H.V.M.; Clokie, M.R.J. A bacteriophage cocktail delivered in feed significantly reduced Salmonella colonization in challenged broiler chickens. *Emerg. Microbes Infect.* **2023**, *12*, 2217947. [CrossRef] [PubMed]
33. Bordet, J.; Ciuca, M. Remarques sur l'historique de recherches concernant la lyse microbienne transmisible. *Compt. Rend. Soc. Biol.* **1921**, 745–747.
34. Wernicki, A.; Nowaczek, A.; Urban-Chmiel, R. Bacteriophage therapy to combat bacterial infections in poultry. *Virol. J.* **2017**, *14*, 179. [CrossRef] [PubMed]
35. Weinbauer, M.G. Ecology of prokaryotic viruses. *FEMS Microbiol. Rev.* **2004**, *28*, 127–181. [CrossRef] [PubMed]
36. Skurnik, M.; Strauch, E. Phage therapy: Facts and fiction. *Int. J. Med. Microbiol.* **2006**, *296*, 5–14. [CrossRef]
37. Favrin, S.J.; Jassim, S.A.; Griffiths, M.W. Development and Optimization of a Novel Immunomagnetic Separation- Bacteriophage Assay for Detection of Salmonella enterica Serovar Enteritidis in Broth. *Appl. Environ. Microbiol.* **2001**, *67*, 217–224. [CrossRef] [PubMed]
38. Fernandes, S.; Proença, D.; Cantante, C.; Silva, F.A.; Leandro, C.; Lourenço, S.; Milheiriço, C.; de Lencastre, H.; Cavaco-Silva, P.; Pimentel, M.; et al. Novel Chimerical Endolysins with Broad Antimicrobial Activity Against Methicillin-Resistant *Staphylococcus aureus*. *Microb. Drug Resist.* **2012**, *18*, 333–343. [CrossRef] [PubMed]

39. Żbikowska, K.; Michalczuk, M.; Dolka, B. The Use of Bacteriophages in the Poultry Industry. *Animals* **2020**, *10*, 872. [CrossRef] [PubMed]
40. O'Flaherty, S.; Ross, R.P.; Coffey, A. Bacteriophage and their lysins for elimination of infectious bacteria. *FEMS Microbiol. Rev.* **2009**, *33*, 801–819. [CrossRef] [PubMed]
41. Kazi, M.; Annapure, U.S. Bacteriophage biocontrol of foodborne pathogens. *J. Food Sci. Technol.* **2016**, *53*, 1355–1362. [CrossRef]
42. Domingo-Calap, P.; Georgel, P.; Bahram, S. Back to the future: Bacteriophages as promising therapeutic tools. *HLA* **2016**, *87*, 133–140. [CrossRef]
43. Cisek, A.A.; Dąbrowska, I.; Gregorczyk, K.P.; Wyżewski, Z. Phage Therapy in Bacterial Infections Treatment: One Hundred Years after the Discovery of Bacteriophages. *Curr. Microbiol.* **2017**, *74*, 277–283. [CrossRef]
44. Grabowski, Ł.; Łepek, K.; Stasiłojć, M.; Kosznik-Kwaśnicka, K.; Zdrojewska, K.; Maciąg-Dorszyńska, M.; Węgrzyn, G.; Węgrzyn, A. Bacteriophage-encoded enzymes destroying bacterial cell membranes and walls, and their potential use as antimicrobial agents. *Microbiol. Res.* **2021**, *248*, 126746. [CrossRef]
45. Huang, Y.; Wang, W.; Zhang, Z.; Gu, Y.; Huang, A.; Wang, J.; Hao, H.; Huang, Y.; Wang, W.; Zhang, Z.; et al. Phage Products for Fighting Antimicrobial Resistance. *Microorganisms* **2022**, *10*, 1324. [CrossRef] [PubMed]
46. Endersen, L.; O'Mahony, J.; Hill, C.; Ross, R.P.; McAuliffe, O.; Coffey, A. Phage therapy in the food industry. *Annu. Rev. Food Sci. Technol.* **2014**, *5*, 327–349. [CrossRef] [PubMed]
47. Hobbs, Z.; Abedon, S.T. Diversity of phage infection types and associated terminology: The problem with 'Lytic or lysogenic'. *FEMS Microbiol. Lett.* **2016**, *363*, fnw047. [CrossRef] [PubMed]
48. Brix, A.; Cafora, M.; Aureli, M.; Pistocchi, A. Animal Models to Translate Phage Therapy to Human Medicine. *Int. J. Mol. Sci.* **2020**, *21*, 3715. [CrossRef] [PubMed]
49. Verheust, C.; Pauwels, K.; Mahillon, J.; Helinski, D.R.; Herman, P. Contained use of Bacteriophages: Risk Assessment and Biosafety Recommendations. *Appl. Biosaf.* **2010**, *15*, 32–44. [CrossRef]
50. Jamal, M.; Bukhari, S.M.A.U.S.; Andleeb, S.; Ali, M.; Raza, S.; Nawaz, M.A.; Hussain, T.; Rahman, S.U.; Shah, S.S.A. Bacteriophages: An overview of the control strategies against multiple bacterial infections in different fields. *J. Basic Microbiol.* **2019**, *59*, 123–133. [CrossRef] [PubMed]
51. Sklar, I.B.; Joerger, R.D. Attempts to utilize bacteriophage to combat Salmonella enterica serovar enteritidis infection in chickens. *J. Food Saf.* **2001**, *21*, 15–29. [CrossRef]
52. Fiorentin, L.; Vieira, N.D.; Barioni, W.J. Oral treatment with bacteriophages reduces the concentration of *Salmonella* Enteritidis PT4 in caecal contents of broilers. *Avian Pathol.* **2005**, *34*, 258–263. [CrossRef]
53. Sorour, H.K.; Gaber, A.F.; Hosny, R.A. Evaluation of the efficiency of using Salmonella Kentucky and *Escherichia coli* O119 bacteriophages in the treatment and prevention of salmonellosis and colibacillosis in broiler chickens. *Lett. Appl. Microbiol.* **2020**, *71*, 345–350. [CrossRef]
54. Loc Carrillo, C.; Atterbury, R.J.; El-Shibiny, A.; Connerton, P.L.; Dillon, E.; Scott, A.; Connerton, I.F. Bacteriophage Therapy to Reduce *Campylobacter jejuni* Colonization of Broiler Chickens. *Appl. Environ. Microbiol.* **2005**, *71*, 6554–6563. [CrossRef]
55. Wagenaar, J.A.; Bergen, M.A.P.V.; Mueller, M.A.; Wassenaar, T.M.; Carlton, R.M. Phage therapy reduces *Campylobacter jejuni* colonization in broilers. *Vet. Microbiol.* **2005**, *109*, 275–283. [CrossRef] [PubMed]
56. Carvalho, C.M.; Gannon, B.W.; Halfhide, D.E.; Santos, S.B.; Hayes, C.M.; Roe, J.M.; Azeredo, J. The in vivo efficacy of two administration routes of a phage cocktail to reduce numbers of *Campylobacter coli* and *Campylobacter jejuni* in chickens. *BMC Microbiol.* **2010**, *10*, 232. [CrossRef] [PubMed]
57. Kittler, S.; Fischer, S.; Abdulmawjood, A.; Glünder, G.; Klein, G. Effect of Bacteriophage Application on *Campylobacter jejuni* Loads in Commercial Broiler Flocks. *Appl. Environ. Microbiol.* **2013**, *79*, 7525–7533. [CrossRef] [PubMed]
58. Richards, P.J.; Connerton, P.L.; Connerton, I.F. Phage Biocontrol of *Campylobacter jejuni* in Chickens Does Not Produce Collateral Effects on the Gut Microbiota. *Front. Microbiol.* **2019**, *10*, 476. [CrossRef]
59. Barrow, P.; Lovell, M.; Berchieri, J. A Use of Lytic Bacteriophage for Control of Experimental *Escherichia coli* Septicemia and Meningitis in Chickens and Calves. *Clin. Diagn. Lab. Immunol.* **1998**, *5*, 294–298. [CrossRef]
60. Huff, W.E.; Huff, G.R.; Rath, N.C.; Balog, J.M.; Xie, H.; Moore, P.A., Jr.; Donoghue, A.M. Prevention of *Escherichia coli* Respiratory Infection in Broiler Chickens with Bacteriophage (SPR02). *Poult. Sci.* **2002**, *81*, 437–441. [CrossRef]
61. El-Gohary, F.A.; Huff, W.E.; Huff, G.R.; Rath, N.C.; Zhou, Z.Y.; Donoghue, A.M. Environmental augmentation with bacteriophage prevents colibacillosis in broiler chickens. *Poult. Sci.* **2014**, *93*, 2788–2792. [CrossRef]
62. Miller, R.W.; Skinner, J.; Sulakvelidze, A.; Mathis, G.F.; Hofacre, C.L. Bacteriophage Therapy for Control of Necrotic Enteritis of Broiler Chickens Experimentally Infected with Clostridium perfringens. *Avian Dis.* **2010**, *54*, 33–40. [CrossRef]
63. Bae, D.; Lee, J.W.; Chae, J.P.; Kim, J.W.; Eun, J.S.; Lee, K.W.; Seo, K.H. Characterization of a novel bacteriophage φCJ22 and its prophylactic and inhibitory effects on necrotic enteritis and *Clostridium perfringens* in broilers. *Poult. Sci.* **2021**, *100*, 302. [CrossRef]
64. Keerqin, C.; McGlashan, K.; Van, T.T.H.; Chinivasagam, H.N.; Moore, R.J.; Choct, M.; Wu, S.B. A lytic bacteriophage isolate reduced *Clostridium perfringens* induced lesions in necrotic enteritis challenged broilers. *Front. Vet. Sci.* **2022**, *9*. [CrossRef]
65. Adhikari, P.A.; Cosby, D.E.; Cox, N.A.; Lee, J.H.; Kim, W.K. Effect of dietary bacteriophage supplementation on internal organs, fecal excretion, and ileal immune response in laying hens challenged by Salmonella Enteritidis. *Poult. Sci.* **2017**, *96*, 3264–3271. [CrossRef]

66. Sevilla-Navarro, S.; Marín, C.; Cortés, V.; García, C.; Vega, S.; Catalá-Gregori, P. Autophage as a control measure for Salmonella in laying hens. *Poult. Sci.* **2018**, *97*, 4367–4373. [CrossRef] [PubMed]
67. Lee, M.J.; Hosseindoust, A.; Oh, S.M.; Ko, H.S.; Cho, E.S.; Sa, S.J.; Kim, Y.I.; Choi, J.W.; Kim, J.S. Impact of an anti-Salmonella. Typhimurium Bacteriophage on intestinal microbiota and immunity status of laying hens. *J. Anim. Physiol. Anim. Nutr.* **2021**, *105*, 952–959. [CrossRef] [PubMed]
68. Evran, S.; Tayyarcan, E.K.; Acar-Soykut, E.; Boyaci, I.H. Applications of Bacteriophage Cocktails to Reduce Salmonella Contamination in Poultry Farms. *Food Environ. Virol.* **2022**, *14*. [CrossRef]
69. Duc, H.M.; Son, H.M.; Honjoh, K.; Miyamoto, T. Isolation and application of bacteriophages to reduce *Salmonella contamination* in raw chicken meat. *Food Sci. Technol.* **2018**, *91*, 353–360. [CrossRef]
70. Kim, J.H.; Kim, H.J.; Jung, S.J.; Mizan, M.F.R.; Park, S.H.; Ha, S.D. Characterization of *Salmonella* spp.-specific bacteriophages and their biocontrol application in chicken breast meat. *J. Food Sci.* **2020**, *85*, 526–534. [CrossRef]
71. Noor Mohammadi, T.; Shen, C.; Li, Y.; Zayda, M.G.; Sato, J.; Masuda, Y.; Honjoh, K.i.; Miyamoto, T. Characterization of Clostridium perfringens bacteriophages and their application in chicken meat and milk. *Int. J. Food Microbiol.* **2022**, *361*. [CrossRef]
72. Tian, R.; Xu, S.; Li, P.; Li, M.; Liu, Y.; Wang, K.; Liu, G.; Li, Y.; Dai, L.; Zhang, W. Characterization of G-type Clostridium perfringens bacteriophages and their disinfection effect on chicken meat. *Anaerobe* **2023**, *81*, 102736. [CrossRef]
73. Sharma, C.S.; Dhakal, J.; Nannapaneni, R. Efficacy of Lytic Bacteriophage Preparation in Reducing Salmonella In Vitro, on Turkey Breast Cutlets, and on Ground Turkey. *J. Food Prot.* **2015**, *78*, 1357–1362. [CrossRef]
74. Sonalika, J.; Srujana, A.S.; Akhila, D.S.; Juliet, M.R.; Santhosh, K.S. Application of bacteriophages to control *Salmonella* Enteritidis in raw eggs. *Iran. J. Vet. Res.* **2020**, *21*, 221. [CrossRef]
75. Azari, R.; Yousefi, M.H.; Taghipour, Z.; Wagemans, J.; Lavigne, R.; Hosseinzadeh, S.; Mazloomi, S.M.; Vallino, M.; Khalatbari-Limaki, S.; Berizi, E. Application of the lytic bacteriophage Rostam to control *Salmonella enteritidis* in eggs. *Int. J. Food Microbiol.* **2023**, *389*, 110097. [CrossRef] [PubMed]
76. Abd-El Wahab, A.; Basiouni, S.; El-Seedi, H.R.; Ahmed, M.F.E.; Bielke, L.R.; Hargis, B.; Tellez-Isaias, G.; Eisenreich, W.; Lehnherr, H.; Kittler, S.; et al. An overview of the use of bacteriophages in the poultry industry: Successes, challenges, and possibilities for overcoming breakdowns. *Front. Microbiol.* **2023**, *14*, 1136638. [CrossRef] [PubMed]
77. Verbeken, G.; Pirnay, J.P.; De Vos, D.; Jennes, S.; Zizi, M.; Lavigne, R.; Casteels, M.; Huys, I. Optimizing the European regulatory framework for sustainable bacteriophage therapy in human medicine. *Arch. Immunol. Ther. Exp.* **2012**, *60*, 161–172. [CrossRef] [PubMed]
78. Naureen, Z.; Malacarne, D.; Anpilogov, K.; Dautaj, A.; Camilleri, G.; Cecchin, S.; Bressan, S.; Casadei, A.; Albion, E.; Sorrentino, E.; et al. Comparison between American and European legislation in the therapeutical and alimentary bacteriophage usage. *Acta Bio Medica Atenei Parm.* **2020**, *91*, 2020023. [CrossRef]
79. Diallo, K.; Dublanchet, A. A Century of Clinical Use of Phages: A Literature Review. *Antibiotics* **2023**, *12*, 751. [CrossRef] [PubMed]
80. Costa, M.J.; Pastrana, L.M.; Teixeira, J.A.; Sillankorva, S.M.; Cerqueira, M.A. Bacteriophage Delivery Systems for Food Applications: Opportunities and Perspectives. *Viruses* **2023**, *15*, 1271. [CrossRef]
81. Guo, Z.; Lin, H.; Ji, X.; Yan, G.; Lei, L.; Han, W.; Gu, J.; Huang, J. Therapeutic applications of lytic phages in human medicine. *Microb. Pathog.* **2020**, *142*, 104048. [CrossRef]
82. Efsa, P.B.H. Evaluation of the safety and efficacy of Listex™ P100 for reduction of pathogens on different ready-to-eat (RTE) food products. *EFSA J.* **2016**, *14*, e04565. [CrossRef]
83. Winfield, M.D.; Groisman, E.A. Role of Nonhost Environments in the Lifestyles of Salmonella and *Escherichia coli*. *Appl. Environ. Microbiol.* **2003**, *69*, 3687–3694. [CrossRef]
84. European Food Safety Authority; European Centre for Disease Prevention and Control. The European Union One Health 2021 Zoonoses Report. *EFSA J.* **2022**, *20*, e07666. [CrossRef]
85. European Food Safety Authority; European Centre for Disease Prevention and Control. The European Union summary report on antimicrobial resistance in zoonotic and indicator bacteria from humans, animals and food in 2016. *EFSA J.* **2018**, *16*, e05182. [CrossRef]
86. Bezek, K.; Avberšek, J.; Zorman Rojs, O.; Barlič-Maganja, D. Antimicrobial and Antibiofilm Effect of Commonly Used Disinfectants on Salmonella Infantis Isolates. *Microorganisms* **2023**, *11*, 301. [CrossRef] [PubMed]
87. Aviv, G.; Tsyba, K.; Steck, N.; Salmon-Divon, M.; Cornelius, A.; Rahav, G.; Grassl, G.A.; Gal-Mor, O. Unique megaplasmid contributes to stress tolerance and pathogenicity of an emergent Salmonella enterica serovar Infantis strain. *Environ. Microbiol.* **2014**, *16*, 977–994. [CrossRef] [PubMed]
88. Aviv, G.; Rahav, G.; Gal-Mor, O. Horizontal Transfer of the Salmonella enterica Serovar Infantis Resistance and Virulence Plasmid pESI to the Gut Microbiota of Warm-Blooded Hosts. *mBio* **2016**, *7*. [CrossRef] [PubMed]
89. Franco, A.; Leekitcharoenphon, P.; Feltrin, F.; Alba, P.; Cordaro, G.; Iurescia, M.; Tolli, R.; D'Incau, M.; Staffolani, M.; Di Giannatale, E.; et al. Emergence of a clonal lineage of multidrug-resistant ESBL-producing Salmonella infantis transmitted from broilers and broiler meat to humans in Italy between 2011 and 2014. *PLoS ONE* **2015**, *10*, e0144802. [CrossRef] [PubMed]
90. Bogomazova, A.N.; Gordeeva, V.D.; Krylova, E.V.; Soltynskaya, I.V.; Davydova, E.E.; Ivanova, O.E.; Komarov, A.A. Mega-plasmid found worldwide confers multiple antimicrobial resistance in Salmonella Infantis of broiler origin in Russia. *Int. J. Food Microbiol.* **2020**, *319*, 108497. [CrossRef]

91. Davies, R.; Breslin, M. Observations on Salmonella contamination of commercial laying farms before and after cleaning and disinfection. *Vet. Rec.* **2003**, *152*, 283–287. [CrossRef] [PubMed]
92. Hernández, A.; Ruiz, F.M.; Romero, A.; Martínez, J.L. The Binding of Triclosan to SmeT, the Repressor of the Multidrug Efflux Pump SmeDEF, Induces Antibiotic Resistance in *Stenotrophomonas maltophilia*. *PLoS Pathog.* **2011**, *7*, e1002103. [CrossRef]
93. Rodriguez, A.; Gutierrez, D.; Garcia, P.; Fernandez, L. Application of Bacteriophages in the Agro-Food Sector: A Long Way toward Approval. *Front. Cell. Infect. Microbiol.* **2018**, *8*, 296. [CrossRef]
94. Fotheringham, V.J.C. Disinfection of livestock production premises. *Rev. Sci. Tech. (Int. Off. Epizoot.)* **1995**, *14*, 191–205. [CrossRef]
95. Münster, P.; Pöppel, L.; Antakli, A.; Müller-Doblies, D.; Radko, D.; Kemper, N. The Detection of *Salmonella* Enteritidis on German Layer Farms after Cleaning and Disinfection. *Animals* **2023**, *13*, 2588. [CrossRef]
96. Gray, J.A.; Chandry, P.S.; Kaur, M.; Kocharunchitt, C.; Bowman, J.P.; Fox, E.M. Novel Biocontrol Methods for Listeria monocytogenes Biofilms in Food Production Facilities. *Front. Microbiol.* **2018**, *9*, 605. [CrossRef] [PubMed]
97. Martínez-Suárez, J.V.; Ortiz, S.; López-Alonso, V. Potential Impact of the Resistance to Quaternary Ammonium Disinfectants on the Persistence of Listeria monocytogenes in Food Processing Environments. *Front. Microbiol.* **2016**, *7*, 638. [CrossRef] [PubMed]
98. Møretrø, T.; Schirmer, B.C.T.; Heir, E.; Fagerlund, A.; Hjemli, P.; Langsrud, S. Tolerance to quaternary ammonium compound disinfectants may enhance growth of *Listeria monocytogenes* in the food industry. *Int. J. Food Microbiol.* **2017**, *241*, 215–224. [CrossRef] [PubMed]
99. Overney, A.; Jacques-André-Coquin, J.; Ng, P.; Carpentier, B.; Guillier, L.; Firmesse, O. Impact of environmental factors on the culturability and viability of *Listeria monocytogenes* under conditions encountered in food processing plants. *Int. J. Food Microbiol.* **2017**, *244*, 74–81. [CrossRef]
100. Gottardi, W.; Nagl, M. Chlorine covers on living bacteria: The initial step in antimicrobial action of active chlorine compounds. *J. Antimicrob. Chemother.* **2005**, *55*, 475–482. [CrossRef] [PubMed]
101. Zhang, L.; Ren, T.; Qiao, M.; Huang, T.; Xia, X. The reduction of Salmonella on chicken skin by the combination of sodium dodecyl sulfate with antimicrobial chemicals and coating wax microemulsions. *Poult. Sci.* **2019**, *98*, 2615–2621. [CrossRef] [PubMed]
102. Byun, K.; Han, S.H.; Yoon, J.; Park, S.H.; Ha, S. Efficacy of chlorine-based disinfectants (sodium hypochlorite and chlorine dioxide) on *Salmonella* Enteritidis planktonic cells, biofilms on food contact surfaces and chicken skin. *Food Control* **2021**, *123*, 107838. [CrossRef]
103. LeChevallier, M.W.; Au, K. *Water Treatment and Pathogen Control*; IWA Publishing: Geneva, Switzerland, 2004.
104. Kim, J.; Pitts, B.; Stewart, P.S.; Camper, A.; Yoon, J. Comparison of the Antimicrobial Effects of Chlorine, Silver Ion, and Tobramycin on Biofilm. *Antimicrob. Agents Chemother.* **2008**, *52*, 1446–1453. [CrossRef]
105. Ran, Y.; Qingmin, C.; Maorun, F. Chlorine Dioxide Generation Method and Its Action Mechanism for Removing Harmful Substances and Maintaining Quality Attributes of Agricultural Products. *Food Bioprocess Technol.* **2019**, *12*, 1110–1122. [CrossRef]
106. Villarreal, M.E.; Baker, R.C.; Regenstein, J.M. The incidence of Salmonella on poultry carcasses following the use of slow release chlorine dioxide (Alcide). *J. Food Prot.* **1990**, *53*, 465–467. [CrossRef] [PubMed]
107. Purnell, G.; James, C.; James, S.J.; Howell, M.; Corry, J.E.L. Comparison of Acidified Sodium Chlorite, Chlorine Dioxide, Peroxyacetic Acid and Tri-Sodium Phosphate Spray Washes for Decontamination of Chicken Carcasses. *Food Bioprocess Technol.* **2014**, *7*, 2093–2101. [CrossRef]
108. Mani-Lopez, E.; Garcia, H.S.; Lopez-Malo, A. Organic acids as antimicrobials to control *Salmonella* in meat and poultry products. *Food Res. Int.* **2012**, *45*, 713–721. [CrossRef]
109. Ricke, S.C. Perspectives on the use of organic acids and short chain fatty acids as antimicrobials. *Poult. Sci.* **2003**, *82*, 632–639. [CrossRef]
110. Madushanka, D.N.N.; Jayaweera, T.S.P.; Jayasinghe, J.M.C.S.; Yasawathie, D.G.; Ruwandeepika, H.A.D. Decontaminating Effect of Organic Acids and Natural Compounds on Broiler Chicken Meat Contaminated with *Salmonella typhimurium*. *Asian Food Sci. J.* **2018**, *3*, 1–9. [CrossRef]
111. Yeh, Y.; de Moura, F.H.; Van Den Broek, K.; de Mello, A.S. Effect of ultraviolet light, organic acids, and bacteriophage on Salmonella populations in ground beef. *Meat Sci.* **2018**, *139*, 44–48. [CrossRef]
112. Fernández, M.; Rodríguez, A.; Fulco, M.; Soteras, T.; Mozgovoj, M.; Cap, M. Effects of lactic, malic and fumaric acids on Salmonella spp. counts and on chicken meat quality and sensory characteristics. *J. Food Sci. Technol.* **2021**, *58*, 3817–3824. [CrossRef]
113. Radkowski, M.; Zdrodowska, B.; Gomółka-Pawlicka, M. Effect of Succinic Acid on Elimination of Salmonella in Chicken Meat. *J. Food Prot.* **2018**, *81*, 1491–1495. [CrossRef]
114. Cosansu, S.; Ayhan, K. Effects of Lactic and Acetic Acid on Survival of *Salmonella enteritidis* during Refrigerated and Frozen Storage of Chicken Meats. *Food Bioprocess Technol.* **2012**, *5*, 372–377. [CrossRef]
115. Abdul-Rahiman, U.A.; Nordin, N.; Abdul-Mutalib, N.A.; Sanny, M. Holistic Approaches to Reducing Salmonella Contamination in Poultry Industry. *Pertanika J. Trop. Agric. Sci.* **2021**, *44*. [CrossRef]
116. Bilgill, S.F.; Conner, D.E.; Pinion, J.L.; Tamblyn, K.C. Broiler skin color as affected by organic acids: Influence of concentration and method of application. *Poult. Sci.* **1998**, *77*, 752–757. [CrossRef]
117. Hajati, H. Application of organic acids in poultry nutrition. *Int. J. Avian Wildl. Biol.* **2018**, *3*, 324–329. [CrossRef]
118. Ben Braïek, O.; Smaoui, S. Chemistry, Safety, and Challenges of the Use of Organic Acids and Their Derivative Salts in Meat Preservation. *J. Food Qual.* **2021**, *2021*, 6653190. [CrossRef]

119. Ramirez-Hernandez, A.; Brashears, M.M.; Sanchez-Plata, M.X. Efficacy of Lactic Acid, Lactic Acid–Acetic Acid Blends, and Peracetic Acid To Reduce Salmonella on Chicken Parts under Simulated Commercial Processing Conditions. *J. Food Prot.* **2018**, *81*, 17–24. [CrossRef]
120. Wessels, K.; Rip, D.; Gouws, P. Salmonella in Chicken Meat: Consumption, Outbreaks, Characteristics, Current Control Methods and the Potential of Bacteriophage Use. *Foods* **2021**, *10*, 1742. [CrossRef]
121. Møretrø, T.; Langsrud, S. Listeria monocytogenes: Biofilm formation and persistence in food-processing environments. *Biofilms* **2004**, *1*, 107–121. [CrossRef]
122. Bridier, A.; Sanchez-Vizuete, P.; Guilbaud, M.; Piard, J.-C.; Naïtali, M.; Briandet, R. Biofilm-associated persistence of food-borne pathogens. *Food Microbiol.* **2015**, *45*, 167–178. [CrossRef]
123. Newell, D.G.; Koopmans, M.; Verhoef, L.; Duizer, E.; Aidara-Kane, A.; Sprong, H.; Opsteegh, M.; Langelaar, M.; Threfall, J.; Scheutz, F.; et al. Food-borne diseases—The challenges of 20years ago still persist while new ones continue to emerge. *Int. J. Food Microbiol.* **2010**, *139*, S3–S15. [CrossRef]
124. Abdallah, M.; Benoliel, C.; Drider, D.; Dhulster, P.; Chihib, N. Biofilm formation and persistence on abiotic surfaces in the context of food and medical environments. *Arch. Microbiol.* **2014**, *196*, 453–472. [CrossRef] [PubMed]
125. Lazazzera, B.A. Lessons from DNA microarray analysis: The gene expression profile of biofilms. *Curr. Opin. Microbiol.* **2005**, *8*, 222–227. [CrossRef] [PubMed]
126. Donlan, R.M.; Costerton, J.W. Biofilms: Survival Mechanisms of Clinically Relevant Microorganisms. *Clin. Microbiol. Rev.* **2002**, *15*, 148–167. [CrossRef] [PubMed]
127. Gutierrez, D.; Rodriguez-Rubio, L.; Martinez, B.; Rodriguez, A.; Garcia, P. Bacteriophages as Weapons Against Bacterial Biofilms in the Food Industry. *Front. Microbiol.* **2016**, *7*, 825. [CrossRef] [PubMed]
128. Chang, C.; Yu, X.; Guo, W.; Guo, C.; Guo, X.; Li, Q.; Zhu, Y. Bacteriophage-Mediated Control of Biofilm: A Promising New Dawn for the Future. *Front. Microbiol.* **2022**, *13*, 825828. [CrossRef] [PubMed]
129. Ioannou, P.; Baliou, S.; Samonis, G. Bacteriophages in Infectious Diseases and Beyond-A Narrative Review. *Antibiotics* **2023**, *12*, 1012. [CrossRef] [PubMed]
130. Hosny, R.A.; Shalaby, A.G.; Nasef, S.A.; Sorour, H.K. Antibiofilm activity of a lytic Salmonella phage on different Salmonella enterica serovars isolated from broiler farms. *Int. Microbiol.* **2023**, *26*, 205–217. [CrossRef]
131. Marin, C.; Hernandiz, A.; Lainez, M. Biofilm development capacity of Salmonella strains isolated in poultry risk factors and their resistance against disinfectants. *Poult. Sci.* **2009**, *88*, 424–431. [CrossRef] [PubMed]
132. Soro, A.B.; Whyte, P.; Bolton, D.J.; Tiwari, B.K. Strategies and novel technologies to control Campylobacter in the poultry chain: A review. *Compr. Rev. Food Sci. Food Saf.* **2020**, *19*, 1353–1377. [CrossRef]
133. Schmelcher, M.; Donovan, D.M.; Loessner, M.J. Bacteriophage endolysins as novel antimicrobials. *Future Microbiol.* **2012**, *7*, 1147–1171. [CrossRef]
134. Yan, J.; Mao, J.; Xie, J. Bacteriophage Polysaccharide Depolymerases and Biomedical Applications. *BioDrugs* **2014**, *28*, 265–274. [CrossRef]
135. Harper, D.; Parracho, H.; Walker, J.; Sharp, R.; Hughes, G.; Werthén, M.; Lehman, S.; Morales, S. Bacteriophages and Biofilms. *Antibiotics* **2014**, *3*, 270–284. [CrossRef]
136. Domingo-Calap, P.; Delgado-Martínez, J. Bacteriophages: Protagonists of a Post-Antibiotic Era. *Antibiotics* **2018**, *7*, 66. [CrossRef] [PubMed]
137. Chan, B.K.; Abedon, S.T. Bacteriophages and their enzymes in biofilm control. *Curr. Pharm. Des.* **2015**, *21*, 85–99. [CrossRef] [PubMed]
138. Young, R. Bacteriophage lysis: Mechanism and regulation. *Microbiol. Rev.* **1992**, *56*, 430–481. [CrossRef]
139. Fernandes, S.; São-José, C. Enzymes and Mechanisms Employed by Tailed Bacteriophages to Breach the Bacterial Cell Barriers. *Viruses* **2018**, *10*, 396. [CrossRef]
140. Davies, R.H.; Breslin, M. Persistence of *Salmonella* Enteritidis Phage Type 4 in the environment and arthropod vectors on an empty free-range chicken farm. *Environ. Microbiol.* **2003**, *5*, 79–84. [CrossRef] [PubMed]
141. Newton, K.; Gosling, B.; Rabie, A.; Davies, R. Field investigations of multidrug-resistant Salmonella Infantis epidemic strain incursions into broiler flocks in England and Wales. *Avian Pathol.* **2020**, *49*, 631–641. [CrossRef] [PubMed]
142. Wales, A.; Breslin, M.; Carter, B.; Sayers, R.; Davies, R. A longitudinal study of environmental salmonella contamination in caged and free-range layer flocks. *Avian Pathol.* **2007**, *36*, 187–197. [CrossRef]
143. Ethèves, M.A.; Choisis, N.; Alvarez, S.; Dalleau, F.; Hascoat, J.; Gallard, V.; Cardinale, E. Risk factors for *Salmonella enterica* subsp. enterica persistence in broiler-chicken flocks on Reunion Island. *Heliyon* **2021**, *7*, e06278. [CrossRef]
144. Korzeniowski, P.; Śliwka, P.; Kuczkowski, M.; Mišić, D.; Milcarz, A.; Kuźmińska-Bajor, M. Bacteriophage Cocktail Can Effectively Control Salmonella Biofilm in Poultry Housing. *Front. Microbiol.* **2022**, *13*, 901770. [CrossRef]
145. Ning, Z.; Zhang, L.; Cai, L.; Xu, X.; Chen, Y.; Wang, H. Biofilm removal mediated by Salmonella phages from chicken-related sources. *Food Sci. Hum. Wellness* **2023**, *12*, 1799–1808. [CrossRef]
146. Ge, H.; Lin, C.; Xu, Y.; Hu, M.; Xu, Z.; Geng, S.; Jiao, X.; Chen, X. A phage for the controlling of Salmonella in poultry and reducing biofilms. *Vet. Microbiol.* **2022**, *269*, 109432. [CrossRef] [PubMed]
147. Gong, C.; Jiang, X.; Wang, J. Application of bacteriophages to reduce Salmonella contamination on workers' boots in rendering-processing environment. *Poult. Sci.* **2017**, *96*, 3700–3708. [CrossRef] [PubMed]

148. Al Kandari, S. Characterization and Comparison of Campylobacter Bacteriophages. Doctoral Dissertation, University of Nottingham, Nottingham, UK, 2013.
149. Zeng, H.; De Reu, K.; Gabriël, S.; Mattheus, W.; De Zutter, L.; Rasschaert, G. Salmonella prevalence and persistence in industrialized poultry slaughterhouses. *Poult. Sci.* **2021**, *100*, 100991. [CrossRef] [PubMed]
150. Pintar, K.; Cook, A.; Pollari, F.; Ravel, A.; Lee, S.; Odumeru, J.A. Quantitative Effect of Refrigerated Storage Time on the Enumeration of Campylobacter, Listeria, and Salmonella on Artificially Inoculated Raw Chicken Meat. *J. Food Prot.* **2007**, *70*, 739–743. [CrossRef]
151. Maziero, M.T.; Oliveira, T.C.R.M.d. Effect of refrigeration and frozen storage on the *Campylobacter jejuni* recovery from naturally contaminated broiler carcasses. *Braz. J. Microbiol.* **2010**, *41*, 501–505. [CrossRef]
152. Blankenship, L.C.; Craven, S.E. *Campylobacter jejuni* survival in chicken meat as a function of temperature [Gastroenteritis]. *Appl. Environ. Microbiol.* **1982**, *44*, 88–92. [CrossRef] [PubMed]
153. Whiley, H.; Ross, K. Salmonella and Eggs: From Production to Plate. *Int. J. Environ. Res. Public Health* **2015**, *12*, 2543–2556. [CrossRef]
154. de Ornellas Dutka Garcia, K.C.; de Oliveira Corrêa, I.M.; Pereira, L.Q.; Silva, T.M.; de Souza Ribeiro Mioni, M.; de Moraes Izidoro, A.C.; Vellano Bastos, I.H.; Marietto Gonçalves, G.A.; Okamoto, A.S.; Andreatti Filho, R.L. Bacteriophage use to control Salmonella biofilm on surfaces present in chicken slaughterhouses. *Poult. Sci.* **2017**, *96*, 3392–3398. [CrossRef]
155. Harada, L.K.; Silva, E.C.; Campos, W.F.; Del Fiol, F.S.; Vila, M.; Dąbrowska, K.; Krylov, V.N.; Balcão, V.M. Biotechnological applications of bacteriophages: State of the art. *Microbiol. Res.* **2018**, *212–213*, 38–58. [CrossRef]
156. Galié, S.; García-Gutiérrez, C.; Miguélez, E.M.; Villar, C.J.; Lombó, F. Biofilms in the Food Industry: Health Aspects and Control Methods. *Front. Microbiol.* **2018**, *9*, 898. [CrossRef]
157. Merino, L.; Procura, F.; Trejo, F.M.; Bueno, D.J.; Golowczyc, M.A. Biofilm formation by *Salmonella* sp. in the poultry industry: Detection, control and eradication strategies. *Food Res. Int.* **2019**, *119*, 530–540. [CrossRef] [PubMed]
158. Siringan, P.; Connerton, P.L.; Cummings, N.J.; Connerton, I.F. Alternative bacteriophage life cycles: The carrier state of *Campylobacter jejuni*. *Open Biol.* **2014**, *4*, 130200. [CrossRef] [PubMed]
159. D'Accolti, M.; Soffritti, I.; Mazzacane, S.; Caselli, E. Bacteriophages as a Potential 360-Degree Pathogen Control Strategy. *Microorganisms* **2021**, *9*, 261. [CrossRef] [PubMed]
160. Hagens, S.; Loessner, M.J. Bacteriophage for Biocontrol of Foodborne Pathogens: Calculations and Considerations. *Curr. Pharm. Biotechnol.* **2010**, *11*, 58–68. [CrossRef] [PubMed]
161. Wu, R.; Davison, M.R.; Nelson, W.C.; Smith, M.L.; Lipton, M.S.; Jansson, J.K.; McClure, R.S.; McDermott, J.E.; Hofmockel, K.S. Hi-C metagenome sequencing reveals soil phage–host interactions. *Nat. Commun.* **2023**, *14*, 7666. [CrossRef] [PubMed]
162. Drulis-Kawa, Z.; Majkowska-Skrobek, G.; Maciejewska, B.; Delattre, A.; Lavigne, R. Learning from Bacteriophages-Advantages and Limitations of Phage and Phage-Encoded Protein Applications. *Curr. Protein Pept. Sci.* **2012**, *13*, 699–722. [CrossRef]
163. Goh, S. Phage Transduction. In *Clostridium Difficile*; Springer: New York, NY, USA, 2016; Volume 1476, pp. 177–185.
164. Chan, B.K.; Abedon, S.T.; Loc-Carrillo, C. Phage cocktails and the future of phage therapy. *Future Microbiol.* **2013**, *8*, 769–783. [CrossRef]
165. Brown, P.; Chen, Y.; Parsons, C.; Brown, E.; Loessner, M.J.; Shen, Y.; Kathariou, S. Whole Genome Sequence Analysis of Phage-Resistant *Listeria monocytogenes* Serotype 1/2a Strains from Turkey Processing Plants. *Pathogens* **2021**, *10*, 199. [CrossRef]
166. Kawacka, I.; Olejnik-Schmidt, A.; Schmidt, M.; Sip, A. Effectiveness of Phage-Based Inhibition of *Listeria monocytogenes* in Food Products and Food Processing Environments. *Microorganisms* **2020**, *8*, 1764. [CrossRef]
167. Aprea, G.; Zocchi, L.; Di Fabio, M.; De Santis, S.; Prencipe, V.A.; Migliorati, G. The applications of bacteriophages and their lysins as biocontrol agents against the foodborne pathogens *Listeria monocytogenes* and Campylobacter: An updated look. *Vet. Ital.* **2018**, *54*, 293–303. [CrossRef]
168. EFSA BIOHAZ Panel; Koutsoumanis, K.; Allende, A.; Alvarez-Ordonez, A.; Bolton, D.; Bover-Cid, S.; Chemaly, M.; De Cesare, A.; Hilbert, F.; Lindqvist, R.; et al. Updated list of QPS-recommended microorganisms for safety risk assessments carried out by EFSA. 2023. Available online: https://zenodo.org/records/8124409 (accessed on 1 August 2023).
169. Boulanger, P. Purification of Bacteriophages and SDS-PAGE Analysis of Phage Structural Proteins from Ghost Particles. In *Bacteriophages*; Humana Press: Totowa, NJ, USA, 2009; Volume 502, pp. 227–238.
170. Gill, J.J.; Hyman, P. Phage Choice, Isolation, and Preparation for Phage Therapy. *Curr. Pharm. Biotechnol.* **2010**, *11*, 2–14. [CrossRef]
171. Manohar, P.; Ramesh, N. Improved lyophilization conditions for long-term storage of bacteriophages. *Sci. Rep.* **2019**, *9*, 15242. [CrossRef] [PubMed]
172. Zimmer, M.; Scherer, S.; Loessner, M.J. Genomic Analysis of Clostridium perfringens Bacteriophage φ3626, Which Integrates into guaA and Possibly Affects Sporulation. *J. Bacteriol.* **2002**, *184*, 4359–4368. [CrossRef] [PubMed]
173. Lim, T.; Kim, M.; Lee, D.; Lee, Y.; Park, J.; Youn, H.; Lee, H.; Yang, S.; Cho, Y.; Lee, J.; et al. Use of bacteriophage for biological control of *Salmonella* Enteritidis infection in chicken. *Res. Vet. Sci.* **2012**, *93*, 1173–1178. [CrossRef]
174. Berchieri, A.; Lovell, M.A.; Barrow, P.A. The activity in the chicken alimentary tract of bacteriophages lytic for *Salmonella typhimurium*. *Res. Microbiol.* **1991**, *142*, 541–549. [CrossRef] [PubMed]
175. Bigot, B.; Lee, W.-J.; McIntyre, L.; Wilson, T.; Hudson, J.A.; Billington, C.; Heinemann, J.A. Control of Listeria monocytogenes growth in a ready-to-eat poultry product using a bacteriophage. *Food Microbiol.* **2011**, *28*, 1448–1452. [CrossRef]

176. EFSA. Scientific Opinion on the maintenance of the list of QPS microorganisms intentionally added to food or feed (2009 update). *EFSA J.* **2009**, *7*, 1431. [CrossRef]
177. Goodridge, L.D.; Bisha, B. Phage-based biocontrol strategies to reduce foodborne pathogens in foods. *Bacteriophage* **2011**, *1*, 130–137. [CrossRef]
178. Moineau, S.; Labrie, S.J.; Samson, J.E. Bacteriophage resistance mechanisms. *Nat. Rev. Microbiol.* **2010**, *8*, 317–327. [CrossRef]
179. Shabbir, M.A.B.; Hao, H.; Shabbir, M.Z.; Wu, Q.; Sattar, A.; Yuan, Z. Bacteria vs. Bacteriophages: Parallel Evolution of Immune Arsenals. *Front. Microbiol.* **2016**, *7*, 1292. [CrossRef]
180. Sorek, R.; Kunin, V.; Hugenholtz, P. CRISPR—A widespread system that provides acquired resistance against phages in bacteria and archaea. *Nat. Rev. Microbiol.* **2008**, *6*, 181–186. [CrossRef]
181. Li, K.; Barksdale, L.; Garmise, L. Phenotypic Alterations Associated with the Bacteriophage Carrier State of *Shigella dysenteriae*. *J. Gen. Microbiol.* **1961**, *24*, 355–367. [CrossRef] [PubMed]
182. Jarling, M.; Bartkowiak, K.; Robenek, H.; Pape, H.; Meinhardt, F. Isolation of phages infecting Actinoplanes SN223 and characterization of two of these viruses. *Appl. Microbiol. Biotechnol.* **2004**, *64*, 250–254. [CrossRef]
183. Bastías, R.; Higuera, G.; Sierralta, W.; Espejo, R.T. new group of cosmopolitan bacteriophages induce a carrier state in the pandemic strain of *Vibrio parahaemolyticus*. *Environ. Microbiol.* **2010**, *12*, 990–1000. [CrossRef] [PubMed]
184. Scott, A.E.; Timms, A.R.; Connerton, P.L.; Loc Carrillo, C.; Adzfa Radzum, K.; Connerton, I.F. Genome Dynamics of *Campylobacter jejuni* in Response to Bacteriophage Predation. *PLoS Pathog.* **2007**, *3*, e119. [CrossRef] [PubMed]
185. Hooton, S.P.T.; Connerton, I.F. *Campylobacter jejuni* acquire new host-derived CRISPR spacers when in association with bacteriophages harboring a CRISPR-like Cas4 protein. *Front. Microbiol.* **2015**, *5*, 744. [CrossRef]
186. Lis, L.; Connerton, I.F. The Minor Flagellin of *Campylobacter jejuni* (FlaB) Confers Defensive Properties against Bacteriophage Infection. *Front. Microbiol.* **2016**, *7*, 1908. [CrossRef]
187. Gencay, Y.E.; Sørensen, M.C.H.; Wenzel, C.Q.; Szymanski, C.M.; Brøndsted, L. Phase Variable Expression of a Single Phage Receptor in *Campylobacter jejuni* NCTC12662 Influences Sensitivity toward Several Diverse CPS-Dependent Phages. *Front. Microbiol.* **2018**, *9*, 82. [CrossRef]
188. Olson, E.G.; Micciche, A.C.; Rothrock, M.J.; Yang, Y.; Ricke, S.C. Application of Bacteriophages to Limit Campylobacter in Poultry Production. *Front. Microbiol.* **2021**, *12*, 458721. [CrossRef]
189. Piel, D.; Bruto, M.; Labreuche, Y.; Blanquart, F.; Goudenège, D.; Barcia-Cruz, R.; Chenivesse, S.; Le Panse, S.; James, A.; Dubert, J.; et al. Phage–host coevolution in natural populations. *Nat. Microbiol.* **2022**, *7*, 1075–1086. [CrossRef]
190. Burmeister, A.R.; Fortier, A.; Roush, C.; Lessing, A.J.; Bender, R.G.; Barahman, R.; Grant, R.; Chan, B.K.; Turner, P.E. Pleiotropy complicates a trade-off between phage resistance and antibiotic resistance. *Proc. Natl. Acad. Sci. USA* **2020**, *117*, 11207–11216.
191. Guenther, S.; Herzig, O.; Fieseler, L.; Klumpp, J.; Loessner, M.J. Biocontrol of Salmonella Typhimurium in RTE foods with the virulent bacteriophage FO1-E2. *Int. J. Food Microbiol.* **2012**, *154*, 66–72. [CrossRef] [PubMed]
192. Oliveira, M.; Viñas, I.; Colàs, P.; Anguera, M.; Usall, J.; Abadias, M. Effectiveness of a bacteriophage in reducing Listeria monocytogenes on fresh-cut fruits and fruit juices. *Food Microbiol.* **2014**, *38*, 137–142. [CrossRef] [PubMed]
193. Guenther, S.; Huwyler, D.; Richard, S.; Loessner, M.J. Virulent Bacteriophage for Efficient Biocontrol of Listeria monocytogenes in Ready-To-Eat Foods. *Appl. Environ. Microbiol.* **2009**, *75*, 93–100. [CrossRef] [PubMed]
194. Anany, H.; Chen, W.; Pelton, R.; Griffiths, M.W. Biocontrol of Listeria monocytogenes and *Escherichia coli* O157:H7 in Meat by Using Phages Immobilized on Modified Cellulose Membranes. *Appl. Environ. Microbiol.* **2011**, *77*, 6379–6387. [CrossRef]
195. Wójcicki, M.; Świder, O.; Gientka, I.; Błażejak, S.; Średnicka, P.; Shymialevich, D.; Cieślak, H.; Wardaszka, A.; Emanowicz, P.; Sokołowska, B.; et al. Effectiveness of a Phage Cocktail as a Potential Biocontrol Agent against Saprophytic Bacteria in Ready-To-Eat Plant-Based Food. *Viruses* **2023**, *15*, 172. [CrossRef]
196. Hudson, J.A.; Billington, C.; Carey-Smith, G.; Greening, G. Bacteriophages as biocontrol agents in food. *J. Food Prot.* **2005**, *68*, 426–437. [CrossRef] [PubMed]
197. Leverentz, B.; Conway, W.S.; Alavidze, Z.; Janisiewicz, W.J.; Fuchs, Y.; Camp, M.J.; Chighladze, E.; Sulakvelidze, A. Examination of Bacteriophage as a Biocontrol Method for Salmonella on Fresh-Cut Fruit: A Model Study. *J. Food Prot.* **2001**, *64*, 1116–1121. [CrossRef]
198. Sharma, M.; Patel, J.R.; Conway, W.S.; Ferguson, S.; Sulakvelidze, A. Effectiveness of Bacteriophages in Reducing *Escherichia coli* O157:H7 on Fresh-Cut Cantaloupes and Lettuce. *J. Food Prot.* **2009**, *72*, 1481–1485. [CrossRef]
199. Atterbury, R.J.; Connerton, P.L.; Dodd, C.E.R.; Rees, C.E.D.; Connerton, I.F. Application of host-specific bacteriophages to the surface of chicken skin leads to a reduction in recovery of *Campylobacter jejuni*. *Appl. Environ. Microbiol.* **2003**, *69*, 6302–6306. [CrossRef]
200. Goode, D.; Allen, V.M.; Barrow, P.A. Reduction of experimental *Salmonella* and *Campylobacter* contamination of chicken skin by application of lytic bacteriophages. *Appl. Environ. Microbiol.* **2003**, *69*, 5032–5036. [CrossRef]
201. Rendueles, O.; de Sousa, J.A.M.; Rocha, E.P.C. Competition between lysogenic and sensitive bacteria is determined by the fitness costs of the different emerging phage-resistance strategies. *eLife* **2023**, *12*. [CrossRef]
202. Jorquera, D.; Galarce, N.; Borie, C. The challenge of controlling foodborne diseases: Bacteriophages as a new biotechnological tool. *Rev. Chil. Infectol.* **2015**, *32*, 678–688. [CrossRef]

animals

Review

Salmonella Control in Swine: A Thoughtful Discussion of the Pre- and Post-Harvest Control Approaches in Industrialized Countries

Ana Carvajal [1], Melvin Kramer [2] and Héctor Argüello [1,*]

[1] Departamento de Sanidad Animal, Campus Vegazana, 2007 Leon, Spain; ana.carvajal@unileon.es
[2] EHA Consulting Group, Fort Lauderdale, FL 33306, USA; mkramer@ehagroup.com
* Correspondence: hector.arguello@unileon.es

Simple Summary: Zoonoses, particularly foodborne zoonoses, are a major problem worldwide, jeopardizing human and animal health and negatively impacting other aspects such as the economy, food systems, and human and animal wellness. This review focuses on the control of the foodborne pathogen *Salmonella* in pigs and pork, revising past and ongoing control programs, the control options in the food chain, and the impact of policies implemented worldwide. The objective is to take a closer look at the efforts made in the last 2–3 decades in *Salmonella* control in pigs, evidencing their strengths, pitfalls, and limitations and determining future directions for the efficient surveillance and control of this pathogen in the future.

Abstract: Pork is among the major sources of human salmonellosis in developed countries. Since the 1990s, different surveys and cross-sectional studies, both national and international (i.e., the baseline studies performed in the European Union), have revealed and confirmed the widespread non-typhoidal *Salmonella* serotypes in pigs. A number of countries have implemented control programs with different approaches and degrees of success. The efforts could be implemented either at farms, in post-harvest stages, or both. The current review revises the current state of the art in *Salmonella* in swine, the control programs ongoing or conducted in the past, and their strengths and failures, with particular attention to the weight of pre- and post-harvest control and the implications that both have for the success of interventions or mitigation after outbreaks. This review provides a novel perspective on *Salmonella* control in swine, a matter that still includes uncertainties and room for improvement as a question of public health and One Health.

Keywords: pig; zoonosis; public health; foodborne pathogen; mitigation

1. Introduction Scope and Aim of This Review

As a pathogen, *Salmonella* is a genus in which species and serovars have evolved to colonize different hosts and environments [1]. Among the serovars included in *Salmonella enterica*, subspecies *enterica*, a few have adapted to specific hosts, causing severe disease. For instance, host-specific serovars include *Salmonella* Typhi and *S.* Paratyphi in humans and the serovar *S.* Cholerasuis in pigs [2–4]. In contrast to these host-adapted serovars, the vast majority of *Salmonella* serovars have chosen a different evolutionary strategy, allowing them to colonize a wide range of hosts, including warm- and cold-blooded species. They are generically known as non-typhoidal serovars, a definition that refers to their tendency to restrict colonization to the intestine without causing systemic disease in the host, unlike the aforementioned serovars [3].

Over the last half-century, non-typhoidal serovars have successfully spread into food production animals, particularly in pig and poultry production. Industrialized production systems with high stock densities, coupled with subclinical outcomes of infection characterized by large pathogen shedding in feces with mild or no pathogenic signs, have facilitated the spread of *Salmonella* in poultry, turkey, pig, and calf productions [5–9].

Consequently, *Salmonella* is a major foodborne pathogen, causing approximately 150 million illnesses and 60,000 deaths globally each year [10]. Major sources of *Salmonella* infection include chicken, turkey meat, eggs, pork, and derived products [11,12]. Human salmonellosis, attributed to pork consumption, ranks among the most highly reported foodborne illnesses [12]. Since the 1990s, many countries have implemented control and surveillance programs aimed at reducing the risk of *Salmonella* transmission by pork into the food chain [13–16]. These programs target *Salmonella* control at different production stages, including feed mills, farms, and post-harvest control. Each strategy or approach has its strengths and shortcomings, leading to different perspectives among the main actors involved in pathogen control. This review aims to bring the readers closer to the control of *Salmonella* in pig production tackling aspects of the origin of control programs, their progression or even discontinuation, main aspects for pre and post-harvest control and legal aspects of *Salmonella* outbreaks in humans in different countries. This review provides a novel perspective on reviews focusing on *Salmonella* control in swine, an issue that still includes uncertainties and room for improvement in public health and One Health.

2. Identifying Relevant Literature

This study did not apply a systematic review or meta-analysis. Instead, the relevant literature on the different topics covered in the review was searched using different scientific databases (www.pubmed.com; accessed the last ie on 20 January 2024; www.sciencedirect.com; accessed the last ie on 20 January 2024) and the website search tool (www.google.es; accessed the last ie on 20 January 2024). Relevant scientific papers published in peer-reviewed English journals were identified using the following keyword combinations: (pig OR Salmonella OR farm OR slaughterhouse OR control OR program) AND (livestock OR swine OR farrow OR weaner OR finisher OR sow OR carcass) AND (acid OR vaccine OR antimic* OR HACCP, OR risk), among others. The search did not have any restriction on dates, so any document in the databases to date was searched to capture up-to-date data. To ensure a wide range of articles from different sources, additional searches were conducted using the reference lists of key articles.

3. *Salmonella* Control in Swine Production: Origins

Salmonella control in swine production was initiated in the 1990s and early 2000s by the establishment of national initiatives to reduce the burden of this pathogen in the food chain, encouraged in part by the results of programs in poultry production.

The pioneering program was the Danish *Salmonella* control program, established in 1993, with the development and implementation of a surveillance program for Danish pork and for slaughter pig herds in 1995 [13]. The program was a response to human cases of salmonellosis linked to pork, with a peak incidence in 1993 [17,18]. Other programs in Europe followed the Danish initiative. Chronologically, Ireland established a surveillance program for pig herds [14] in 1997, which remains in force today [19]. The German *Salmonella* Monitoring Program was established in 2002 through the German Quality Assurance System for the food chain, the so-called "QS-System" [20,21]. In the same year, the pig industry in the UK launched the Zoonosis Action Plan (ZAP) to categorize pig herds based on their *Salmonella* prevalence. The program was revised in 2008 under a new name, the "Zoonoses Control Program" [16], and was finally suspended in 2012 [22]. Finally, the Netherlands and Belgium established their respective programs in 2005. In the Netherlands, compulsory *Salmonella* monitoring in fattening pigs was initiated by the Product Boards for Livestock, Meat, and Eggs, whereas the Belgian program [23], named the National *Salmonella* Action Plan (SAP), originally targeted herds with more than 30 pigs and has since changed several times, currently operating on a voluntary basis [24]. For further information about control programs in European countries, specific reviews are accessible elsewhere [25].

Outside of European countries, few have undertaken actions to mitigate the pathogen in primary production. Despite the extensive literature on *Salmonella* epidemiology in Asian countries [26–28] and North and South American regions [29–32], only specific actions have been implemented in the United States [33]. In this country, the Food Safety and Inspection Service (FSIS) published a final rule on pathogen reduction (PR) and hazard analysis and critical control point (HACCP) systems in 1996. The final rule required meat and poultry establishments under federal inspection to take responsibility for preventing and reducing physical, chemical, and biological hazards throughout the food production process by implementing a system of science-based preventive controls, known as HACCP. Establishments must have an effective HACCP food safety system to comply with regulatory requirements, focusing on controlling hazards to prevent product adulteration. This approach relies on post-harvest decontamination for *Salmonella* control, with no surveillance in primary production. Undoubtedly, the approach taken to perform *Salmonella* control impacts the results and approaches to dealing with the pathogen, factors which will be introduced and discussed in subsequent sections.

4. Rationale of *Salmonella* Surveillance in Control Programs

Salmonella control can be approached from various perspectives or strategies, all valid but each with its weaknesses. Not previously mentioned, the most successful strategy, followed in Sweden, Norway, and Finland, is based on pre-harvest surveillance programs combined with an eradication strategy [34–36]. This strategy maintains *Salmonella* prevalence at 0% by identifying infected animals that are then condemned. However, this ideal strategy is only feasible when sporadic infections in production animals occur. In countries with a significant prevalence or larger production, characterized by higher movement and pig imports, this approach would fail, and the cost of eradication would be impractical. Therefore, in countries with a non-negligible herd and within-herd prevalence of *Salmonella*, more realistic options are needed to mitigate the risk of human infections.

All other European control programs rely on farm categorization based on serological *Salmonella* surveillance (Table 1). This on-farm categorization is based on the presence and quantification of antibodies against *Salmonella*-LPS, which demonstrate the on-farm contact between the pathogen and the animal [37]. Serology through Enzyme-Linked Immunosorbent Assays (ELISA) has been the technique of choice for several reasons. Firstly, it is less expensive than microbiological methods, can be easily automated, and *Salmonella* antibodies persist for long periods, overcoming the intermittent fecal shedding that may lead to false-negative results. Additionally, samples can be easily collected at the slaughterhouse (either sampling blood or meat juice samples). Thus, this technique has become the gold standard for *Salmonella* monitoring, with only the Danish control program conducting complementary bacteriological analyses [18].

Table 1. Summary of surveillance programs in place since the 1990s in *Salmonella* control in pigs.

Program (by Country)	Status	Farm Monitoring	Carcass Monitoring	Penalties	Demonstrated Impact	References
Denmark	Ongoing	Yes	Yes	Yes	Yes	[13,17,18]
Germany	Ongoing, voluntarily	Yes	No	Yes	No	[15,21]
Ireland	Ongoing, under animal health program	Yes	No	No	No	[14]
United Kingdom	Discontinued	Yes	No	No	No	[16,22]
Belgium	Ongoing, voluntarily	Yes	No	No	No	[24]
The Netherlands	Ongoing	Yes	No	No	No	[23]

Despite herd surveillance and on-farm categorization based on *Salmonella* indirect burden estimation, not all programs act based on prevalence reduction or penalty systems considering their *Salmonella* serological results (Table 1). Indeed, only the Danish and German programs have established penalty systems that devalue carcass refunds from animals from infected farms [18,25]. Furthermore, some programs (such as in Denmark) also monitor carcass prevalence. Danish prevalence targets are based on carcass prevalence rather than herd prevalence [18]. *Salmonella* carcass prevalence provides complementary information to serological surveillance on the farm, reflecting the potential introduction of *Salmonella* into the food chain (cutting plants and retailers). Moreover, this relevant information can be used to assess the effectiveness of interventions taken to reduce prevalence, not only on the farm but also in post-harvest stages [38]. Carcass monitoring is also performed in EU countries, regardless if they have a control program or not, by the Competent Authorities and Food Business Operators, either by official requirements or internal audits. Although this carcass monitoring is not part of the control programs, the results may help to inform about the success of the interventions put in place. More information about carcass contamination and results from scientific studies can be found in specific reviews on the topic [38,39]. The national control programs currently underway do not establish penalties for either the supplier (farm) or processor (abattoir) associated with carcass prevalence, although purchasers or importers may establish regulations that include the absence of *Salmonella* in the products, thus establishing an indirect firewall against *Salmonella*-contaminated pork.

The aim of these programs is not just limited to detecting *Salmonella* carriers in the food chain but also to reducing the *Salmonella* burden over time. Have the current control programs and their policies achieved any reduction? This question will be discussed in the next section.

5. Outcomes of over Two Decades of *Salmonella* Surveillance

The ongoing programs mentioned above imply a significant economic investment and efforts by public bodies involved in official control, as well as industries, practitioners, and farmers participating in the surveillance. Indirectly, concern about *Salmonella* also generates costs in research about *Salmonella* etiology, epidemiology, and control. With this information in hand, the question to be asked is probably "is all this effort worthwhile?".

A case study published by Alban and colleagues in 2012 [18] described the lessons learned in the Danish *Salmonella* surveillance and control program for pigs and pork. In 2001, its prevalence in pork produced by members of the Danish Agriculture & Food Council was 1.7%, according to that study. A decade later, in 2011 and onwards, carcass prevalence ranged between 1 and 1.4% [40]. According to the authors of the last reference, in 2015, 24 years after the beginning of the program, the Danish *Salmonella* program reduced the number of human cases associated with Danish pork by more than 95% [40]. Undoubtedly, these efforts show figures that highlight the benefits of the interventions. But what are the costs? According to Danish researchers, the initial expenses of the program were approximately EUR 13 million per year. Adjusting the program, costs were significantly reduced to approximately EUR 3 million per year. To be a pioneer implies risks and failures. Not everything in the Danish program is a successful story, and some interventions or decisions did not provide the desired results. Between 1996 and 2010, an eradication policy was pursued for a highly concerning strain from a human health perspective, the multi-resistant *Salmonella* Typhimurium DT104 [41]. The plan did not achieve the established targets, and the total expenses spent were enormous, reaching more than EUR 14 million [18]. The nature of the design made *Salmonella* TyphimuriumDT104 vanish from pig farms in the decade of the 2010s, replaced by other emerging strains [42,43].

The demonstrated success of the Danish control program, both in carcass prevalence and human-related cases, contrasts with the partial figures reported by other European control programs. The ZAP program in the UK, as mentioned above, was discontinued in 2015. The German control program aimed at reducing herds with high seroprevalence of

Salmonella. According to data provided by Blaha and colleagues [44], although herds falling into the highest level category of *Salmonella* prevalence were able to reduce prevalence figures, thus lowering their intra-herd prevalence of *Salmonella*-antibody positive pigs and herds with mild or low prevalence used to increase in prevalence. As a result, the overall national frequency of *Salmonella*-antibody-positive pigs supplied to the slaughter plants did not show any sort of improvement [45]. Other programs such as the Irish or Belgian programs also did not observe any improvement in *Salmonella* figures [21]. Consequently, the Irish NSCP is under review by the Pig HealthCheck program of Animal Health Ireland [25].

Why have these programs failed? Probably, the question requires a complex analysis, but several reasons can be identified as major motives. Firstly, as mentioned above, different programs such as the Belgian, Irish, or Dutch do not include penalties for high-prevalence herds. Without that penalty, the motivation to reduce on-farm prevalence may not be as crucial. A recent study performed in Denmark observed differences among farmers' perceptions of costs and actions toward *Salmonella* control, which might hamper the effectiveness of the penalty scheme as a regulatory instrument to influence farmers' behavior [46].

Other potential challenges that may hamper the interventions taken could be related to the complex control of the pathogen [47–50], as well as human factors [51]. For instance, Blaha identified a "lean-back" attitude in farmers that succeeded in achieving a reduction in *Salmonella* prevalence, linking low prevalence serology with *Salmonella*-free herds, while the pathogen was still circulating there. That relaxation in the interventions taken to reduce *Salmonella* prevalence ended in an increase in prevalence finally [44]. A similar idea is reflected by Marier and colleagues in a study performed in the UK [52]. This study confirmed that farmers accepted their responsibility for controlling *Salmonella* in pork, even though their confidence in their ability to control *Salmonella* decreased over time, and believed that responsibility should be shared with the rest of the production chain, a fact that matches the perception of German studies [45].

The control of the pathogen seems to be complex according to the information gained so far in this article. However, complexity is not synonymous with impossibility. The next sections summarize the experience gained in *Salmonella* control both on farms and in post-farm stages.

6. *Salmonella* Control in Pig Production

To assist surveillance programs in pig production, actions must be taken to mitigate *Salmonella* risk throughout the food chain. *Salmonella* mitigation strategies can be approached from different angles, meaning different production stages. Each intervention will have benefits but also weaknesses and risks of reinfection or recontamination. In this subsection, we aim to discuss the various alternatives available for *Salmonella* control (Figure 1). A thorough review of potential interventions at each production stage is not the objective of this review. Detailed and extensive reviews on this topic can be found elsewhere [39,44,47,48,51–57]. Nonetheless, several notes about the two main stages of pathogen control will be described in the subsequent paragraphs. This information will allow us to discuss the impact of on-farm and post-harvest control strategies for the control programs and the perception of the actors involved in *Salmonella* control.

6.1. The Control of Salmonella on the Farm

The study of control options to mitigate *Salmonella* prevalence and infection transmission on swine farms has been and is the subject of scientific research. Numerous published scientific peer-reviewed articles have evaluated in vivo control strategies or risk factors for *Salmonella* prevalence on swine farms, all aiming to extend the current knowledge in *Salmonella* control and identify tools and strategies to mitigate the risk of infection or the burden of this pathogen in live pigs. Despite the extensive research, variability in results is common, thus not providing absolute certainty about the formula to mitigate *Salmonella* on the farm.

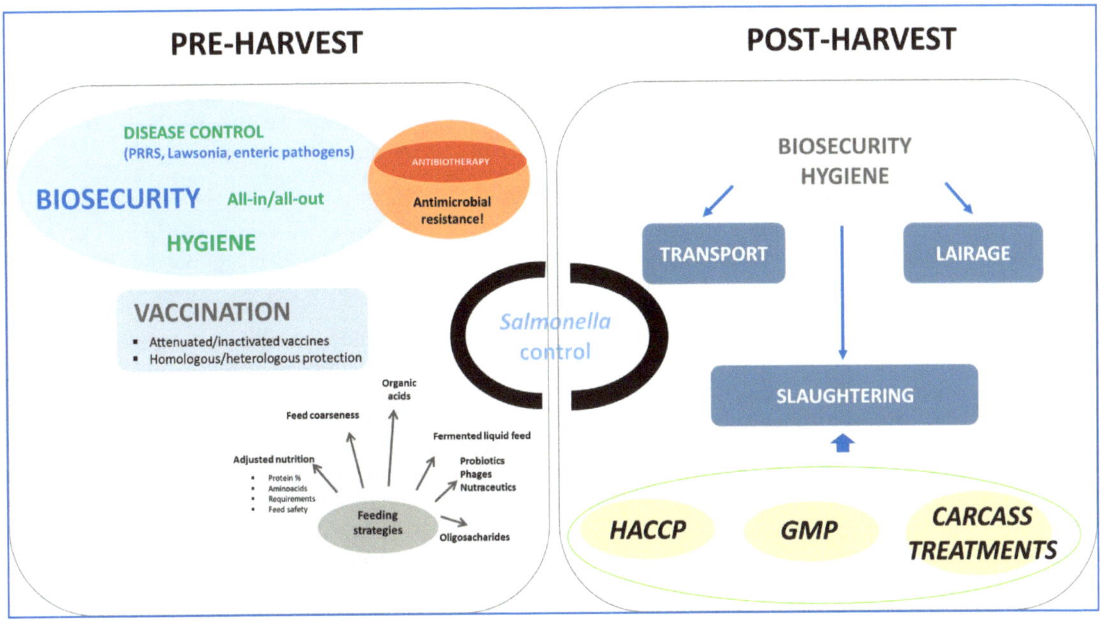

Figure 1. Summary of *Salmonella* control options in pre-harvest and post-harvest swine production.

In 2015, the FAO published a systematic review of *Salmonella* control in beef and pork, covering different steps in the food chain [58]. The FAO book report performed a thorough review of the options to control the pathogen on farms. Among all the interventions mentioned, experts highlighted several as the most efficient or useful. Biosecurity was highlighted as a good farm practice limiting the likelihood of pathogen introduction on the farm. Several studies have demonstrated that improved on-farm biosecurity limits the risk of on-farm *Salmonella* prevalence [59–61]. It seems logical to think that it is more plausible to introduce *Salmonella* by "four legs" than other potential sources such as feed, wild animals, or through water or semen. Nonetheless, all can be prevented with appropriate measures of external biosecurity on the farm, particularly restrictive policies for vehicles and visitors who come into contact with other herds. Measures of internal biosecurity, meaning limiting the transmission or circulation of the pathogen within the farm, seem to be less efficient than external biosecurity measures, probably due to lower or poorer internal biosecurity [32,62]. All-in/all-out husbandry, which includes emptying and cleaning barns and disinfection of facilities before introducing new animals, is the most useful tool to break infection cycles between animal batches [63]. Despite this activity being common among herds, risk factor studies have not demonstrated any clear benefit for *Salmonella* mitigation when this intervention is applied [59,64,65]. Different aspects like inefficient cleaning or commingling of animals from different batches may be behind the lack of benefits [66]. Thus, while introducing the pathogen on a farm can be prevented, once inside, breaking transmission is highly difficult.

The FAO document also mentions other interventions associated with feed or vaccination [58]. Feed serves as a vehicle for different strategies to mitigate *Salmonella* on the farm. Feed form [49,67,68], feed coarseness [69], feed additives [70–72], or probiotics administered through feed [73] have demonstrated their potential beneficial effect either by impacting the pathogen directly [74] or by improving intestinal health, thus limiting the opportunity for *Salmonella* to colonize the gut. Apart from the mentioned strategies, only vaccination is a strategy that can be performed on a large scale [58]. Again, different vaccines have been tested using different vaccination approaches [75–78]. *Salmonella* vaccines are not among the commonly used vaccines on farms, probably for different reasons; firstly, because they are not available in all countries, and secondly, because their efficacy is limited to the serotype infecting the animal [77]. While countries with serological surveillance may be reluctant to the use of vaccination due to interference with serological tests, they are extensively used in countries with clinical problems of salmonellosis like the USA (Fernando Leite, personal communication). Current policies to reduce antimicrobial use, together with the therapeutic use of ZnO in the EU [79], may be a reason to increase the use of vaccines in *Salmonella* control.

Apart from these interventions, other options are currently at a lower technology readiness level. For instance, phages, new nutraceuticals, or antimicrobial peptides are strategies with potential but with low field applicability right now [80–82]. There is not a magic formula for *Salmonella* control on the farm. Indeed, none of the options mentioned in this section exhibits enough robustness to successfully control the pathogen on its own. Controlling *Salmonella* on farms involves a combination of strategies, and their efficacy can vary based on farm-specific factors, management practices, and the *Salmonella* serotype involved. In a recent study evaluating factors associated with high and low seroprevalence on Irish farms, based on surveillance data from the National Control program, we demonstrated that low prevalence herds were related to farms in which aspects associated with biosecurity, feed coarseness, and herd health influenced the on-farm *Salmonella* prevalence [49].

In summary, a comprehensive and multifaceted approach, incorporating various control measures, is most effective in managing *Salmonella* on farms. The combination of biosecurity, sanitation, water and feed management, pest control, vaccination, monitoring, and employee training is key to reducing the prevalence and impact of *Salmonella* infections on pig farms.

6.2. *Salmonella on Post Farm Stages, Epidemiology and Control*

Post-harvest is the term that defines the production stages after the farm and which include transport of the animals to the slaughterhouse, their resting, slaughter and carcass fabrication and the latter stages of pork meat processing at the cutting plants and retailers. Again, the number of studies that have focused their aims on disclosing information about the dynamics of *Salmonella* in these stages and the potential mitigation strategies to be put in place is almost overwhelming. If readers want to extend their knowledge about *Salmonella* and post-harvest in pig production, we recommend a few review studies which go into depth into this topic [39,56,83–85].

There are a few studies that exemplify what happens in the *Salmonella* epidemiology after the farm. In 2001, Hurd and colleagues [86] performed a study in which 50 pigs were slaughtered on the farm and another 567 market-weight pigs were transported (mean distance, 169 km) to the slaughterhouse with 2 h to 3 h of holding in ante-mortem pens before slaughtering, as usual. Lymph nodes from both groups were collected and *Salmonella* prevalence was determined by microbiological detection of the pathogen. Interestingly, the prevalence at the slaughterhouse was five times higher than on the farm. In 2013, the European Food Safety Agency (EFSA) published the results of the analysis performed on carcasses from finisher pigs, as part of the cross-sectional study performed in member states to determine the basal prevalence of *Salmonella* in slaughter pigs [87]. From the results and conclusions obtained in the study, it is noteworthy to mention here that carcass prevalence was positively associated with on-farm prevalence, i.e., the higher the prevalence on the farm, the larger the number of contaminated carcasses detected. However, when farms with similar prevalence supplied different abattoirs, differences in carcass prevalence were observed among slaughterhouses. The third study to mention is a study in which the authors aimed at organizing the slaughtering by means of on-farm seroprevalence [88]. First, pigs from low prevalence farms were slaughtered followed by farms with high prevalence. Unexpectedly, when carcass prevalence was analyzed, a higher prevalence was obtained in carcasses from *Salmonella*-free and low *Salmonella* seroprevalence herds compared to high seroprevalence herds. Further typing of *Salmonella* isolates obtained from concomitant samples collected in the slaughterhouse environment revealed the link between carcasses from *Salmonella*-free herds and lairage environment, thus revealing new infections or contamination occurring in the slaughterhouse facilities.

We have chosen these three examples as they exemplify with accuracy the role of post-harvest in *Salmonella* epidemiology and control after the farm. The study of Hurd and colleagues, supported by results from subsequent studies [89–92], clearly highlights the potential hotspots in *Salmonella* new infections or re-infections after the farm. Indeed, the studies referenced and others [39] point out both transport and lairage as hotspots where new infections can occur and also where re-activation of infections can happen. The arrival of pigs from infected herds which spread *Salmonella* in these environments, sometimes in large concentrations, together with the rapid infection onset and fecal shedding in infected animals [93], explains the results observed by Hurd and colleagues in their study performed two decades ago. Is this information asserting that post-harvest is more important in *Salmonella* control than the on-farm prevalence?

The baseline *Salmonella* prevalence study performed in the EU and already mentioned [87,94] is the largest study performed so far, all over the world, involving a total of 19,159 slaughter pigs with validated results from 26 countries and monitored 5736 carcass swab samples in 146 slaughterhouses (from 13 Member States). The weighted prevalence of *Salmonella* contamination of carcasses was greater for slaughter pigs with *Salmonella* infection in lymph nodes compared to the pigs with un-infected lymph nodes. This result strongly evidences the link between on-farm status and carcass contamination risk. Nevertheless, when carcass contamination was compared in slaughterhouses with similar inputs of infected pig lymph nodes, differences were also observed, a fact which was repeated by other contemporary studies involving different slaughterhouses [89,92,95–97]. The result demonstrates that the impact of *Salmonella* inputs by infected animals in carcass contamination varies among establishments, indisputably associated with hygiene in the carcass fabrication.

And what about the third and last example? Well, on the one hand, it demonstrates that if we fail in *Salmonella* control in post-harvest, non-infected pigs become positive, but also that correct carcass processing, from a hygienic perspective, can mitigate external carcass contamination, even in highly infected batches.

The preceding section provided a brief overview of various control options that can be implemented on farms. While interventions on farms lack standardization, slaughterhouses have adhered to a standardized procedure for decades to uphold hygiene and ensure meat safety through the hazard analysis critical control points (HACCP) system. This system serves as a framework for monitoring the food system, aimed at reducing the risk of foodborne illness. Different points in the slaughter process or carcass fabrication can be incorporated into the HACCP plan of the abattoir. The control of endemic bacteria, including *Salmonella* can be performed through proper cleaning and disinfection with good manufacturing practices rules. According to Borch and colleagues [98], the following affiliation to CCPs made for specific steps during slaughter and dressing may serve as a guidance: (i) lairage (CP), (ii) killing (CP), (iii) scalding (CP), (iv) dehairing (CP), (v) singeing/flaming (CP), (vi) polishing (CP), (vii) circumoral incision and removal of the intestines (CCP), (viii) excision of the tongue, pharynx, and in particular the tonsils (CCP), (ix) splitting (CP), (x) post mortem inspection procedures (CCP) and (xi) deboning of the head (CCP). Most, if not all, of these CCPs have been highlighted by *Salmonella* post-harvest studies, but the most important CCPs associated with *Salmonella* contamination and spread, both between carcasses and within the slaughter environment, include carcass singeing, carcass scalding (considering water temperature and duration of scalding), carcass inspection (to detect the presence of fecal material), lacerations at evisceration [39,58] or carcass chilling [99,100]. Additionally, strategies such as hot water showers [18] and showers with antibacterial compounds (e.g., organic acids or disinfectants like peroxides) aid in mitigating carcass contamination at the conclusion of the fabrication process [48,101]. Individual plans in the slaughterhouse should identify and correct their CCPs and guarantee that strategies that help in *Salmonella* mitigation (singeing, potential hot washing or chilling) are run accordingly.

Apart from the strategies integrated into the HACCP plan, other measures contributing to *Salmonella* control involve enhancing hygiene in the slaughter process, encompassing both the environment and equipment. Moreover, mitigating the risk of cross-contamination during human carcass dressing activities is imperative, emphasizing strict hygiene practices such as tool sterilization (e.g., knives) and frequent glove changes [39,48]. Furthermore, interventions targeting the processing stage should be complemented by actions to limit the risk of new infections during transport and lairage, as discussed in previous reviews [39]. Enhancing cleaning protocols in both transport and lairage [64] and restricting transport duration or distance and minimizing lairage resting time are well-known practices effective in reducing *Salmonella* transmission risk.

7. *Salmonella* Control and Legal Aspects

The main objective of *Salmonella* control in livestock species is mitigating the risk for humans. The current legislation, actions taken or lack thereof have consequences in the aim just exposed. For the first time in this review, let us talk about *Salmonella* control in poultry within the EU. There are currently compulsory programs to control the bacteria in chickens, laying hens and breeders both in poultry and turkey meat production [102]. These control programs include microbiological samplings with *Salmonella* isolation, a fact that together with the zoonoses monitoring in humans [103] offers accurate information on the serotypes and strains circulating both in humans and animals and their associations. In the hypothetical case of human outbreaks, molecular typing methods allow us to trace back the case to the infection source [104,105]. The lack of bacteriological analyses in control programs, or even worse, the lack of control programs and involvement of some of the production stages in the actions taken to control *Salmonella* in swine, hampers clarifying and effectively cutting the outbreaks from swine origin. As an example, in 2015 and 2016, two outbreaks of the emerging monophasic variant of *Salmonella* Typhimurium occurred in the state of Washington in the United States [106,107]. Both outbreaks occurred by the consumption of roasted infected pork and involved over 40 human infections with hospitalizations [107]. According to the scientist involved in the investigations of the first outbreak in 2015 "*Our investigation could not determine the relative importance of specific points in the pork production process that contributed to this outbreak*" and "*we were unable to assess practices or conduct environmental or animal testing at establishment A's source farms because farms were reluctant to participate, and unclear jurisdictional authority of state agriculture agencies did not require farms to comply with our request*". Indeed, Salmonellosis is not listed as an animal reportable disease in the US despite its proven importance, which can involve large morbidity and relevant mortality rates in humans [108], a fact that limits potential investigations in live animals. Another recent outbreak in another country with no control program does not offer information about the source of infection either [109]. While in the references provided above [104,105], outbreaks could be traced back, the lack of official involvement of producers in official *Salmonella* surveillance and control in pigs [106,109] was an obstacle to disclosing the origin of both outbreaks. Furthermore, in the outbreak detected in the US, an investigation could prevent a second outbreak with pigs from the same origin occurring in 2016. The examples provided here clearly point out the importance of control programs and the involvement of all production stages in the mitigation of the first foodborne zoonosis worldwide [110]. When authorities excuse any of the actors from the *Salmonella* equation, for instance, primary production on the farm, despite their relevance in the equation, demonstrated by the information gathered here or elsewhere [47,48,51,56], commitment in the mitigation of the human risk fails, as pointed out by Kawami and colleagues [106].

8. Conclusions

Over the last 20 years, actions have been put in place to mitigate the main foodborne zoonosis in industrialized countries, which, together with antimicrobial resistance and slurry environmental pollution, are the major One Health problems to be tackled by the swine industry in the forthcoming years. The European Union led the initiative with consolidated compulsory programs in avian species which have ended up in half of the human Salmonellosis reported two decades ago. The success and progress in avian production contrast with the slow implementation of actions in the pig industry. A higher complexity of production systems, infection epidemiology, and lack of clear strategies to mitigate *Salmonella* have discouraged its control, with countries that have not implemented any control and others that have discontinued it. Unfortunately, outside the EU, there is not any leading country in *Salmonella* mitigation in pigs and pork, despite the frequent outbreaks observed and the evidenced risk for humans. Monitoring and surveillance programs offer valuable information to mitigate the pre-harvest and post-harvest *Salmonella* hazards. Countries with programs in place should, in our opinion, put in place actions to reduce prevalence both in animals and pork but avoid penalty systems as much as possible, which limit the profitability of the production and envisage a negative reaction to deal with *Salmonella* but also finding a strategy which does not waste the economic efforts of monitoring and control programs. In addition, the scientific knowledge and experience running control programs acquired through the last decades are useful to re-think and design new, where not in place, efficient and reliable programs to mitigate the pathogen, always considering economic constraints. As stated above, it is highly important to involve and encourage the participation of all the production stages and build awareness of the impact that health actions taken in primary production have on human and environmental health, named One Health.

Author Contributions: H.A. and M.K. prepared the draft of the scope of the article. A.C., M.K. and H.A. contributed to the information search, discussions, writing and revision of the sections included in this article. H.A. conducted a final review of the information and A.C. prepared the tables and figures. M.K. also revised the English grammar of the article. All authors have read and agreed to the published version of the manuscript.

Funding: This research received no external funding.

Institutional Review Board Statement: Not applicable.

Informed Consent Statement: Not applicable.

Acknowledgments: We want to acknowledge the reviewers who participated in the review process. Their ideas im-proved the quality of the manuscript and the information provided here.

Conflicts of Interest: Author Melvin Kramer is CEO in the company EHA consulting group. The remaining authors declare that the research was conducted in the absence of any commercial or financial relationships that could be construed as a potential conflict of interest.

References

1. Wray, C. Mammalian salmonellosis. In *Handbook of Zoonoses—Section A: Bacterial, Rickettsial, Chlamydial, and Mycotic*, 2nd ed.; Beran, G.W., Steele, J.H., Eds.; CRC Press: Boca Raton, FL, USA, 1999; pp. 289–302.
2. Raffatellu, M.; Wilson, R.P.; Winter, S.E.; Bäumler, A.J. Clinical pathogenesis of typhoid fever. *J. Infect. Dev. Ctries.* **2008**, *2*, 260–266.
3. Hiyoshi, H.; Tiffany, C.R.; Bronner, D.N.; Bäumler, A.J. Typhoidal *Salmonella* serovars: Ecological opportunity and the evolution of a new pathovar. *FEMS Microbiol. Rev.* **2018**, *42*, 527–541. [CrossRef]
4. Pedersen, K.; Sørensen, G.; Löfström, C.; Leekitcharoenphon, P.; Nielsen, B.; Wingstrand, A.; Aarestrup, F.M.; Hendriksen, R.S.; Baggesen, D.L. Reappearance of *Salmonella* serovar Choleraesuis var. Kunzendorf in Danish pig herds. *Vet. Microbiol.* **2015**, *176*, 282–291. [CrossRef] [PubMed]
5. Wales, A.D.; Davies, R.H. A critical review of *Salmonella* Typhimurium infection in laying hens. *Avian Pathol.* **2011**, *40*, 429–436. [CrossRef] [PubMed]
6. Porter, S.; Strain, S.A.J.; Bagdonaite, G.; McDowell, S.W.; Bronckaers, T.; Sherrey, M.; Devine, P.; Pascual-Linaza, A.V.; Spence, N.; Porter, R.; et al. Trends in *Salmonella* serovars and antimicrobial resistance in pigs and poultry in Northern Ireland between 1997 and 2016. *Vet. Rec.* **2020**, *186*, 156. [CrossRef] [PubMed]

7. Papadopoulou, C.; Davies, R.H.; Carrique-Mas, J.J.; Evans, S.J. *Salmonella* serovars and their antimicrobial resistance in British turkey flocks in 1995 to 2006. *Avian Pathol.* **2009**, *38*, 349–357. [CrossRef] [PubMed]
8. Arguello, H.; Sørensen, G.; Carvajal, A.; Baggesen, D.L.; Rubio, P.; Pedersen, K. Prevalence, serotypes and resistance patterns of *Salmonella* in Danish pig production. *Res. Vet. Sci.* **2013**, *95*, 334–342. [CrossRef] [PubMed]
9. Mohler, V.L.; Izzo, M.M.; House, J.K. *Salmonella* in calves. *Vet. Clin. N. Am. Food Anim. Pract.* **2009**, *25*, 37–54. [CrossRef] [PubMed]
10. Nemhauser, J.B. (Ed.) *CDC Yellow Book 2024*; Oxford University press: Oxford, UK, 2023; ISBN 9780197570944.
11. Pires, S.M.; Vieira, A.R.; Hald, T.; Cole, D. Source attribution of human salmonellosis: An overview of methods and estimates. *Foodborne Pathog. Dis.* **2014**, *1*, 667–676. [CrossRef]
12. Kirk, M.D.; Pires, S.M.; Black, R.E.; Caipo, M.; Crump, J.A.; Devleesschauwer, B.; Döpfer, D.; Fazil, A.; Fischer-Walker, C.L.; Hald, T.; et al. World Health Organization Estimates of the Global and Regional Disease Burden of 22 Foodborne Bacterial, Protozoal, and Viral Diseases, 2010: A Data Synthesis. *PLoS Med.* **2015**, *12*, e1001921.
13. Mousing, J.; Jensen, P.T.; Halgaard, C.; Bager, F.; Feld, N.; Nielsen, B.; Nielsen, J.P.; Bech-Nielsen, S. Nation-wide *Salmonella enterica* surveillance and control in Danish slaughter swine herds. *Prev. Vet. Med.* **1997**, *29*, 247–261. [CrossRef] [PubMed]
14. Quirke, A.M.; Leonard, N.; Kelly, G.; Egan, J.; Lynch, P.B.; Rowe, T.; Quinn, P.J. Prevalence of *Salmonella* serotypes on pig carcasses from high- and low-risk herds slaughtered in three abattoirs. *Berl. Munch. Tierarztl. Wochenschr.* **2001**, *114*, 360–362. [PubMed]
15. Osterkorn, K.; Czerny, C.P.; Wittkowski, G.; Huber, M. Sampling plan for the establishment of a serologic *Salmonella* surveillance for slaughter pigs with meat juice ELISA. *Berl. Munch. Tierarztl. Wochenschr.* **2001**, *114*, 30–34. [PubMed]
16. Snary, E.L.; Munday, D.K.; Arnold, M.E.; Cook, A.J. Zoonoses action plan *Salmonella* monitoring programme: An investigation of the sampling protocol. *J. Food Prot.* **2010**, *73*, 488–494. [CrossRef] [PubMed]
17. Baggesen, D.L.; Wegener, H.C.; Bager, F.; Stege, H.; Christensen, J. Herd prevalence of *Salmonella* enterica infections in Danish slaughter pigs determined by microbiological testing. *Prev. Vet. Med.* **1996**, *26*, 201–213. [CrossRef]
18. Alban, L.; Baptista, F.M.; Møgelmose, V.; Sørensen, L.L.; Christensen, H.; Aabo, S.; Dahl, J. *Salmonella* surveillance and control for finisher pigs and pork in Denmark—A case study. *Food Res. Int.* **2012**, *45*, 656–665. [CrossRef]
19. Deane, A.; Murphy, D.; Leonard, F.C.; Byrne, W.; Clegg, T.; Madigan, G.; Griffin, M.; Egan, J.; Prendergast, D.M. Prevalence of *Salmonella* spp. in slaughter pigs and carcasses in Irish abattoirs and their antimicrobial resistance. *Ir. Vet. J.* **2022**, *75*, 4. [CrossRef]
20. Blaha, T. The present state of the German QS *Salmonella* monitoring and reduction programme. *Dtsch. Tierärztl. Wschr.* **2004**, *111*, 324–326.
21. Merle, R.; Kösters, S.; May, T.; Portsch, U.; Blaha, T.; Kreienbrock, L. Serological *Salmonella* monitoring in German pig herds: Results of the years 2003–2008. *Prev. Vet. Med.* **2011**, *99*, 229–233. [CrossRef]
22. Anonymous. UK: New Direction for Zoonoses National Control Programme (ZNCP). 2012. Available online: https://www.pigprogress.net/Health-Diseases/Health/2012/6/UK-New-direction-for-Zoonoses-National-Control-Programme-ZNCP-PP008961W/ (accessed on 7 January 2024).
23. Van Der Wolf, P.J. Monitoring for Salmonella in Swine in The Netherlands. Pig333. Available online: https://www.pig333.com/articles/monitoring-for-salmonella-in-swine-in-the-netherlands_12866/ (accessed on 7 January 2024).
24. Méroc, E.; Strubbe, M.; Vangroenweghe, F.; Czaplicki, G.; Vermeersch, K.; Hooyberghs, J.; Van der Stede, Y. Evaluation of the *Salmonella* surveillance program in Belgian pig farms. *Prev. Vet. Med.* **2012**, *105*, 309–314. [CrossRef]
25. Correia-Gomes, C.; Leonard, F.; Graham, D. Description of control programmes for *Salmonella* in pigs in Europe. *Progress to date? J. Food Saf.* **2021**, *41*, e12916. [CrossRef]
26. Kishima, M.; Uchida, I.; Namimatsu, T.; Osumi, T.; Takahashi, S.; Tanaka, K.; Aoki, H.; Matsuura, K.; Yamamoto, K. Nationwide surveillance of *Salmonella* in the faeces of pigs in Japan. *Zonoses Pub. Health* **2008**, *55*, 139–144. [CrossRef] [PubMed]
27. Tian, Y.; Gu, D.; Wang, F.; Liu, B.; Li, J.; Kang, X.; Meng, C.; Jiao, X.; Pan, Z. Prevalence and Characteristics of *Salmonella* spp. from a Pig Farm in Shanghai, China. *Foodborne Pathog. Dis.* **2021**, *18*, 477–488. [CrossRef]
28. Ntakiyisumba, E.; Lee, S.; Won, G. Identification of risk profiles for *Salmonella* prevalence in pig supply chains in South Korea using meta-analysis and a quantitative microbial risk assessment model. *Food Res. Int.* **2023**, *170*, 112999. [CrossRef] [PubMed]
29. Haley, C.A.; Dargatz, D.A.; Bush, E.J.; Erdman, M.M.; Fedorka-Cray, P.J. *Salmonella* prevalence and antimicrobial susceptibility from the National Animal Health Monitoring System Swine 2000 and 2006 studies. *J. Food Prot.* **2012**, *75*, 428–436. [CrossRef] [PubMed]
30. Sanchez-Maldonado, A.F.; Aslam, M.; Service, C.; Narváez-Bravo, C.; Avery, B.P.; Johnson, R.; Jones, T.H. Prevalence and antimicrobial resistance of *Salmonella* isolated from two pork processing plants in Alberta, Canada. *Int. J. Food Microbiol.* **2017**, *241*, 49–59. [CrossRef] [PubMed]
31. Lopes, G.V.; Pissetti, C.; da Cruz Payão Pellegrini, D.; da Silva, L.E.; Cardoso, M. Resistance phenotypes and genotypes of *Salmonella enterica* subsp. *enterica* isolates from feed, pigs, and carcasses in Brazil. *J. Food. Prot.* **2015**, *78*, 407–413. [PubMed]
32. Vico, J.P.; Lorenzutti, A.M.; Zogbi, A.P.; Aleu, G.; Sánchez, I.C.; Caffer, M.I.; Rosmini, M.R.; Mainar-Jaime, R.C. Prevalence, associated risk factors, and antimicrobial resistance profiles of non-typhoidal Salmonella in large scale swine production in Cordoba, Argentina. *Res. Vet. Sci.* **2020**, *130*, 161–169. [CrossRef]
33. Anon 2023. FSIS Guideline to Control *Salmonella* in Swine Slaughter and Pork Processing Establishments June 2023 USDA-FSIS. Available online: https://www.fsis.usda.gov/contact-us/askFSIS (accessed on 8 January 2024).
34. Carlsson, U.; Elvander, M. Surveillance and control programs. In *Domestic and Wild Animals in Sweden*; SVA National Veterinary Institute: Uppsala, Sweden, 2008; pp. 13–17.

35. Hofshagen, M.; Nygard, K.; Hauge, K. *The Surveillance and Control Programme for Salmonella in Live Animals, Eggs and Meat in Norway—Annual Report*; National Veterinary Institute: Oslo, Norway, 2008; 10p.
36. Huttunen, A.; Johansson, T.; Kostamo, P.; Kuronen, P.; Kuronen, H.; Laaksonen, T. *Salmonella Control and Occurence of Salmonella from 1995 to 2004*; Finnish Food Safety Authority Evira: Helsinki, Finland, 2006; Available online: https://www.ruokavirasto.fi/en/about-us/publications/finnish-food-authority-publications/ (accessed on 8 January 2024).
37. Nielsen, B.; Baggesen, D.; Bager, F.; Haugegaard, J.; Lind, P. The serological response to *Salmonella* serovars typhimurium and infantis in experimentally infected pigs. The time course followed with an indirect anti-LPS ELISA and bacteriological examinations. *Vet. Microbiol.* **1995**, *47*, 205–218. [CrossRef]
38. O'Connor, A.M.; Wang, B.; Denagamage, T.; McKean, J. Process mapping the prevalence of *Salmonella* contamination on pork carcass from slaughter to chilling: A systematic review approach. *Foodborne Pathog. Dis.* **2012**, *9*, 386–395. [CrossRef]
39. Arguello, H.; Alvarez-Ordoñez, A.; Carvajal, A.; Rubio, P.; Prieto, M. Role of slaughtering in *Salmonella* spreading and control in pork production. *J. Food Prot.* **2013**, *76*, 899–911. [CrossRef]
40. Salmonella Control in the Pig and Pork Production in Denmark—Lessons Learnt. Available online: https://www.pig333.com/articles/salmonella-control-in-the-pig-and-pork-production-in-denmark-%E2%80%93-lessons_12329/#:~:text=Salmonella%20control%20in%20the%20pig%20and%20pork%20production%20in%20Denmark%20%E2%80%93%20lessons%20learnt&text=The%20success%20of%20the%20Danish,to%20control%20in%20primary%20production (accessed on 20 December 2023).
41. Møgelmose, V.; Bagger, J.; Nielsen, B.; Baggesen, D.L. Reduction of multiresistant Salmonella TyphimuriumDT104 in Danish swineherds—New strategy. In Proceedings of the 4th International Symposium on the Epidemiology and Control of Salmonella and Other Food Borne Pathogens in Pork, Leipzig, Germany, 2–5 September 2001; pp. 69–71.
42. Hopkins, K.L.; de Pinna, E.; Wain, J. Prevalence of *Salmonella* enterica serovar 4,[5],12:i:- in England and Wales, 2010. *Eurosurveillance* **2012**, *13*, 20275. [CrossRef]
43. Argüello, H.; Sørensen, G.; Carvajal, A.; Baggesen, D.L.; Rubio, P.; Pedersen, K. Characterization of the emerging *Salmonella* 4,[5],12:i:- in Danish animal production. *Foodborne Pathog. Dis.* **2014**, *11*, 366–372. [CrossRef]
44. Blaha, T. The German Salmonella Serological Monitoring Programme. Pig333. *Com.* Available online: https://www.pig333.com/articles/the-german-salmonella-serological-monitoring-programme_12395/ (accessed on 20 December 2023).
45. QS. Anzahl Betriebe Mit Erhöhtem Salmonellenrisiko So Gering Wie Noch Nie. 2020. Available online: https://www.q-s.de/pressemeldungen/anzahl-betriebeerhoehtes-salmonellenrisiko-gering.html (accessed on 8 January 2024).
46. Olsen, J.V.; Christensen, T.; Jensen, J.D. Pig Farmers' Perceptions of Economic Incentives to Control *Salmonella* Prevalence at Herd Level. *Front. Vet. Sci.* **2021**, *8*, 647697. [CrossRef] [PubMed]
47. Boyen, F.; Haesebrouck, F.; Maes, D.; Van Immerseel, F.; Ducatelle, R.; Pasmans, F. Non-typhoidal *Salmonella* infections in pigs: A closer look at epidemiology, pathogenesis and control. *Vet. Microbiol.* **2008**, *130*, 1–19. [CrossRef]
48. De Busser, E.V.; De Zutter, L.; Dewulf, J.; Houf, K.; Maes, D. *Salmonella* control in live pigs and at slaughter. *Vet. J.* **2013**, *196*, 20–27. [CrossRef]
49. Argüello, H.; Manzanilla, E.G.; Lynch, H.; Walia, K.; Leonard, F.C.; Egan, J.; Duffy, G.; Gardiner, G.E.; Lawlor, P.G. Surveillance Data Highlights Feed Form, Biosecurity, and Disease Control as Significant Factors Associated with *Salmonella* Infection on Farrow-to-Finish Pig Farms. *Front. Microbiol.* **2018**, *9*, 187. [CrossRef] [PubMed]
50. Galán-Relaño, Á.; Valero Díaz, A.; Huerta Lorenzo, B.; Gómez-Gascón, L.; Mena Rodríguez, M.Á.; Carrasco Jiménez, E.; Pérez Rodríguez, F.; Astorga Márquez, R.J. *Salmonella* and Salmonellosis: An Update on Public Health Implications and Control Strategies. *Animals* **2023**, *13*, 3666. [CrossRef]
51. Marier, E.; Piers Smith, R.; Ellis-Iversen, J.; Watson, E.; Armstrong, D.; Hogeveen, H.; Cook, A.J. Changes in perceptions and motivators that influence the implementation of on-farm *Salmonella* control measures by pig farmers in England. *Prev. Vet. Med.* **2016**, *133*, 22–30. [CrossRef]
52. Laanen, M.; Maes, D.; Hendriksen, C.; Gelaude, P.; De Vliegher, S.; Rosseel, Y.; Dewulf, J. Pig, cattle and poultry farmers with a known interest in research have comparable perspectives on disease prevention and on-farm biosecurity. *Prev. Vet. Med.* **2014**, *115*, 1–9. [CrossRef]
53. Berge, A.C.; Wierup, M. Nutritional strategies to combat *Salmonella* in mono-gastric food animal production. *Animal* **2012**, *6*, 557–564. [CrossRef] [PubMed]
54. Wales, A.D.; Davies, R.H. *Salmonella* Vaccination in Pigs: A Review. *Zoonoses Public Health* **2017**, *64*, 1–13. [CrossRef] [PubMed]
55. de la Cruz, M.L.; Conrado, I.; Nault, A.; Perez, A.; Dominguez, L.; Alvarez, J. Vaccination as a control strategy against *Salmonella* infection in pigs: A systematic review and meta-analysis of the literature. *Res. Vet. Sci.* **2017**, *114*, 86–94. [CrossRef] [PubMed]
56. Rodrigues da Costa, M.; Pessoa, J.; Meemken, D.; Nesbakken, T. A Systematic Review on the Effectiveness of Pre-Harvest Meat Safety Interventions in Pig Herds to Control *Salmonella* and Other Foodborne Pathogens. *Microorganisms* **2021**, *9*, 1825. [CrossRef] [PubMed]
57. Viltrop, A.; Niine, T.; Tobias, T.; Sassu, E.L.; Bartolo, I.D.; Pavoni, E.; Alborali, G.L.; Burow, E.; Smith, R.P. A Review of Slaughter Practices and Their Effectiveness to Control Microbial *Salmonella* spp., contamination of Pig Carcasses. *J. Food Prot.* **2023**, *86*, 100171. [PubMed]
58. FAO. *Interventions for the Control of Non-Typhoidal Salmonella in Beef and Pork*; FAO/WHO: Rome, Italy, 2016; ISBN 978-92-5-109049-7.

59. Beloeil, P.A.; Chauvin, C.; Proux, K.; Fablet, C.; Madec, F.; Alioum, A. Risk factors for *Salmonella* seroconversion of fattening pigs in farrow-to-finish herds. *Vet. Res.* **2007**, *38*, 835–848. [CrossRef]
60. Correia-Gomes, C.; Mendonça, D.; Vieira-Pinto, M.; Niza-Ribeiro, J. Risk factors for *Salmonella* spp. in Portuguese breeding pigs using a multilevel analysis. *Prev. Vet. Med.* **2013**, *108*, 159–166. [CrossRef]
61. Smith, R.P.; May, H.E.; Burow, E.; Meester, M.; Tobias, T.J.; Sassu, E.L.; Pavoni, E.; Di Bartolo, I.; Prigge, C.; Wasyl, D.; et al. Assessing pig farm biosecurity measures for the control of *Salmonella* on European farms. *Epidemiol. Infect.* **2023**, *151*, e130. [CrossRef]
62. Alarcón, L.V.; Allepuz, A.; Mateu, E. Biosecurity in pig farms: A review. *Porc. Health Manag.* **2021**, *7*, 5. [CrossRef]
63. Dahl, J.; Wingstrand, A.; Nielsen, B.; Baggesen, D.L. Eradication of *Salmonella* Typhimurium infection by the strategic movement of pigs. *Vet. Rec.* **1997**, *140*, 679–681. [CrossRef]
64. Beloeil, P.A.; Fravalo, P.; Fablet, C.; Jolly, J.P.; Eveno, E.; Hascoet, Y.; Chauvin, C.; Salvat, G.; Madec, F. Risk factors for *Salmonella* enterica subsp. enterica shedding by market-age pigs in French farrow-to-finish herds. *Prev. Vet. Med.* **2004**, *63*, 103–120. [CrossRef] [PubMed]
65. García-Feliz, C.; Carvajal, A.; Collazos, J.A.; Rubio, P. Herd-level risk factors for faecal shedding of *Salmonella enterica* in Spanish fattening pigs. *Prev. Vet. Med.* **2009**, *91*, 130–136. [CrossRef] [PubMed]
66. Walia, K.; Argüello, H.; Lynch, H.; Grant, J.; Leonard, F.C.; Lawlor, P.G.; Gardiner, G.E.; Duffy, G. The efficacy of different cleaning and disinfection procedures to reduce *Salmonella* and Enterobacteriaceae in the lairage environment of a pig abattoir. *Int. J. Food Microbiol.* **2017**, *246*, 64–71. [CrossRef]
67. Farzan, A.; Friendship, R.M.; Dewey, C.E.; Warriner, K.; Poppe, C.; Klotins, K. Prevalence of *Salmonella* spp. on Canadian pig farms using liquid or dry-feeding. *Prev. Vet. Med.* **2006**, *73*, 241–254. [CrossRef]
68. Højberg, O.; Canibe, N.; Knudsen, B.; Jensen, B.B. Potential rates of fermentation in digesta from the gastrointestinal tract of pigs: Effect of feeding fermented liquid feed. *Appl. Environ. Microbiol.* **2003**, *69*, 408–418. [CrossRef]
69. Van Immerseel, F.; Russell, J.B.; Flythe, M.D.; Gantois, I.; Timbermont, L.; Pasmans, F.; Haesebrouck, F.; Ducatelle, R. The use of organic acids to combat *Salmonella* in poultry: A mechanistic explanation of the efficacy. *Avian Pathol.* **2006**, *35*, 182–188. [CrossRef]
70. Creus, E.; Pérez, J.F.; Peralta, B.; Baucells, F.; Mateu, E. Effect of acidified feed on the prevalence of *Salmonella* in market-age pigs. *Zoonoses Public Health* **2007**, *54*, 314–319. [CrossRef] [PubMed]
71. Argüello, H.; Carvajal, A.; Costillas, S.; Rubio, P. Effect of the addition of organic acids in drinking water or feed during part of the finishing period on the prevalence of *Salmonella* in finishing pigs. *Foodborne Pathog. Dis.* **2013**, *10*, 842–849. [CrossRef]
72. Lynch, H.; Leonard, F.C.; Walia, K.; Lawlor, P.G.; Duffy, G.; Fanning, S.; Markey, B.K.; Brady, C.; Gardiner, G.E.; Argüello, H. Investigation of in-feed organic acids as a low cost strategy to combat *Salmonella* in grower pigs. *Prev. Vet. Med.* **2017**, *139*, 50–57. [CrossRef]
73. Barba-Vidal, E.; Castillejos, L.; Roll, V.F.B.; Cifuentes-Orjuela, G.; Moreno Muñoz, J.A.; Martín-Orúe, S.M. The Probiotic Combination of *Bifidobacterium longum* subsp. *infantis* CECT 7210 and *Bifidobacterium animalis* subsp. *lactis* BPL6 Reduces Pathogen Loads and Improves Gut Health of Weaned Piglets Orally Challenged with *Salmonella* Typhimurium. *Front. Microbiol.* **2017**, *8*, 1570.
74. Boyen, F.; Haesebrouck, F.; Vanparys, A.; Volf, J.; Mahu, M.; Van Immerseel, F.; Rychlik, I.; Dewulf, J.; Ducatelle, R.; Pasmans, F. Coated fatty acids alter virulence properties of *Salmonella* Typhimurium and decrease intestinal colonization of pigs. *Vet. Microbiol.* **2008**, *132*, 319–327. [CrossRef] [PubMed]
75. Haesebrouck, F.; Pasmans, F.; Chiers, K.; Maes, D.; Ducatelle, R.; Decostere, A. Efficacy of vaccines against bacterial diseases in swine: What can we expect? *Vet. Microbiol.* **2004**, *100*, 255–268. [CrossRef] [PubMed]
76. Haneda, T.; Okada, N.; Kikuchi, Y.; Takagi, M.; Kurotaki, T.; Miki, T.; Arai, S.; Danbara, H. Evaluation of *Salmonella enterica* serovar Typhimurium and Choleraesuis slyA mutant strains for use in live attenuated oral vaccines. *Comp. Immunol. Microbiol. Infect. Dis.* **2011**, *34*, 399–409. [CrossRef] [PubMed]
77. Arguello, H.; Carvajal, A.; Naharro, G.; Rubio, P. Evaluation of protection conferred by a *Salmonella* Typhimurium inactivated vaccine in *Salmonella*-infected finishing pig farms. *Comp. Immunol. Microbiol. Infect. Dis.* **2013**, *36*, 489–498. [CrossRef] [PubMed]
78. van der Wolf, P.; Meijerink, M.; Libbrecht, E.; Tacken, G.; Gijsen, E.; Lillie-Jaschniski, K.; Schüller, V. *Salmonella* Typhimurium environmental reduction in a farrow-to-finish pig herd using a live attenuated *Salmonella* Typhimurium vaccine. *Porc. Health Manag.* **2021**, *7*, 43. [CrossRef]
79. Ortiz Sanjuán, J.M.; Manzanilla, E.G.; Cabrera-Rubio, R.; Crispie, F.; Cotter, P.D.; Garrido, J.J.; Argüello, H. Using Shotgun Sequencing to Describe the Changes Induced by In-Feed Zinc Oxide and Apramycin in the Microbiomes of Pigs One Week Postweaning. *Microbiol. Spectr.* **2022**, *10*, e0159722. [CrossRef] [PubMed]
80. Jordá, J.; Lorenzo-Rebenaque, L.; Montoro-Dasi, L.; Marco-Fuertes, A.; Vega, S.; Marin, C. Phage-Based Biosanitation Strategies for Minimizing Persistent *Salmonella* and Campylobacter Bacteria in Poultry. *Animals* **2023**, *13*, 3826. [CrossRef] [PubMed]
81. Walia, K.; Argüello, H.; Lynch, H.; Leonard, F.C.; Grant, J.; Yearsley, D.; Kelly, S.; Duffy, G.; Gardiner, G.E.; Lawlor, P.G. Effect of strategic administration of an encapsulated blend of formic acid, citric acid, and essential oils on *Salmonella* carriage, seroprevalence, and growth of finishing pigs. *Prev. Vet. Med.* **2017**, *137*, 28–35. [CrossRef] [PubMed]

82. Young, I.; Wilhelm, B.J.; Cahill, S.; Nakagawa, R.; Desmarchelier, P.; Rajić, A. A Rapid Systematic Review and Meta-Analysis of the Efficacy of Slaughter and Processing Interventions to Control Non-Typhoidal *Salmonella* in Beef and Pork. *J. Food Prot.* **2016**, *79*, 2196–2210. [CrossRef]
83. Ojha, S.; Kostrzynska, M. Approaches for reducing *Salmonella* in pork production. *J. Food Prot.* **2007**, *70*, 2676–2694. [CrossRef]
84. Totton, S.C.; Glanville, J.M.; Dzikamunhenga, R.S.; Dickson, J.S.; O'Connor, A.M. Systematic review of the magnitude of change in prevalence and quantity of *Salmonella* after administration of pathogen reduction treatments on pork carcasses. *Anim. Health Res. Rev.* **2016**, *17*, 39–59. [CrossRef] [PubMed]
85. Blyth, G.A.D.; Connors, L.; Fodor, C.; Cobo, E.R. The Network of Colonic Host Defense Peptides as an Innate Immune Defense against Enteropathogenic Bacteria. *Front. Immunol.* **2020**, *11*, 965. [CrossRef] [PubMed]
86. Hurd, H.S.; McKean, J.D.; Griffith, R.W.; Wesley, I.V.; Rostagno, M.H. *Salmonella* enterica infections in market swine with and without transport and holding. *Appl. Environ. Microbiol.* **2002**, *68*, 2376–2381. [CrossRef] [PubMed]
87. EFSA. Report of the Task Force on Zoonoses Data Collection on the Analysis of the baseline survey on the prevalence of *Salmonella* in slaughter pigs, Part, B. *EFSA J.* **2008**, *206*, 1–111.
88. Arguello, H.; Carvajal, A.; Alvarez-Ordoñez, A.; Jaramillo-Torres, A.H.; Rubio, P. Effect of logistic slaughter on *Salmonella* contamination on pig carcasses. *Food Res. Int.* **2014**, *55*, 77–82. [CrossRef]
89. Swanenburg, M.; Urlings, H.A.; Snijders, J.M.; Keuzenkamp, D.A.; van Knapen, F. *Salmonella* in slaughter pigs: Prevalence, serotypes and critical control points during slaughter in two slaughterhouses. *Int. J. Food Microbiol.* **2001**, *70*, 243–254. [CrossRef] [PubMed]
90. O'Connor, A.M.; Gailey, J.; McKean, J.D.; Hurd, H.S. Quantity and distribution of *Salmonella* recovered from three swine lairage pens. *J. Food Prot.* **2006**, *69*, 1717–1719. [CrossRef]
91. Boughton, C.; Egan, J.; Kelly, G.; Markey, B.; Leonard, N. Quantitative examination of *Salmonella* spp. in the lairage environment of a pig abattoir. *Foodborne Pathog. Dis.* **2007**, *4*, 26–32. [CrossRef] [PubMed]
92. Arguello, H.; Carvajal, A.; Collazos, J.A.; Garcia-Feliz, C.; Rubio, P. Prevalence and serovars of *Salmonella* enterica on pig carcasses, slaughtered pigs and the environment of four Spanish slaughterhouses. *Food Res. Int.* **2012**, *45*, 905–912. [CrossRef]
93. Fedorka-Cray, P.J.; Whipp, S.C.; Isaacson, R.E.; Nord, N.; Lager, K. Transmission of *Salmonella* Typhimurium to swine. *Vet. Micro.* **1994**, *41*, 333–344. [CrossRef]
94. EFSA. Report of the Task Force on Zoonoses Data Collection on the analysis of the baseline survey on the prevalence of *Salmonella* in slaughter pigs, Part, A. *EFSA J.* **2008**, *135*, 1–111.
95. Hald, T.; Wingstrand, A.; Swanenburg, M.; von Altrock, A.; Thorberg, B.M. The occurrence and epidemiology of *Salmonella* in European pig slaughterhouses. *Epidemiol. Infect.* **2003**, *131*, 1187–1203. [CrossRef] [PubMed]
96. Bonardi, S.; Bassi, L.; Brindani, F.; D'Incau, M.; Barco, L.; Carra, E.; Pongolini, S. Prevalence, characterization and antimicrobial susceptibility of *Salmonella enterica* and *Yersinia enterocolitica* in pigs at slaughter in Italy. *Int. J. Food Microbiol.* **2013**, *163*, 248–257. [CrossRef] [PubMed]
97. Botteldoorn, N.; Herman, L.; Rijpens, N.; Heyndrickx, M. Phenotypic and molecular typing of *Salmonella* strains reveals different contamination sources in two commercial pig slaughterhouses. *Appl. Environ. Microbiol.* **2004**, *70*, 5305–5314. [CrossRef] [PubMed]
98. Borch, E.; Nesbakken, T.; Christensen, H. Hazard identification in swine slaughter with respect to foodborne bacteria. *Int. J. Food Microbiol.* **1996**, *30*, 9–25. [CrossRef] [PubMed]
99. Bolton, D.J.; Pearce, R.A.; Sheridan, J.J.; Blair, I.S.; McDowell, D.A.; Harrington, D. Washing and chilling as critical control points in pork slaughter hazard analysis and critical control point (HACCP) systems. *J. Appl. Microbiol.* **2002**, *92*, 893–902. [CrossRef] [PubMed]
100. Cê, E.R.; Giombelli, A.; Kich, J.D.; Moresco, K.S.; Miranda, A.; Pedrão, M.R.; Johann, G.; Badaró, A.C.L.; Hashimoto, E.H.; Machado-Lunkes, A. Monitoring of Pig Slaughter Stages and Correlation in the Prevalence of Pathogens and Levels of Microorganisms That Indicate Microbiological Quality and Hygiene Using a Predictive Model. *J. Food Prot.* **2023**, *86*, 100034. [CrossRef]
101. Mikołajczyk, A. Evaluation of the effects of a mixture of organic acids and duration of storage on the survival of *salmonella* on turkey carcasses. *J. Food Prot.* **2015**, *78*, 585–589. [CrossRef] [PubMed]
102. European Union. *Directive 2003/99/EC of the European Parliament and of the Council of 17 November 2003 on the Monitoring of Zoonoses and Zoonotic Agents, Amending Council Decision 90/424/EEC and Repealing Council Directive 92/117/EEC*; European Union: Brussels, Belgium, 2003.
103. EFSA; ECDC. The European Union One Health 2022 Zoonoses Report. *EFSA J.* **2023**, *21*, e8442.
104. Kinross, P.; van Alphen, L.; Martinez Urtaza, J.; Struelens, M.; Takkinen, J.; Coulombier, D.; Makela, P.; Bertrand, S.; Mattheus, W.; Schmid, D.; et al. Multidisciplinary investigation of a multicountry outbreak of *Salmonella* Stanley infections associated with turkey meat in the European Union, August 2011 to January 2013. *Eurosurveillance* **2014**, *19*, 20801. [CrossRef]
105. Sarno, E.; Pezzutto, D.; Rossi, M.; Liebana, E.; Rizzi, V. A Review of Significant European Foodborne Outbreaks in the Last Decade. *J. Food Prot.* **2021**, *84*, 2059–2070. [CrossRef] [PubMed]
106. Kawakami, V.; Bottichio, L.; Lloyd, J.; Carleton, H.; Leeper, M.; Olson, G.; Li, Z.; Kissler, B.; Angelo, K.M.; Whitlock, L.; et al. Multidrug-Resistant *Salmonella* I 4,[5],12:i:- and *Salmonella Infantis* Infections Linked to Whole Roasted Pigs from a Single Slaughter and Processing Facility. *J. Food Prot.* **2019**, *82*, 1615–1624. [CrossRef] [PubMed]

107. Bearson, B.L.; Trachsel, J.M.; Holman, D.B.; Brunelle, B.W.; Sivasankaran, S.K.; Simmons, M.; Wasilenko, J.; Tillman, G.; Johnston, J.J.; Bearson, S.M.D. Complete Genome Sequence of Multidrug-Resistant *Salmonella enterica* Serovar I 4,[5],12:i:- 2015 U.S. Pork Outbreak Isolate USDA15WA-1. *Microbiol. Resour. Announc.* **2019**, *8*, e00791-19. [CrossRef] [PubMed]
108. USDA. Available online: https://www.aphis.usda.gov/aphis/ourfocus/animalhealth/nvap/NVAP-Reference-Guide/Animal-Health-Emergency-Management/Notifiable-Diseases-and-Conditions (accessed on 19 January 2024).
109. Napoleoni, M.; Villa, L.; Barco, L.; Lucarelli, C.; Tiengo, A.; Baggio, G.; Dionisi, A.M.; Angellotti, A.; Ferretti, E.; Ruggeri, S.; et al. Monophasic Variant of *Salmonella* Typhimurium 4,[5],12:i:- (ACSSuGmTmpSxt Type) Outbreak in Central Italy Linked to the Consumption of a Roasted Pork Product (Porchetta). *Microorganisms* **2023**, *15*, 2567. [CrossRef]
110. Kuchenmüller, T.; Abela-Ridder, B.; Corrigan, T.; Tritscher, A. World Health Organization initiative to estimate the global burden of foodborne diseases. *Rev. Sci. Tech.* **2013**, *32*, 459–467. [CrossRef]

animals

MDPI

Review

Scoping Review About *Salmonella* spp. in Colombian Pig Farms from 2009 to Mid-2024

Adriana Pulido-Villamarín [1,*], Iliana Chamorro-Tobar [2,3], Ana K. Carrascal-Camacho [2], Fernando Sampedro [4], Marcela Rodríguez-Moreno [3], Fernando Rojas-Bermúdez [3], Mónica Pérez-Vargas [3], Ivonne Hernández-Toro [5], Alejandra Camacho-Carrillo [6] and Raúl A. Poutou-Piñales [2]

1 Unidad de Investigaciones Agropecuarias (UNIDIA), Departamento de Microbiología, Facultad de Ciencias, Pontificia Universidad Javeriana, Bogotá 110231, Colombia
2 Grupo de Biotecnología Ambiental e Industrial (GBAI), Departamento de Microbiología, Facultad de Ciencias, Pontificia Universidad Javeriana, Bogotá 110231, Colombia; rpoutou@javeriana.edu.co (R.A.P.-P.)
3 Asociación Porkcolombia—Fondo Nacional de la Porcicultura, Centro de Investigación y Transferencia de Tecnología del Sector Porcícola (CENIPORCINO), Bogotá 110231, Colombia
4 School of Public Health, Environmental Health Sciences, University of Minnesota, Minneapolis, MN 55455, USA
5 Área de Bacteriología, Laboratorio Nacional de Diagnóstico Veterinario (LNDV), Instituto Colombiano Agropecuario—ICA, Bogotá 110221, Colombia
6 Programa de Microbiología Industrial, Facultad de Ciencias, Pontificia Universidad Javeriana, Bogotá 110231, Colombia
* Correspondence: adriana.pulido@javeriana.edu.co

Simple Summary: The global increase in pig production has been economically significant. However, a critical issue is the pathogens affecting the production chain, particularly *Salmonella* spp. This microorganism compromises animal health and growth, resulting in economic losses and posing a public health risk due to potential transmission to humans via contaminated pork product consumption. The lack of epidemiological data about *Salmonella* spp. in Colombian pig farms has become a challenge to effective risk mitigation measures, highlighting an urgent need to improve surveillance and disease control within the production chain.

Abstract: In Colombia, research on Salmonella concerning animal health, veterinary diagnostics, and epidemiology within the primary production chain is limited. This study aimed to analyze the published data about Salmonella in the Colombian primary pig production chain from 2009 to mid-2024. This involved an exploratory literature review using systematic search strategies, including articles, graduate studies, conference presentations, and technical reports from the selected period. Of the 35 studies reviewed, 30 met the inclusion criteria, with eleven being from the grey literature. The pooled prevalence of *Salmonella* spp. on Colombian farms was 8.9%, while the seroprevalence ranged from 27 to 40%. Risk factors associated with the presence of this bacterium on farms included aspects such as water sources, pest control, the farm type, and management practices. Few scientific publications address the presence of this pathogen in primary pig production in Colombia, underscoring the need to raise awareness within the academic and production communities about the importance of conducting and reporting new studies and cases.

Keywords: Colombia; pig farms; pig production; primary production chain; prevalence; risk factors; *Salmonella* spp.; seroprevalence

1. Introduction

The worldwide pork production and trade have increased recently, especially in the European Union, Russia, and the United States. Over the last two decades, the global pork production has increased by 111%, with technological innovation being a fundamental factor [1]. Colombian pork production represents 1.4% of the agricultural gross domestic product (GDP) and 4.8% of the livestock sector's GDP, generating approximately USD 900 million/year [2]. Its production has continued to grow: in 2015, there were 285,280 tons of pork, and, by 2019, there were nearly four million slaughtered pigs, amounting to 406,085 tons of meat [3]. In 2019, the pork production in Colombia amounted to approximately 6,473,525 animals, concentrated mainly in the departments of Antioquia (29.68%), Valle del Cauca (14.36%), Cundinamarca (8.94%), Meta (6.03%), Córdoba (6.01%), Magdalena (4.21%), Sucre (3.33%), Boyacá (3.12%), Atlántico (2.74%), and Bolívar (2.27%) [4].

Pork production is susceptible to various pathogens, including viruses, parasites, and bacteria, which cause serious health issues and economic losses by slowing growth rates, reducing animals' weight, decreasing the meat quality, and causing some cases of animal mortality. Additionally, the diagnosis and treatment of diseases incur additional expenses that can be large [5]. Pathogens also pose public health risks, particularly foodborne zoonotic pathogens, which can increase human morbidity and mortality. *Salmonella* spp. is one of the primary pathogens that is responsible for around 94 million gastroenteritis cases and 155,000 deaths yearly worldwide [6].

Pigs are a significant reservoir of *Salmonella* spp., being a source of infection for humans when they consume contaminated pork products or come into contact with the animals [7]. This transmission leads to salmonellosis, presenting symptoms such as diarrhea, fever, and abdominal cramps, with severe cases causing arthritis, pneumonia, meningitis, and septicemia [8]. The European Food Safety Authority's (EFSA) studies reported that 8.9% of salmonellosis cases were due to consuming contaminated pork or pork by-products [9]. In the United States, the reported values were between 3 and 10% in 2015–2017 [10,11], in which there were 281 outbreaks related to the intake of contaminated pork and its by-products [12]. Similarly, in Colombia, there were 867 outbreaks of foodborne diseases in 2017, with *Salmonella* spp. as the etiological agents in 15 cases [13]. The INS clarified that the FBD incidence has a direct relationship with food quality, especially when the contamination occurs with the same organic matter of the animal, as many human pathogens are considered normal microbiota in protein-source animals. The INS reported that, of the 881 outbreaks detected in 2018, beef and pork meat were involved in 6% of them [14].

Salmonella is a genus within the Enterobacteriaceae family, consisting of Gram-negative bacilli that are facultative anaerobes and grow optimally at 37 °C. It has been found in reptiles, birds, and rodents, among other wildlife specimens, which behave as natural reservoirs of the pathogen and can sporadically enter farms [15]; on the other hand, the bacterium can persist in the environment for extended periods, contaminating water, feed, and surfaces [16–18]. Once Salmonella infects pigs, it causes asymptomatic infections, allowing subclinical carrier animals to serve as a contamination source for the healthy populations of farms [19] and processing plants [20]. Worldwide, the average on-farm prevalence has oscillated between 17 and 59% [21,22]. However, in Latin America, and specifically in Colombia, epidemiological data regarding the presence of this microorganism in swine production farms are scarce, which hinders the implementation of plans to reduce the zoonotic risk in the primary chain.

Given the economic and public health significance of pig production, the present research team aimed to analyze publications on *Salmonella* spp. in the pig production chain of Colombia between 2009 and mid-2024. This study followed the PICoR scheme (Patient/Problem, Intervention, Comparison/Alternative vs. Standard, Reports), addressing the following question: what has been published about *Salmonella* spp., within the primary chain of Colombian pig production, from 2009 to mid-2024? However, the answer may not fully capture the reality of the situation regarding *Salmonella* spp. in Colombian pig farms due to underdiagnosis and a lack of official reporting. The team hopes this exploratory review will promote awareness and potentially encourage more rigorous surveillance and reporting practices within Colombia and similar contexts.

2. Materials and Methods

2.1. Protocol

This analysis was conducted according to the methods outlined by previously published strategies [23,24].

2.2. Search Strategy

A literature review was conducted by searching databases such as ScienceDirect, SciELO, EBSCOhost, Redalyc, Google Scholar, and all grey literature derived from associations and governmental institutions, corresponding to the period between 2009 and mid-2024. The MeSH search terms were Salmonella "AND" Colombia "AND" farm "AND" pig/swine/pork and Salmonellosis "AND" Colombia "AND" pig/swine/pork "AND" farm, in Spanish, English, and Portuguese.

2.3. Inclusion and Eligibility Criteria

Based on the acronym PICoR (Table 1), the inclusion and eligibility criteria concerned all publication types/documents between 2009 and mid-2024 related to *Salmonella* spp. in the Colombian primary pig production chain. There was no restriction on the age of the animals, the place or region, or the type of study reported.

Table 1. Eligibility criteria definitions according to PICoR approach.

Acronym	Criteria
P (Population/Patient/Problem)	Studies and reports obtained from the pig production chain at the primary scale in Colombia.
I (Intervention)	Studies and reports related to *Salmonella* spp. and salmonellosis.
Co (Comparison, Outcomes, Alternative vs. Standard)	Studies and reports related to prevalence, seroprevalence, pathology, risk factors, and farm management.
R (Reports)	Studies and reports published in English or Spanish throughout the country during the years 2009–2024.

2.4. Data Extraction

The Microsoft Excel software allowed for the collection of information such as the author, year, production stage, age, place or region, type and number of samples, technique, methodology, seroprevalence, prevalence, serotypes, and other results.

The prevalence analysis reported in all papers considered in this study referred to the traditional microbiological method of the pre-enrichment and handling of selective and differential media, with confirmation by molecular methods (MDS 3M® and PCR) in three of the reports. The methodology used for seroprevalence determination was ELISA.

Once the data were tabulated, we carefully reviewed the abstracts, procedures, and results from all documents that met the inclusion criteria. Two researchers independently reviewed the abstracts and selected those relevant for this review. Figure 1 illustrates the procedure performed to select and discard bibliographic sources. No duplicated documents were included.

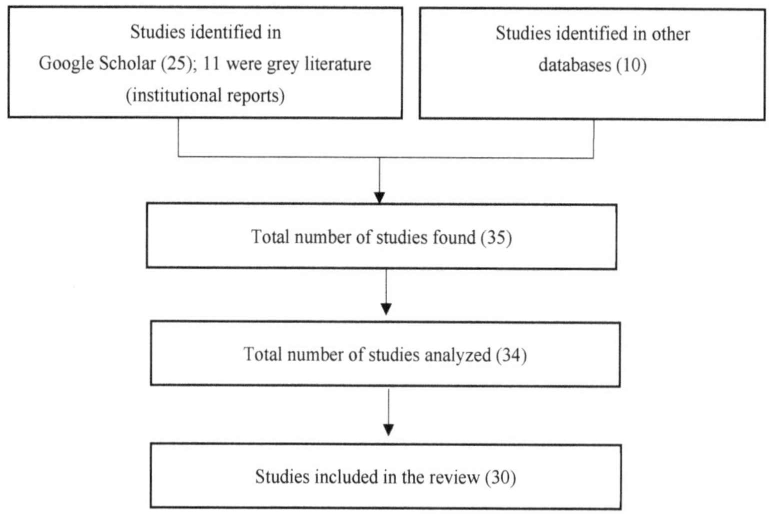

Figure 1. PRISMA flowchart of the procedure for the selection of the studies analyzed.

2.5. *Analysis*

Fifteen selected articles (50%) containing prevalence and seroprevalence data served for analysis. The meta-analysis methodology previously described, using the MetaXL V5.3 software (Epigear International, Sunrise Beach, Australia), was used to calculate the combined prevalence of all of the studies from the literature search in Instituto Colombiano Agropecuario [20]. To this end, we used three models from the meta-analysis methodology, i.e., fixed effects, random effects, and inverse variance heterogeneity models, to estimate the data heterogeneity and select the most appropriate model to estimate the combined prevalence and seroprevalence of *Salmonella* spp. in Colombian pig farms.

3. Results

Related to the searches performed in the scientific databases, only Scielo and Google Scholar yielded results for the keywords used. Additionally, we included information from other sources, such as technical reports, undergraduate theses, etc., corresponding to 11 grey literature documents.

Based on the information analyzed, we found that clinical, serological, microbiological, and molecular studies on the detection of *Salmonella* spp. have been conducted throughout the primary production cycle in different departments in the national territory, which we will discuss later in the corresponding section.

3.1. *Pig Salmonellosis*

In Colombia, the ICA (animal health regulator) has reported salmonellosis as an infectious pathological condition. Figure 2 shows the data on the affected farms and consolidated epidemiological data from 2009 to 2016 [25–32].

On the other hand, one of the clinical reports on the disease in Antioquia showed weak animals with evidence of anoxia ("purple ears and extremities") in the rearing stage and animals with "liquid" diarrhea, with the subsequent mortality of three piglets within 24 h of this clinical evidence; the symptomatology increased with time, necessitating the administration of trimethoprim sulfa (sulfatrim). The mentioned study reported a morbidity rate of 90% and a mortality rate of 10%, with the diagnosis of salmonellosis confirmed by the isolation of the pathogen from the spleen, cecum, mesenteric lymph nodes, intestinal abscesses, and feces [33]. In the same department, animals with clinical signs of *Salmonella* spp. were treated with trimethoprim sulfa and enrofloxacin; however, there were repeated cases, suggesting resistance to these antibiotics, especially to trimethoprim sulfa [34]. In the Atlantic region of the country, the microorganism's isolation was positive in 33% of the farms, and the authors found weaned piglets with symptoms compatible with salmonellosis (diarrhoea). All isolates were susceptible to the antibiotics tested (ampicillin, cefotaxime, ceftazidime, ceftriaxone, ciprofloxacin, and trimethoprim/sulfamethoxazole) [35].

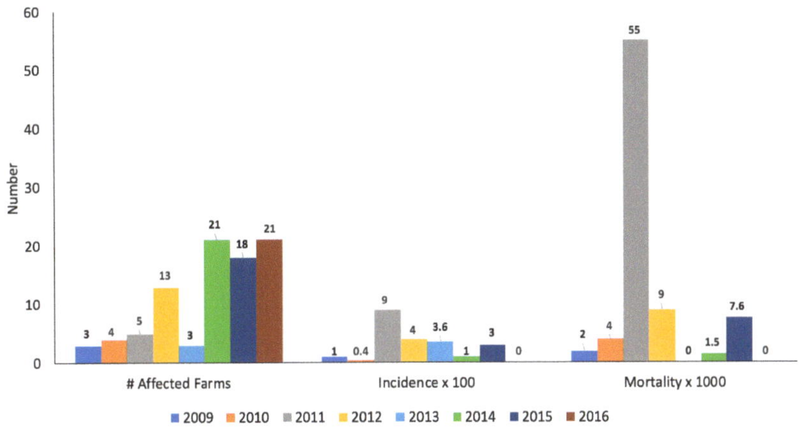

Figure 2. Swine salmonellosis is an infectious pathological condition reported in Colombia, according to the parameters of several affected farms and its incidence and mortality.

3.2. Epidemiological Parameters

Prevalences (P) and seroprevalences (SP) determined by microbiological, serological, and molecular tests have been reported throughout the national territory (Tables 2–4). The P data ranged from 0.58 to 28% in several departments. These included unpublished data from the Colombian national casuistry and a report for internal circulation only [36]. Moreover, the SP averaged 29% [16,37,38] (internal circulation report (Porkcolombia)) (Figures 3–5).

Table 2. *Salmonella* spp. prevalence (P) reported in Colombian pig farms.

Reference	Year of Origin	Geographic Location	No. Samples Analyzed	No. Positive Samples	% P
[16] *		Colombia	504	45	8.0
[35]		Valledupar, Atlántico	90	30	33.0
[36]	(Data from 2018 *)	Colombia	385	108	28.0
	(Data from 2018 *)	Colombia	203	18	8.9
	(Data from 2017 *)	Colombia	147	7	4.8
	(Data from 2016)	Cundinamarca, Antioquia	93	3	3.2
	(Data from 2015 *)	Colombia	273	15	5.5
	(Data from 2014 *)	Colombia	238	12	5.0
	(Data from 2013 *)	Colombia	88	10	11.4
	(Data from 2012)	Valle del Cauca	344	2	0.58
[17]		Antioquia	653	149	23

* More than three Colombian Departments analyzed.

The above depicts each study's estimated prevalence, ranging from 1 to 33%. In turn, the work conducted by the ICA had the highest relative weight concerning the combined prevalence calculation, with 55.6%. However, the sensitivity analysis showed that no study significantly affected the estimation of the combined prevalence [16] (Figure 3).

Table 3. Pooled prevalence and confidence intervals for the fixed effects model, random effect model, and inverse variance heterogeneity model.

Model Type	Combined Prevalence	LCI *	HCI *	Range **
Fixed effects	0.120	0.109	0.131	0.022
Random effect	0.172	0.075	0.295	0.22
Inverse variance heterogeneity	0.120	0.029	0.289	0.25

* LCI and HCI are low (2.5%) and high confidence intervals (97.5%). ** The range is the difference between the minimum and maximum values.

Table 4. Seroprevalence (SP) of *Salmonella* spp. reported in Colombian pig farms.

Reference	Geographic Location	No. Samples Analyzed	% SP
[16]	Colombia	231	38.1
(Internal circulation report (Porkcolombia))	Colombia	7140	26.7
[38]	Cundinamarca	89	40.0
[37]	Tolima	420	36.1

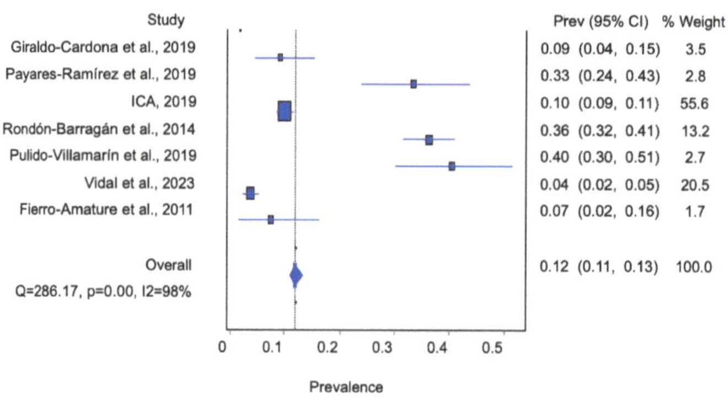

Figure 3. Estimation of *Salmonella* spp. prevalence (P) in Colombian pig farms (the percentage weight is in the order 3.5; 2.8; 55.6; 13.2; 2.7, and 1.7 and an overall value of 100.0) [16,17,35–39].

Table 3 shows the combined probability estimated with the three models. As can be seen, the value of Cochran's Q test for heterogeneity (Q = 286.17) and the inverse variance index (I2 = 98) was greater than 75%, indicating high heterogeneity among the studies. According to Barendregt et al. (2013), the inverse variance heterogeneity model should be used to analyze high-heterogeneity data. In this study, the model yielded a pooled prevalence value of 12.0% (confidence interval (CI), 10.9–13.1) [40].

The inverse variance heterogeneity model reported a pooled seroprevalence value of 27.6% (CI, 15.8–40.3). Figure 4 shows the estimated seroprevalence of each study, which ranged between 27 and 40% (unpublished data from the Colombian national casuistry). The report for internal circulation only [36] showed the highest relative weight concerning the calculation of the combined seroprevalence (90.6%) due to the number of samples analyzed. It is crucial to remark that the studies that reported P and SP data were conducted using validated detection techniques for both the bacterium and the antibodies against it.

The report from the Colombian Association of Pig Farmers—National Fund for Pig Farming (Asociación Colombiana de Porcicultores—Porkcolombia—Fondo Nacional de la Porcicultura (FNP)) indicates that, for the period between 2014 and 2018, when employing voluntary routine diagnosis in Colombian farms, the national average seroprevalence was 26.7%. The results by department evidenced that the highest value was found for Valle del Cauca, with 66.2%, followed by Caldas at 14.7%, Cundinamarca at 9%, Antioquia and

Cauca at 3%, Quindío at 2.7%, and Boyacá at 1.5%; 2018 was the year with the highest seroprevalence (53.9%). The high seroprevalence was due to the producers' awareness of animal health and public health problems through research conferences. In 2018, more samples were sent to diagnostic laboratories, with 3406 samples received (samples)—a value that increased over time (2014: 30 samples, 2015: 726 samples, 2016: 893 samples, 2017: 2055 samples) [36].

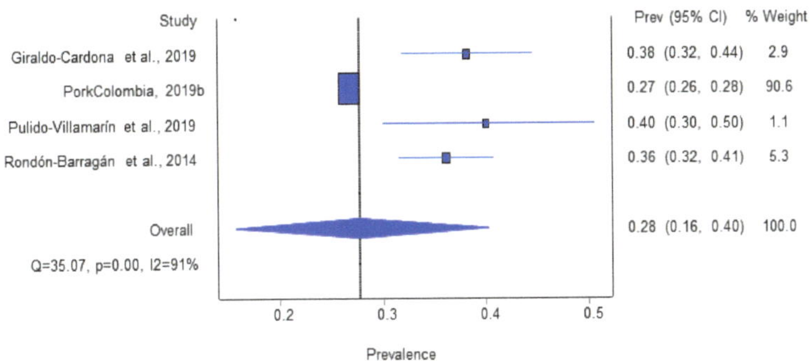

Figure 4. Estimation of *Salmonella* spp. seroprevalence (SP) in Colombian pig farms [3,16,37,38] 2014.

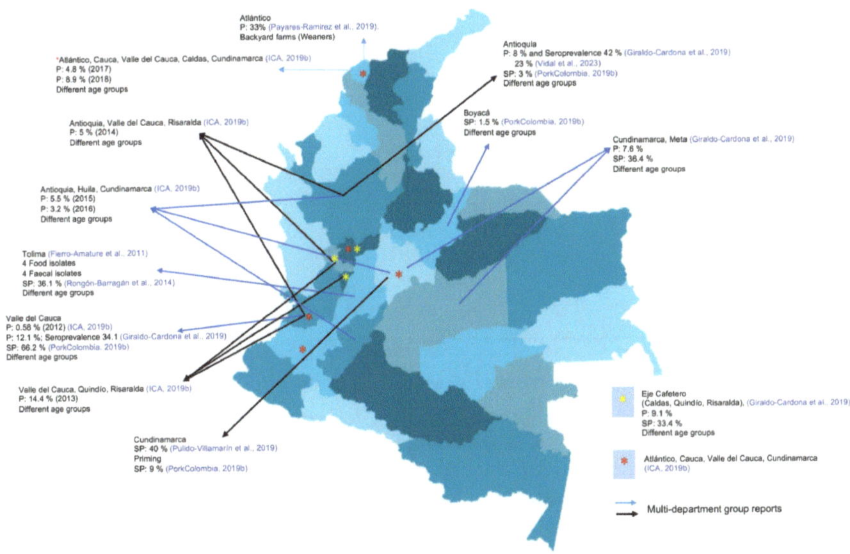

Figure 5. Map of prevalence (P) and seroprevalence (SP) against *Salmonella* spp. as reported in primary production throughout the national territory from 2009 to mid-2024 [3,16,17,35–39].

Concerning seropositivity according to the age group, 40.9% of the positivity occurred in the growing-finishing stages, 24.8% in the nursery, 20.1% in sows, and 6.1% in replacement gilts [37].

At the national level, there are diverse production systems and different types of farms. According to the parameters established by Porkcolombia, following the sanitary design submitted to the World Organization for Animal Health (WOAH) for notifiable diseases, and based on the 2018 census, the findings for the farms by type of exploitation definition were as follows: (i) backyard—there were 23,539 holdings, ≤3 female breeding piglets (FPB) or ≤15 fat; (ii) commercial family—with 92,572 holdings, >3 and <10 HC and or >15 and <100 fat; (iii) industrial commercial—with 7659, owning between ≥10 and <100

HC and/or ≥100 and <600 fat; and (iv) commercial technical (Colombian denomination) —with 9122, those with ≥100 HC and/or ≥600 fat.

The farms by the type of production system and production site are named as follows: full-cycle farms (FC); multi-site farms (MSF); farms with mothers for piglet rearing (piglet); and farms where animals are raised and fattened (fattening).

On the other hand, according to the flow of animals on the farm, there is the continuous flow (CF) and the all-in/all-out (AIAO) production system. There are two production lines: the first is the production, which refers to the nursery, growing, and finishing stages, and the reproduction line refers to sows, piglets, barrows, and replacement gilts. According to these production parameters, the percentages of SP obtained through the routine diagnosis carried out by Porkcolombia were different (Table 5).

Table 5. Percentages of seroprevalence (SP) according to production parameters, through routine diagnosis [36].

Production Parameters	% SP
Type of Operation	
Full cycle	97.7
Multiple sites	0.0
Breeding	0.9
Fattening	1.4
On-Farm Flow	
Continuous flow—FC	100.0
All in/all out—AIAO	0.0
Property Type	
Commercially technified	95.7
Commercial industrial	4.3
Production Line	
Production	64.0
Reproduction	36.0

3.3. Serovars and Antimicrobial Resistance (AMR)

The reports found in Colombia indicate the following the predominant serovars: *S.* Typhimurium, *S.* Brancaster, *S.* Derby, *S.* Typhimurium variant Copenhagen, *S.* Heidelberg, *S.* Group B, *S.* Group E1, *S.* Group D2, and *S.* Group F. [36,39], with the presence of *S.* Typhimurium as a monophasic variant in the last few years [17].

The isolates obtained from Colombian farms show different antimicrobial susceptibility profiles, with evidence of sensitivity to amikacin (100%) and resistance close to 95% for penicillin, lincomycin, and tetracycline [16,39], among other antibiotics. On the other hand, in recent years, multi-resistance rates (resistance to three or more antibiotic classes) have been reported between 44 and 70.3% [17,37,41].

3.4. Risk Factors

In Colombia, there are few studies carried out on issues other than sanitary and epidemiological aspects. In 2012, a study reported the risk factors associated with the presence of the bacterium on farms in Tolima; in this study, some parameters were established, such as the control of rodents, insects, and birds. New animals and self-replacements were significantly associated with the presence of the pathogen on farms [42]; however, no further specific mention was made of the characteristics of the variables analyzed. Furthermore, a consolidated report for the four regions of Colombia where pig production is predominant (Antioquia, Valle del Cauca, Eje Cafetero, Cundinamarca-Meta) concluded that the factors associated with the presence of the bacterium on farms are related to the source of water supplied to the pigs [43]; rough flooring in the animal pens; the feed supplied in hoppers; and workers' boots [16].

On the other hand, analyzing the data related to the risk factors, Porkcolombia indicates the possibility of increasing the presence of antibodies against *Salmonella* spp. with a statistically significant association. These factors include (a) the type of farm, where those with a complete cycle exhibited a 27.9 times higher likelihood of presenting antibodies against *Salmonella* spp. (odds ratio (OR) 27.9; 95% CI 5.79–315.6) concerning fattening farms; (b) the flow of the farm, as those farms with an "all in/all out" system showed a 24 times higher likelihood (OR 24.9; 95% CI 5.22–272. 39) of seropositivity concerning continuous-flow farms; and (c) the type of farm, where "commercial technical" farms (farms with ≥100 breeding females and or ≥600 heads) had an OR of 1.92 (95% CI 1.18–3.13), which, although relatively low, is a relevant factor when establishing prevention and control measures [36].

3.5. Water and Organic Waste Management

Some studies have reported pathogen studies related to the environmental management of waste from pig production and water. The first one was related to the management and quality of pig manure in Caldas, in which the fermentation process favored the reduction of the Salmonella load, ensuring the safety of silage supplemented with pig manure [44]. In Antioquia, the process of composting pig manure and pig mortality using a bioinoculant accelerator favors the production of compost free of bacterial pathogens (including Salmonella) and parasites in a shorter time (30 days) compared to the traditional method (45 days) [45].

Regarding water management, a study in Cundinamarca showed the effectiveness of photolysis and photolysis with hydrogen peroxide in pig farm water for the inactivation of the microorganism [43]. In Antioquia, the additives (Selko 0.15–0.3 mg/L) used in the water showed promising results in reducing the symptoms [34]; similarly, in 2023, it was found that the maintenance and cleaning of water pipes and the administration of organic acids in pigs' drinking water favored the productivity of the animals and delayed their exposure to the bacterium [46].

4. Discussion

Clinically, salmonellosis in pigs generates gastrointestinal, systemic, and occasionally subclinical symptoms, making the detection of carrier animals difficult and facilitating the pathogen's spread on farms, because Salmonella persists in the animal's spleen and mesenteric lymph nodes [19,47]. This situation has been observed in several departments in Colombia, where subclinical carrier animals are likely to maintain the pathogen within the farm population [16,36,37].

The disease may appear at any stage of the production cycle, although its severity is generally more pronounced in younger animals [48], such as suckling piglets, where diagnosis is limited to obtaining rectal swabs, thus complicating accurate diagnosis and allowing these animals to become risk factors for the maintenance of bacteria on farms [49].

The information reported confirms the importance of piglets in the maintenance of the microorganism within the farm, facilitating its rotation in nursing mothers [26,36]; however, in the data reported for Colombia, the pre-feeding and rearing–weaning stages presented the highest seropositivity values [44].

On the other hand, although the treatment of salmonellosis by antibiotics is recognized worldwide, as reported in Colombian studies, an Argentinian work evaluated extracts obtained from lactic acid bacteria (LAB) isolated from the colostrum of mothers as a treatment alternative, demonstrating inhibitory effects "In Vitro" [39]. On the other hand, with the preventive management of antibiotics and the use of pH-modifying additives, the symptomatology decreased, a situation that was achieved by the addition of organic acids such as propionic acid (0.8%), formic acid (0.2%), or lactic acid (0.4%) to feed or drinking water [42,50]. This is a practice that has yielded good results as a prevention and control strategy against salmonellosis worldwide.

While Colombian studies have not yet explored the efficacy of animal vaccination against Salmonella, international findings suggest that the vaccine's effectiveness varies depending on the products and protocols used. Live vaccines using S. Typhimurium and S. Choleraesuis serovars are prevalent, while inactivated vaccines are less common; however, both have been considered effective in reducing infection rates and minimizing the contamination of pork products intended for human consumption [42,44].

Epidemiological data related to P and SP were calculated and analyzed based on the laboratory tests performed for their detection, either because the bacterium was recovered/isolated by microbiological methods or if using serological methods, the antibody response was positive. Such positivity was detected in pigs of different age groups; however, the highest proportion of positivity was reported in adult animals, as corroborated by some studies [51,52].

In Colombia, the combined prevalence was 21.3%, and the seroprevalence was 27.6%. Worldwide, P and SP are usually high [19]; however, reports for farms in different countries have been variable, e.g., Germany reported 7.9% (SP) [53], Romania 35.8% (SP) [53], 36% (P) for Canada [51], and 41.5% (P) in Argentina [54]. These data suggest that the management conditions and prevention and control programs are different in each country.

Although more than 2000 serovars are recognized within the *Salmonella enterica* species [6], many of them tend to be host-specific. Thus, *S. enterica* serovar Cholerasuis is usually associated with disease in pigs; however, *S. enterica* serovar Typhimurium cannot be overlooked as it can also be clinically relevant [19], and, being a non-host-specific serovar, it represents a major public health concern due to its zoonotic transmission.

Internationally, S. Typhimurium and S. Derby are the most frequently reported serovars in studies related to animals and farms [21,53]; compared to this, the data found for Colombia do not differ from the serovars that have predominated, such as S. Typhimurium, S. Brancaster, S. Derby, S. Typhimurium variant Copenhagen, and S. Heidelberg, as well as a few others [38,51]. This information also coincides with the serovars found along the primary chain in Canadian farms (S. Derby, S. Typhimurium var. Copenhagen, S. Putten, S. Infantis, and S. Mbandaka) [52,55]; for Latin America, specifically in Argentina, S. Anatum and S. Typhimurium have predominated, with the detection of the monophasic strain S. 4, 5, 12:i: [54]. Although, for Colombia, specific data on the serotypes found by age group and the presence of the monophasic serovar of S. Typhimurium (S. 4, 5, 12:i:) have been reported, they highlight the necessity to expand field studies. In Spanish farms, its presence has been detected together with S. Rissen, S. Derby, and S. Bovismorbificans in weaned piglets and their mothers, leading to the establishment of the bacterium in these farms [26]. Furthermore, the detection of the mentioned serovar indicates a high degree of antimicrobial resistance [53,56], either by intrinsic genetic factors or acquired through plasmids; thus, in Colombia, it is necessary to be alert at the time of its appearance to take control and prevention measures.

The resistance to penicillin, lincomycin, and tetracycline [16,17,39,41] could be similar to that found in Brazil [51], Argentina [52], Portugal [53], and, in general, the European Union [54], suggesting that the indiscriminate use of antibiotics throughout the primary production chain could favor this species, enabling it to become a reservoir of multi-resistant strains [55].

According to Davies [11,57], one- to three-site AIAO production systems represent higher protection and better control for pig health than FC farms and can prevent cross-contamination between production cycles by facilitating thorough cleaning and disinfection, reducing the potential for exposure to Salmonella infection [11,57]. In Colombia, AIAO systems and practices are frequent; however, in routine diagnosis, a statistically significant association ($p = 0.05$) was found between AIAO/FC and the presence of the bacterium. In other countries, some of the risk factors associated with *Salmonella* spp. on farms include the inadequate handling of operators' boots, the contamination of water drains [55], and nose-to-nose contact between animals in the same pen [51], as well as feed prepared with contaminated raw materials and possibly "pelleted" food [43,52]. Therefore, the risk factors described in national and international scientific publications agree with those analyzed by Porkcolombia [38].

5. Conclusions

In summary, the estimated combined prevalence for *Salmonella* spp. in Colombian pig farms is 21.3%, while the seroprevalence is 27.6%, with the production systems and animal flow significantly influencing the pathogens' presence. Data on *Salmonella* spp. in Colombia's primary pork production chain remain scarce, with most reports categorized as grey literature.

Although *Salmonella* spp. are leading zoonotic pathogen responsible for foodborne outbreaks traced to animal products, including pork, there is a lack of official reporting for these pathogens in Colombia's primary production chain. It is essential to raise awareness within the government and academic circles regarding the importance of reporting epidemiological data on pathogens that affect animal production and pose risks to human health.

Author Contributions: Conceptualization, A.P.-V.; literature review and formal analysis, A.K.C.-C., M.R.-M., F.R.-B., M.P.-V., I.H.-T. and A.C.-C.; methodology, F.S., I.C.-T. and R.A.P.-P.; writing—original draft, A.P.-V., F.S. and R.A.P.-P.; translation and English revision, R.A.P.-P. All authors have read and agreed to the published version of the manuscript.

Funding: This collaborative work was part of the project entitled "Salmonella Control in the Colombian Pig Industry—SALPORK", which received financial support from the Ministry of Foreign Affairs (Danida), Denmark, through grant number 18-M07-KU, and Pontificia Universidad Javeriana VRI, Bogotá, D.C., Colombia, through grants ID 0008825 and ID 21274.

Data Availability Statement: Data are available upon request.

Conflicts of Interest: The authors declare that they have no competing interests. The financing entity had no role in the study design, data collection or analysis, decision to publish or preparation of the manuscript.

References

1. Organización para la Cooperación y el Desarrollo Económicos (OCDE)/Organización de las Naciones Unidas para la Alimentación y la Agricultura (FAO), Perspectivas Agrícolas 2019–2028. 2019. Available online: https://doi.org/10.1787/7b2e8ba3-es (accessed on 16 August 2022).
2. Maya, C. Sector Porcicultor, uno de los más Productivos del Momento. 2018. Available online: https://www.semana.com/edicion-impresa/negocios/articulo/balance-del-sector-porcicultor-en-colombia/255321/ (accessed on 16 August 2022).
3. Porkcolombia, Boletín Económico Diciembre 2019. 2019. Available online: https://porkcolombia.co/wp-content/uploads/2019/12/Bol_Dic_19.pdf (accessed on 16 August 2022).
4. ICA, Censo Pecuario año 2018. 2022. Available online: https://www.ica.gov.co/areas/pecuaria/servicios/epidemiologia-veterinaria/censos-2016/censo-2018 (accessed on 16 August 2022).
5. Hurd, H.S.; McKean, J.D.; Griffith, R.W.; Wesley, I.V.; Rostagno, M.H. Salmonella enterica infections in market swine with and without transport and holding. *Appl. Environ. Microbiol.* **2002**, *68*, 2376–2381. [CrossRef] [PubMed]
6. Majowicz, S.E.; Musto, J.; Scallan, E.; Angulo, F.J.; Kirk, M.; O'Brien, S.J.; Jones, T.F.; Fazil, A.; Hoekstra, R.M. The Global Burden of Nontyphoidal Salmonella Gastroenteritis. *Clin. Infect. Dis.* **2010**, *50*, 882–889. [CrossRef] [PubMed]
7. Simpson, K.M.J.; Hill-Cawthorne, G.A.; Ward, M.P.; Mor, S.M. Diversity of Salmonella serotypes from humans, food, domestic animals and wildlife in New South Wales, Australia. *BMC Infect. Dis.* **2018**, *18*, 623. [CrossRef] [PubMed]
8. Organización Mundial de la Salud (OMS), Salmonella (No Tifoidea). 2019. Available online: https://www.who.int/es/news-room/fact-sheets/detail/salmonella-(non-typhoidal) (accessed on 16 August 2022).
9. European Food Safety Authority; European Centre for Disease Prevention and Control. The European Union summary report on trends and sources of zoonoses, zoonotic agents and food-borne outbreaks in 2015. *EFSA J.* **2016**, *14*, 4634.
10. Funk, J. Pre-harvest food safety diagnostics for Salmonella serovars. Part 1: Microbiological culture. *J. Swine Health Prod.* **2003**, *11*, 87–90. [CrossRef]
11. Davies, P.R. Intensive Swine Production and Pork Safety. *Foodborne Pathog. Dis.* **2011**, *8*, 189–201. [CrossRef]
12. Centers of Disease Control and Prevention (CDC), National Outbreak Reporting System (NORS). 2019. Available online: https://wwwn.cdc.gov/norsdashboard/ (accessed on 16 August 2022).

13. Ministerio de Salud Pública de Colombia, Instituto Nacional de Salud (MSPS-INS) República de Colombia. Informe de Evento Enfermedades Transmitidas por Alimentos, Colombia, 2017. 2017. Available online: https://www.ins.gov.co/buscador-eventos/Informesdeevento/ETA%202017.pdf (accessed on 16 August 2022).
14. Instituto Nacional de Salud, (INS). Boletín epidemiológico Semanal 52. 2018. Available online: https://www.ins.gov.co/buscador-eventos/BoletinEpidemiologico/2018%20Bolet%C3%ADn%20epidemiol%C3%B3gico%20semana%2052.pdf (accessed on 16 August 2022).
15. Guillier, L.; Thébault, A.; Fravalo, P.; Mughini-Gras, L.; Jourdan-da Silva, N.; David, J.; Kooh, P.; Cadavez, V.; Gonzales-Barron, U. Risk factors for sporadic salmonellosis: A systematic review and meta-analysis. *Microb. Risk Anal.* **2021**, *17*, 100138. [CrossRef]
16. Giraldo-Cardona, J.P.; Gualdrón-Ramírez, D.; Chamorro-Tobar, I.; Pulido-Villamarín, A.; Santamaría-Durán, N.; Castañeda-Salazar, R.; Zambrano-Moreno, C.; Carrascal-Camacho, A.K. *Salmonella* spp. prevalence, antimicrobial resistance and risk factor determination in Colombian swine farms. *Pesqui. Veterinária Bras.* **2019**, *39*, 816–822. [CrossRef]
17. Vidal, J.L.; Clavijo, V.; Castellanos, L.R.; Kathiresan, J.; Kumar, A.M.V.; Mehta, K.; Chaparro-Gutierrez, J.J. Multidrug-resistant *Salmonella* spp. in fecal samples of pigs with suspected salmonellosis in Antioquia, Colombia, 2019-2021. *Rev. Panam. Salud Pública* **2023**, *47*, e46. [CrossRef]
18. Hurst, M.; Nesbitt, A.; Kadykalo, S.; Dougherty, B.; Arango-Sabogal, J.C.; Ravel, A. Attributing salmonellosis cases to foodborne, animal contact and waterborne routes using the microbial subtyping approach and exposure weights. *Food Control* **2023**, *148*, 109636. [CrossRef]
19. Patterson, S.K.; Kim, H.B.; Borewicz, K.; Isaacson, R.E. Towards an understanding of Salmonella enterica serovar Typhimurium persistence in swine. *Anim. Health Res. Rev.* **2016**, *17*, 159–168. [CrossRef] [PubMed]
20. Instituto Colombiano Agropecuario, (ICA). Resolución 20148. Por la cual se Establecen los Requisitos para Obtener la Autorización Sanitaria y de Inocuidad en los Predios Pecuarios Productores de Animales Destinados al Sacrificio para Consumo Humano. 2016. Available online: http://acovez.org/index.php/noticias/gremiales/27-expedicion-resolucion-20148-del-8-de-agosto-de-2016.html (accessed on 20 May 2017).
21. Sanchez, J.; Dohoo, I.R.; Christensen, J.; Rajic, A. Factors influencing the prevalence of *Salmonella* spp. in swine farms: A meta-analysis approach. *Prev. Vet. Med.* **2007**, *81*, 148–177. [CrossRef] [PubMed]
22. Dors, A.; Pomorska-Mol, M.; Czyzewska, E.; Wasyl, D.; Pejsak, Z. Prevalence and risk factors for Lawsonia intracellularis, Brachyspira hyodysenteriae and *Salmonella* spp. in finishing pigs in Polish farrow-to-finish swine herds. *Pol. J. Vet. Sci.* **2015**, *18*, 825–831. [CrossRef] [PubMed]
23. Tricco, A.C.; Lillie, E.; Zarin, W.; O'Brien, K.K.; Colquhoun, H.; Levac, D.; Moher, D.; Peters, M.D.J.; Horsley, T.; Weeks, L.; et al. PRISMA Extension for Scoping Reviews (PRISMA-ScR): Checklist and Explanation. *Ann. Intern. Med.* **2018**, *169*, 467–473. [CrossRef]
24. Munn, Z.; Peters, M.D.J.; Stern, C.; Tufanaru, C.; McArthur, A.; Aromataris, E. Systematic review or scoping review? Guidance for authors when choosing between a systematic or scoping review approach. *BMC Med. Res. Methodol.* **2018**, *18*, 143. [CrossRef]
25. ICA. Sanidad Animal. 2016. Available online: https://www.ica.gov.co/getattachment/6d2f08b5-da5d-49a2-ad3c-ef3ccfe06df7/Boletin-2016-Sanidad-Animal.aspx (accessed on 16 August 2022).
26. ICA. Sanidad Animal. 2015. Available online: https://www.ica.gov.co/getattachment/4d163775-d3d8-47ab-92ba-fa5a0140bfdc/2015.aspx (accessed on 16 August 2022).
27. ICA. Sanidad Animal. 2014. Available online: https://www.ica.gov.co/getattachment/986dd783-8f37-4ab3-bc33-39995bd8c065/2014.aspx (accessed on 16 August 2022).
28. ICA. Sanidad Animal. 2013. Available online: https://www.ica.gov.co/getattachment/0b099ac3-d670-4c11-be1b-02e50db63047/2013.aspx (accessed on 18 August 2022).
29. ICA. Sanidad Animal. 2012. Available online: https://www.ica.gov.co/getattachment/bce28fb3-c2c7-4f46-99fc-6bae850353fc/2012.aspx (accessed on 16 August 2022).
30. ICA. Sanidad Animal. 2011. Available online: https://www.ica.gov.co/getattachment/b66f6f33-43bb-4c2c-a8f6-e66ab31194e0/2011.aspx (accessed on 16 August 2022).
31. ICA. Sanidad Animal. 2010. Available online: https://www.ica.gov.co/getattachment/5822cada-667f-4541-8b86-c0258be04b64/2010.aspx (accessed on 16 August 2022).
32. ICA. Sanidad Animal. 2009. Available online: https://www.ica.gov.co/getattachment/7b0e7f6a-4c43-4dff-be39-ab9f4d1b0e0a/2009.aspx (accessed on 16 August 2022).
33. Escobar Cardona, X. *Caso Clínico Salmonella en Precebo*; Corporación Universitaria Lasallista: Caldas-Antioquia, Colombia, 2017; p. 49.
34. Betancur Vélez, E. *Control de Salmonellosis en Cerdos con uso de Aditivos y sin el uso de Antibióticos en el Alimento, en Cinco Granjas Porcícolas del Municipio de Santa Rosa de Osos (Antioquia)*; Corporación Universitaria Lasallista: Caldas-Antioquia, Colombia, 2018; p. 39.
35. Payares-Ramírez, K.; Cañate-González, A.; Polo-Gutiérrez, W.; Prada-Herrera, J.; Esquea-Larios, K. Salmonelosis en lechones destetados de granjas de traspatio en Valledupar, Colombia. *Rev. Vet. Zootec.* **2019**, *13*, 57–67.
36. ICA. *Unpublished Data from the Colombian National Casuistry*; Report for internal circulation only; The Colombian Agricultural Institute (ICA): Bogotá, Colombia, 2019.
37. Rondón-Barragán, I.S.; Rodríguez, G.A.; Marín, M.G.A. Determinación de la seroprevalencia de *Salmonella* spp. en granjas porcinas del departamento del Tolima. *Orinoquia* **2014**, *18*, 60–67. [CrossRef]

38. Pulido-Villamarín, A.; Castañeda-Salazar, R.; Mendoza-Gómez, M.F.; Vivas-Díaz, L. Presence of antibodies against some pathogens of zoonotic interest in four pig farms of Cundinamarca, Colombia. *Rev. Investig. Vet. Perú* **2019**, *30*, 446–454. [CrossRef]
39. Fierro-Amature, M.A.; Osorio-Amortegui, C.A.; de Rubio, L.C.F.; Rondón-Barragán, I.S. Resistencia antibiótica en *Salmonella enterica* serovar Typhimurium aisladas de granjas porcícolas en el Departamento del Tolima. *Orinoquia* **2011**, *15*, 71–78. [CrossRef]
40. Barendregt, J.J.; Doi, S.A.; Lee, Y.Y.; Norman, R.E.; Vos, T. Meta-analysis of prevalence. *J. Epidemiol. Commun. Health* **2013**, *67*, 974–978. [CrossRef]
41. Mantilla, J.F.; Villar, D.; Gómez-Beltrán, D.A.; Vidal, J.L.; Chaparro-Gutiérrez, J.J. High antimicrobial resistance in *Salmonella* spp. and Escherichia coli isolates from swine fecal samples submitted to a veterinary diagnostic laboratory in Colombia. *Rev. Colomb. Cienc. Pecu.* **2022**, *35*, 26–35. [CrossRef]
42. Henao Beltrán, J.S.; Ramírez Aguirre, E.; Rondón-Barragán, I.S. Analysis of Good Production Practices in pig farms on the departamento del Tolima and risk factors associated with presence of *Salmonella* spp. *Rev. CES Med. Vet. Zootec.* **2012**, *7*, 11–20.
43. Valencia Acosta, C.L. *Evaluación de la Fotólisis UV/254 nm como Tratamiento para la Eliminación de Salmonella spp. en aguas de Granjas Porcícolas*; Pontificia Universidad Javeriana: Bogotá, Colombia, 2019; p. 71.
44. Estrada, J.; Aranda, E.M.; Pichard, G.; Henao, F.J. Efecto de la fermentación en estado sólido de la porcinaza sobre la persistencia de patógenos en el ensilaje. *Boletín Científico Mus. Hist. Nat.* **2011**, *15*, 71–80.
45. Matiz-Villamil, A.; Chamorro-Tobar, I.C.; Saenz-Aponte, A.; Pulido-Villamarin, A.; Carrascal-Camacho, A.K.; Gutierrez-Rojas, I.S.; Sanchez-Garibello, A.M.; Barrientos-Anzola, I.A.; Zambrano-Moreno, D.C.; Poutou-Pinales, R.A. Management of swine mortalities through the use of a mixed composting-accelerating bio-inoculant. *Heliyon* **2021**, *7*, e05884. [CrossRef]
46. Roldan-Henao, M.; Dalsgaard, A.; Cardona-Castro, N.; Restrepo-Rivera, L.; Veloza-Angulo, L.C.; Alban, L. Pilot study of the productivity and Salmonella seroprevalence in pigs administered organic acids. *Front. Vet. Sci.* **2023**, *10*, 1123137. [CrossRef]
47. Harris, I.T. Serologic basis for assessment of subclinical Salmonella infection in swine: Part 2. *J. Swine Health Prod.* **2003**, *11*, 300–303. [CrossRef]
48. Hoelzer, K.; Moreno Switt, A.I.; Wiedmann, M. Animal contact as a source of human non-typhoidal salmonellosis. *Vet. Res.* **2011**, *42*, 34. [CrossRef]
49. Casanova-Higes, A.; Marin-Alcala, C.M.; Andres-Barranco, S.; Cebollada-Solanas, A.; Alvarez, J.; Mainar-Jaime, R.C. Weaned piglets: Another factor to be considered for the control of Salmonella infection in breeding pig farms. *Vet. Res.* **2019**, *50*, 45. [CrossRef]
50. Martin-Pelaez, S.; Costabile, A.; Hoyles, L.; Rastall, R.A.; Gibson, G.R.; La Ragione, R.M.; Woodward, M.J.; Mateu, E.; Martin-Orue, S.M. Evaluation of the inclusion of a mixture of organic acids or lactulose into the feed of pigs experimentally challenged with *Salmonella typhimurium*. *Vet. Microbiol.* **2010**, *142*, 337–345. [CrossRef]
51. Bessa, M.C.; Michael, G.B.; Canu, N.; Canal, C.W.; Cardoso, M.; Rabsch, W.; Rubino, S. Phenotypic and genetic characterization of Salmonella enterica subsp. enterica serovar Typhimurium isolated from pigs in Rio Grande do Sul, Brazil. *Res. Vet. Sci.* **2007**, *83*, 302–310. [CrossRef]
52. Vico, J.P.; Lorenzutti, A.M.; Zogbi, A.P.; Aleu, G.; Sanchez, I.C.; Caffer, M.I.; Rosmini, M.R.; Mainar-Jaime, R.C. Prevalence, associated risk factors, and antimicrobial resistance profiles of non-typhoidal Salmonella in large scale swine production in Cordoba, Argentina. *Res. Vet. Sci.* **2020**, *130*, 161–169. [CrossRef] [PubMed]
53. Fernandes, L.; Centeno, M.M.; Couto, N.; Nunes, T.; Almeida, V.; Alban, L.; Pomba, C. Longitudinal characterization of monophasic *Salmonella typhimurium* throughout the pig's life cycle. *Vet. Microbiol.* **2016**, *192*, 231–237. [CrossRef] [PubMed]
54. European Food Safety Authority; European Centre for Disease Prevention and Control. The European Union Summary Report on Antimicrobial Resistance in zoonotic and indicator bacteria from humans, animals and food in 2017/2018. *EFSA J* **2020**, *18*, e06007.
55. Mora-Gamboa, M.P.C.; Rincón-Gamboa, S.M.; Ardila-Leal, L.D.; Poutou-Piñales, R.A.; Pedroza-Rodríguez, A.M.; Quevedo-Hidalgo, B.E. Impact of antibiotics as waste, physical, chemical, and enzymatical degradation: Use of laccases. *Molecules* **2022**, *27*, 4436. [CrossRef] [PubMed]
56. Wierup, M.; Haggblom, P. An assessment of soybeans and other vegetable proteins as source of Salmonella contamination in pig production. *Acta Vet. Scand.* **2010**, *52*, 15. [CrossRef]
57. Davies, P.R. Pork Safety: Achievements and Challenges. *Zoonoses Public Health* **2010**, *57*, 1–5. [CrossRef]

animals

MDPI

Review

Pre-Harvest Non-Typhoidal *Salmonella* Control Strategies in Commercial Layer Chickens

Roshen N. Neelawala, Lekshmi K. Edison and Subhashinie Kariyawasam *

Department of Comparative, Diagnostic, and Population Medicine, College of Veterinary Medicine, University of Florida, Gainesville, FL 32608, USA; roshen.neelawala@ufl.edu (R.N.N.); edison.le@ufl.edu (L.K.E.)
* Correspondence: skariyawasam@ufl.edu

Simple Summary: *Salmonella* poses a significant threat to poultry farming, as it can asymptomatically infect flocks, ultimately leading to foodborne illnesses when contaminated poultry products reach consumers. This review focuses on various strategies for controlling *Salmonella* colonization in commercial layer chickens at the pre-harvest level. The approaches discussed include enhancing farm biosecurity, using vaccines, incorporating feed additives, and improving genetic resistance to infection through selective breeding. Each of these approaches offers unique benefits, such as preventing the introduction of pathogens, enhancing immunity against *Salmonella*, and/or improving intestinal health to reduce *Salmonella* colonization. Therefore, a combined approach is essential for minimizing the risk of *Salmonella*, promoting food safety, and supporting sustainable poultry production.

Abstract: Non-typhoidal *Salmonella* (NTS) infections in poultry, particularly in commercial-layer chickens, pose a critical risk to food safety and public health worldwide. NTS bacteria can remain undetected in poultry flocks, contaminating products and potentially leading to gastroenteritis in humans. This review examines pre-harvest control strategies for NTS in layer chickens, including biosecurity protocols, vaccinations, feed additives, genetic selection, and environmental management. These strategies have substantially reduced *Salmonella* colonization and product contamination rates in the commercial layer industry. By evaluating these strategies, this review highlights the importance of integrated control measures to limit NTS colonization, reduce antimicrobial resistance, and improve poultry health. This review aims to provide producers, researchers, and policymakers with insights into effective practices to minimize *Salmonella* contamination and enhance both animal and human health outcomes.

Keywords: antimicrobial resistance; biosecurity; vaccination; feed additives; food safety; layer chickens; non-typhoidal *Salmonella*; pre-harvest control; poultry health

1. Introduction

Non-typhoidal *Salmonella* (NTS) refers to a group of *Salmonella* serovars associated with gastroenteritis in humans. Notable serovars include *S.* Enteritidis, *S.* Typhimurium, *S.* Kentucky, *S.* Newport, *S.* Javiana, *S.* Heidelberg, *S.* Muenchen, *S.* Infantis, *S.* Braenderup, *S.* Saintpaul, *S.* Montevideo, and *S.* Thompson [1]. Some of these serovars pose a significant threat due to their zoonotic transmission from poultry to humans [2]. Poultry products, such as meat and eggs, serve as critical reservoirs of NTS; therefore, they contribute significantly to foodborne disease outbreaks globally [3]. Among NTS serovars, *S.* Enteritidis and *S.* Typhimurium are the most commonly linked to foodborne illnesses worldwide, particularly in the United States and the European Union, stemming from the consumption of eggs and egg products [1,4]. In commercial poultry operations, especially in layer chickens, NTS colonization usually occurs without symptoms, allowing *Salmonella* to persist undetected in flocks [5]. This silent carriage has a direct impact on public health by facilitating contamination of poultry environment and poultry products. Therefore, controlling NTS in poultry has become an urgent global priority to reduce foodborne illnesses and underscores the need for effective pre-harvest interventions. The occurrence and spread of NTS in commercial layer chicken facilities can vary widely depending on the geographical region, farm management practices, and the effectiveness of biosecurity measures [6]. In the United States, *Salmonella* is the leading cause of foodborne infections, and poultry products have consistently been identified as major contributors [5].

In avian hosts, NTS colonization of the gastrointestinal tract (GIT) is a complex process involving the adhesion of bacteria to intestinal epithelial cells and subsequent invasion. During this course of colonization, the NTS evolved several adaptive mechanisms to evade the host immune system. These survival mechanisms include modulation of the local immune environment and the use of virulence factors that enable the pathogen to exist within the host without causing overt clinical manifestations [7–9]. In poultry, the transmission of NTS can occur through vertical or horizontal (Figure 1) pathways, necessitating that the pre-harvest control strategies focus on various farm management aspects. Vertical transmission occurs when *Salmonella* is transmitted from the breeder hens to their progeny through contaminated eggs. This is observed when *Salmonella* contaminates the hen's reproductive organs, particularly the ovaries or oviducts, leading to egg contamination before shell formation. Such infected eggs can hatch into chicks carrying the pathogen throughout their development, thereby perpetuating the infection cycle. Horizontal transmission of NTS implies pathogen dissemination within a flock or between different flocks, often through direct or indirect contact. The fecal-oral route is the main route of infection; wherein other birds ingest *Salmonella* from contaminated droppings by pecking contaminated litter or drinking water. Poor quality feed or water sources may act as reservoirs, bringing the pathogen into contact with healthy birds. Other possible vectors include contaminated farm equipment, clothing, rodents, and footwear from farm workers [8].

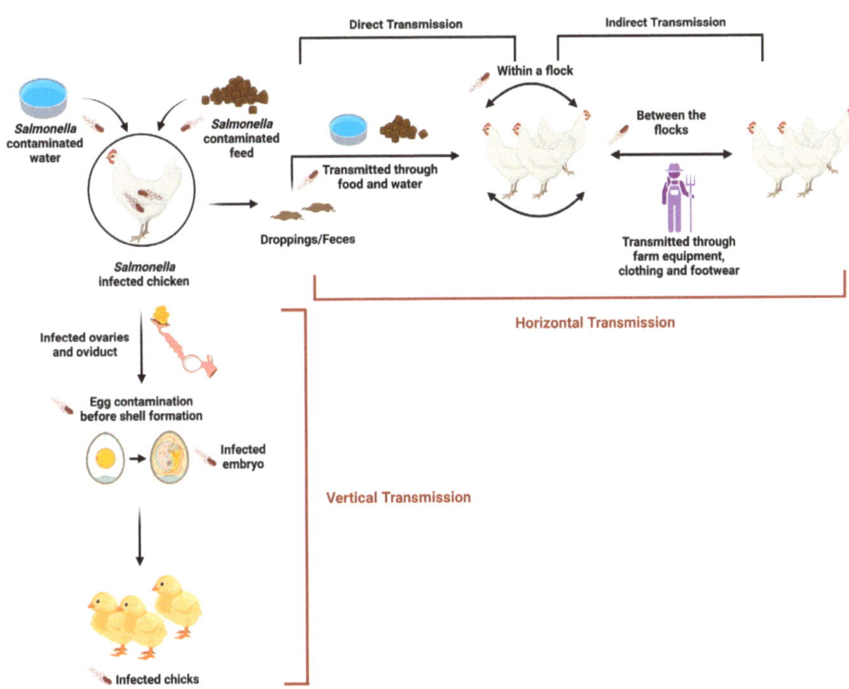

Figure 1. Transmission pathways of *Salmonella* in poultry (created with BioRender.com).

From a production point of view, NTS can lead to economic losses in commercial layer operations by decreasing egg production, increasing mortality in severe cases, and raising the cost of implementing control measures [8]. Although many NTS infections in poultry are subclinical, their presence may still subtly affect the health status, impacting feed conversion efficiency, growth rate, and overall productivity of the flock [10]. From a public health perspective, NTS has a significant impact as a leading cause of foodborne illness worldwide. Contaminated poultry products, particularly eggs, are recognized as a main source of NTS to humans. Human infections generally lead to self-limiting gastroenteritis, which presents typical symptoms of diarrhea, abdominal pain, fever, and vomiting [9]. However, in the elderly, children, and immunocompromised individuals, it can lead to serious extraintestinal infections. The ability of *Salmonella* to acquire antimicrobial resistance presents an additional challenge to veterinary and human health care [11].

Controlling NTS during the pre-harvest stage in commercial-egg production is critical in reducing foodborne diseases due to *Salmonella*, limiting the spread of antimicrobial resistance, and improving public confidence in food safety [12]. Therefore, this review comprehensively evaluates various pre-harvest control strategies that benefit poultry producers, researchers, and policymakers in effectively implementing NTS control measures.

2. Risk Factors Influencing *Salmonella* Contamination in the Pre-Harvest Stage

Understanding the factors influencing *Salmonella* contamination in the pre-harvest stage is essential for developing effective control strategies in commercial layer chicken operations. Various environmental, management, biological, and microbial factors contribute to the risk of *Salmonella* presence in poultry flocks, each playing a unique role in facilitating the spread and persistence of these bacteria [13,14].

2.1. Environmental Factors

Environmental conditions within the poultry housing system significantly influence *Salmonella* contamination [15]. Temperature, humidity, and ventilation all impact *Salmonella* survival and transmission within poultry facilities. Higher temperatures and humidity can increase bacterial survival rates in litter, dust, feces, and surfaces, creating an environment conducive to pathogen persistence [16]. Ventilation in poultry houses significantly influences *Salmonella* contamination levels, as large volumes of air can transport bacteria from litter and dust into the aerosol environment within the sheds. Mechanical ventilation systems, such as tunnel ventilation, move substantial air volumes through poultry sheds, potentially aerosolizing *Salmonella* and enabling its dispersal inside and outside the housing unit. However, external levels are usually lower than internal *Salmonella* concentrations [17]. Nonetheless, inadequate light and ventilation can increase stress among birds and weaken their immune responses, making them more susceptible to infections like *Salmonella* [18–20]. Additionally, the type and quality of litter used in non-cage housing systems play a role; soiled or moist litter can support *Salmonella* proliferation, especially if it is not changed or managed regularly [21,22]. Maintaining optimal environmental conditions through temperature control, adequate ventilation, and regular litter management is crucial to minimizing contamination risks.

2.2. Farm Management Practices

Poor hygiene practices, such as inadequate cleaning and disinfection of poultry houses, equipment, and feed areas, create environments where *Salmonella* can thrive and spread. Farm biosecurity practices are also essential, as inadequate restrictions on personnel movement, ineffective entry points, and equipment disinfection can introduce or spread *Salmonella* within flocks. Farm management includes implementing proper waste disposal systems and rigorous cleaning schedules [23,24]. Additionally, feed and water management play a role; contaminated feed and water are primary sources of *Salmonella*, and failure to routinely monitor and sanitize these inputs increases the risk of introducing the pathogen to chickens [25]. Ensuring regular inspections, enforcing strict biosecurity protocols, and implementing cleaning schedules help mitigate the risk of contamination [23,24].

2.3. Flock Density and Housing Systems

The housing system and flock density directly affect *Salmonella* transmission among birds. In high-density rearing systems, birds are in closer contact, which can facilitate the rapid spread of *Salmonella* through direct and indirect contact, including via droppings, feathers, and feed. Cage-free and free-range systems, while offering benefits for animal welfare, may pose unique challenges for *Salmonella* control [26]. In free-range systems, birds may be exposed to contaminated soil, insects, or wild animals, all of which can serve as reservoirs for *Salmonella*. In addition, free-range environments make it difficult to control environmental contamination sources, adding complexity to management practices [26,27]. However, cage production systems pose a greater risk of *Salmonella* contamination compared to non-cage production systems, such as barn and free-range systems. This increased risk is attributed to several factors inherent to cage systems. Cages usually hold a large number of birds in a confined space, which facilitates the rapid spread of microorganisms, including *Salmonella*. Moreover, the compact environment can cause stress to the hens, weakening their immune system and making them more vulnerable to disease. In addition, the structural layout of cage systems makes effective cleaning and disinfection challenging, allowing *Salmonella* to persist in the environment and further increasing the risk of contamination [15,28–31]. The recovery rate of *Salmonella* from the dust and fecal samples in caged systems is typically higher than in non-cage systems [32]. Additionally, *Salmonella* persists longer in houses with a deep pit (step-cage houses and cages with a scraper manure disposal system) than in non-cage systems [31]. Controlling flock density, regularly rotating pastures for free-range birds, and maintaining clean housing environments can help minimize the risk of contamination in various housing systems.

2.4. Bird Age and Immunity

The age and immune status of birds influence their susceptibility to *Salmonella* infection. Younger chickens with immature immune systems are generally more susceptible to *Salmonella* and are often the focus of preventative strategies such as vaccination. Due to overcrowding, poor nutrition, or suboptimal environmental conditions, birds under stress may experience immune suppression, making them more vulnerable to infections [18–20]. Vaccinating chickens against *Salmonella* at an early age has proven to effectively reduce susceptibility by providing them with specific immunity to certain *Salmonella* serovars [33]. Additionally, selective breeding for resistance to pathogens, including *Salmonella*, is an emerging area of interest. Recent studies have identified genetic variations and specific loci associated with immune and disease-resistance traits, indicating the potential for the use of breeding programs to enhance resistance to infections [34]. Ensuring optimal health, nutritional support, and targeted vaccination schedules can strengthen flock immunity, reducing the likelihood of *Salmonella* establishment and transmission.

2.5. Feed and Water Quality

Contaminated feed and water are among the primary sources of *Salmonella* entry into poultry flocks [8]. Feed contamination can occur at any stage in the supply chain, including during production, storage, or transport. In particular, animal-based feed ingredients, like fishmeal or meat by-products, have higher contamination risks [35]. Ensuring that feed undergoes processing treatments, such as heat-pelleting, can reduce *Salmonella* presence [36]. Water quality also plays a crucial role in influencing *Salmonella* contamination at the pre-harvest stage, as water systems can serve as reservoirs for *Salmonella* biofilms. This protection allows the bacteria to persist over time within the water systems, contributing to the ongoing risk of contamination in poultry environments. Regular water testing and appropriate water sanitizers can help prevent contamination [37].

2.6. Role of Rodents, Wild Birds, and Insects as Vectors

Rodents, wild birds, and insects act as vectors, facilitating the introduction and spread of *Salmonella* in poultry farms. Rodents, in particular, are known carriers of *Salmonella* and can easily transmit the bacteria through droppings, which can contaminate feed, water, and litter [14]. Wild birds that access free-range or cage-free areas can introduce *Salmonella* to poultry through droppings or direct contact. Similarly, flies and other insects can pick up *Salmonella* from contaminated environments and transmit it to birds through feeding or nesting materials [8]. Effective pest control measures are essential in reducing the likelihood of *Salmonella* transmission from these external vectors. Measures such as securing feed storage, installing barriers to restrict wild bird access, and implementing rodent and insect control programs are vital to minimizing contamination risks.

2.7. Antimicrobial Resistance (AMR) and Its Impact on Salmonella Control

The rise in antimicrobial resistance (AMR) among *Salmonella* strains is an increasingly concerning topic in pre-harvest control of *Salmonella*. The use of antibiotics in poultry production for growth promotion and disease prevention has contributed to the development of resistant *Salmonella* strains, making infections more challenging to manage [38,39]. Resistant strains can persist longer in the environment and within birds, reducing the efficacy of traditional control measures. AMR complicates treatment options and poses a risk to human health, as resistant *Salmonella* can be transmitted to humans through contaminated poultry products [39]. AMR can also facilitate the horizontal transfer of resistance genes between different bacterial populations within the poultry environment, compounding the issue by increasing the diversity of resistant pathogens present [40]. This prolonged environmental survival and gene transfer capability amplify the risk of resistant *Salmonella* strains spreading to humans, underscoring the critical need to understand AMR's impact on *Salmonella* dynamics in pre-harvest settings [41,42]

2.8. Human Interaction and Farm Personnel Practices

Human interaction, including the activities of farm personnel, is another significant factor influencing *Salmonella* contamination in the pre-harvest stage. Workers can inadvertently introduce *Salmonella* through contaminated clothing, footwear, and equipment if proper biosecurity measures are not followed. Without strict hygiene protocols, personnel moving between different farm areas or flocks can facilitate cross-contamination [43,44]. Producers must commit to training staff on proper hygiene, providing dedicated clothing and equipment for specific zones, and implementing hand-washing and disinfection protocols in order to minimize human-mediated *Salmonella* transmission. Additionally, the establishment of clear visitor policies can reduce the risk of introducing external contamination sources [45].

3. Pre-Harvest Control Strategies

3.1. Biosecurity Measures

Biosecurity measures are a key component of *Salmonella* control in commercial poultry that have contributed greatly to the current levels of success observed in the industry [46–48]. Implementing comprehensive biosecurity measures is fundamental to infection control, as the effectiveness of concurrent preventive interventions is significantly diminished when biosecurity protocols are compromised. Biosecurity protocols encompass diverse preventive measures designed to minimize pathogen introduction and transmission within poultry flocks. These include physical barriers, sanitation protocols, pest management systems, personal protective equipment requirements, and strict personnel movement controls [15,49].

Establishing physical barriers represents a cornerstone of biosecurity protocols in preventing pathogen introduction to poultry facilities. Implementing perimeter fencing and controlled entry points for personnel and vehicles is essential. Enhanced biosecurity measures, such as perimeter fencing and restricted access, can reduce *Salmonella* contamination rates in poultry facilities by up to 80% [50]. Epidemiological studies, also highlighted the effectiveness of stringent biosecurity measures in reducing *Salmonella* contamination in commercial layer farms [14]. Similarly, Samper-Cativiela et al. (2023) also demonstrated the positive impact of physical barriers and entry restrictions in commercial layer facilities in Spain [6]. These studies underscore the importance of well-defined boundaries and access control measures in minimizing *Salmonella* outbreaks [51].

Moreover, the strategic limitation of equipment and vehicle movement between facilities effectively reduces cross-contamination risks. Specific control measures include the implementation of dedicated facility-specific footwear and protective clothing, alongside the adoption of all-in/all-out production systems, which substantially minimizes cross-contamination between different age cohorts [52].

Environmental decontamination protocols targeting water systems, feed storage, and litter management are crucial in preventing *Salmonella* introduction. These protocols aim to substantially reduce microbial load within poultry housing facilities. Research demonstrates that implementing systematic cleaning and decontamination procedures significantly reduces *Salmonella* contamination in broiler housing environments [53]. Surveillance and monitoring are other essential components of an effective biosecurity framework. Regular microbiological testing of flocks, feed supplies, and environmental samples enables the identification of potential contamination sources and informs evidence-based management decisions. These surveillance systems facilitate early detection of *Salmonella* outbreaks, enabling rapid corrective interventions. Integrating biosecurity measures with systematic surveillance enhances the overall efficacy of *Salmonella* control programs in poultry operations [54].

Feed represents a primary vector for pathogen transmission in poultry flocks, with *Salmonella* identified as the predominant biological hazard across animal feed categories [55]. These risks can be mitigated by sourcing ingredients from verified suppliers who maintain rigorous sampling and testing protocols [56]. At the level of the feed mill, the establishment of segregated clean and dirty zones, coupled with strict access controls, significantly reduces the potential for *Salmonella* and other pathogen cross-contamination between raw materials and finished feed products [57]. Another prominent source of *Salmonella* in poultry farms are rodents [14] and insects, and they function as both mechanical and biological vectors in pathogen transmission. The mitigation of these risks requires the implementation of integrated pest management protocols, incorporating physical barriers, strategic trap placement, vegetation management, and systematic chemical intervention through targeted pesticide applications [58].

3.2. Vaccination

Although numerous vaccine platforms, such as live attenuated vaccines, inactivated vaccines, subunit vaccines, and emerging options like ghost vaccines, have been tested experimentally for *Salmonella* control in layer operations, only a few of these are currently available for commercial use. The available vaccines include inactivated whole-cell preparations, and live attenuated strains, each type providing distinct benefits in terms of immune response and protection [59]. Over the past few decades, these vaccination efforts have significantly reduced *Salmonella* incidence in poultry flocks over recent decades, contributing to improved food safety outcomes [60,61]. Studies have demonstrated that vaccination can reduce *Salmonella* prevalence in flocks by as much as 70% [62]. However, vaccine efficacies can vary depending on the *Salmonella* serovar, environmental conditions, and types of poultry.

3.2.1. Live Attenuated Vaccines

Attenuated vaccines against *Salmonella* comprise live bacteria that have been specifically altered to reduce their virulence while preserving essential immunogenic properties needed to trigger a protective immune response in birds. Several live attenuated vaccines are commonly used in commercial poultry production, including commercial layer production, due to their lower cost, ease of manufacture, and simpler administration than inactivated vaccines (Table 1). These vaccines mimic natural infection by adhering to mucosal surfaces and interacting with mucosa-associated lymphoid tissue, thereby eliciting robust humoral and cell-mediated immune responses [63–65]. They are also known to provide heterologous protection and confer immunity against serovars beyond the serovar of the specific vaccine strain and protection against a broader range of *Salmonella* compared to killed vaccines [66,67]. Furthermore, live vaccines provide an additional protective mechanism through the competitive exclusion of pathogens, particularly benefiting young chickens with a still-developing gut microflora and an immature immune system [68]. However, live vaccines could persist for extended periods post-vaccination, potentially transferring to poultry products and, subsequently, to consumers [69]. Although these strains are attenuated, their persistence in the environment and poultry products raises concerns for food safety officials and the public. Another downside of this persistence is that these vaccine strains may interfere with pre-harvest *Salmonella* surveillance and performance standards during poultry processing, leading to financial losses for the farmer [59]. To address this issue, novel methods that could discriminate between field strains and vaccine strains of *Salmonella* have been widely used [69].

3.2.2. Inactivated Vaccines

Inactivated vaccines typically consist of whole bacterial cells combined with an adjuvant, where cultured pathogens are rendered non-infectious by inactivation with heat, chemicals, or irradiation before being used for immunization. Inactivated vaccines against *Salmonella* have been extensively studied and utilized in poultry (Table 1) [70–72]. These vaccines elicit an immune response without the risk of causing disease, as they contain killed bacteria. In response to the challenges posed by the diversity of *Salmonella* serovars, researchers have developed multivalent inactivated vaccines that demonstrate protective efficacy against an expanded range of serovars, including emerging serovars such as *Salmonella* Infantis [73]. However, inactivated vaccines primarily stimulate humoral immunity rather than mucosal or cell-mediated immunity, which are crucial for controlling intracellular pathogens and may limit the vaccine's ability to control *Salmonella* effectively [74]. Moreover, most inactivated vaccines must be administered via parenteral routes, such as subcutaneous or intramuscular routes, and require at least two doses. These requirements significantly increase the labor and economic costs associated with farming operations [75].

3.2.3. Subunit Vaccines

Subunit vaccines present a promising approach to controlling *Salmonella* infections in poultry. Unlike live attenuated or inactivated whole-cell vaccines, subunit vaccines do not contain the entire pathogen and therefore stimulate c immune responses against the specific immunogenic and protective antigenic components included in the vaccine. Thus, subunit vaccines are safer than live or inactivated whole-cell vaccine types as they minimize potential adverse side effects by generating an immune response to targeted antigens only. Examples of antigens used in experimental poultry *Salmonella* subunit vaccines include outer membrane proteins (OMPs), flagella (FLA) antigens, and outer membrane vesicles (OMVs) [76,77]. Li et al. (2020) reported that recombinant OMP F (rOmpF) mixed with QuilA adjuvant and extracted OMVs alone induced strong antibody and cell-mediated immune responses, protecting vaccinated chickens from a subsequent challenge with *S.* Enteritidis and effectively reducing intestinal *Salmonella* colonization [78].

Subunit vaccines have gained special interest due to their safety and specificity in targeting *Salmonella*. Among these, nanoparticle-based vaccines, such as chitosan nanoparticle (CS-NP) formulations, have emerged as a promising approach for immunizing poultry [79]. CS-NPs have been shown to enhance both mucosal and systemic immune responses due to their biocompatibility, biodegradability, and ability to encapsulate antigens. Recent studies have also investigated alternative delivery methods of CS-NP-formulated vaccines, including gel sprays, which optimize vaccine distribution and uptake. For example,

Han et al. (2020) demonstrated that gel spray delivery of a CS-NP vaccine encapsulating OMP and FLA antigens, either through drinking water or mixed into feed, resulted in significant mucosal and systemic antibody responses as well as cell-mediated immune response, thereby successfully reducing *Salmonella* colonization in vaccinated chickens. Experiments conducted with different dosages of CS-NP revealed the importance of optimal dosing to balance immune response while minimizing adverse effects [80–84]. However, there is a significant gap in knowledge regarding the application of these vaccines in commercial-layer chickens, as most studies have primarily focused on broiler chickens [85]. Layer chickens represent a distinct group of poultry with unique physiological and production characteristics, which could influence vaccine efficacy and outcomes [86]. Addressing these gaps, including the effects of different delivery routes and dosage regimens in layer chickens, is crucial for developing effective nanoparticle-based vaccines to control *Salmonella* in diverse poultry production systems. Investigating these aspects in future research will greatly enhance the understanding and applicability of nanoparticle-based vaccines, ultimately improving poultry health and food safety.

3.2.4. Ghost Vaccines

Bacterial ghost vaccines represent a novel strategy used in developing poultry *Salmonella* vaccines. These vaccines are prepared from genetically modified Gram-negative bacteria that have had their cytoplasmic contents removed while leaving the cellular membranes intact. This is typically achieved by the controlled expression of the lysis gene E from bacteriophage PhiX174 in bacteria, which triggers the formation of transmembrane channels to release intracellular contents [87,88]. The remaining empty cellular envelopes, so-called "ghosts", would still maintain the native antigenic structures of bacteria, thereby retaining the capability to induce strong immune responses with a reduced risk of infection [89]. One significant application of this technology is the construction of a ghost vaccine against *Salmonella* Gallinarum to combat fowl typhoid [89]. Chaudhari et al. (2012) reported that the *S.* Gallinarum ghost vaccine was safe and elicited strong antibody- and cell-mediated immune responses, providing protection against *S.* Gallinarum [90]. Recent studies on ghost vaccines have incorporated adjuvants to enhance immunogenicity. For example, Jawale et al. (2014) constructed an *S.* Typhimurium-derived ghost vaccine expressing the heat-labile enterotoxin B subunit (LTB) of *Escherichia coli*, which acted as an adjuvant. Incorporation of LTB significantly enhanced systemic and mucosal antibody responses in vaccinated chickens [91]. When these chickens were subsequently challenged with *S.* Typhimurium, vaccinated chickens exhibited reduced *Salmonella* counts in the internal organs compared to unvaccinated chickens [91]. Similarly, adding surface-displayed FliC as an adjuvant elicited robust antigen-specific immune responses, both humoral and cell-mediated immunity, and significantly reduced bacterial loads in target organs after a challenge with virulent *Salmonella* [92]. Despite the advantages of ghost vaccines—such as increased safety due to the absence of live pathogens and enhanced immunogenicity from the retention of native antigenic structures that provoke a robust immune response— there are still some challenges with regard to their optimization, stability, and consistent immunogenicity across different *Salmonella* serovars [93]. Ongoing research aims to address these challenges and develop improved ghost vaccines that can be integrated into overall *Salmonella* control programs in poultry production.

Table 1. Overview of currently used vaccines and their efficacy.

Vaccine Name	Constituents	Outcomes	Routes and Frequency of Administration
Live attenuated			
Nobilis® SG 9R, (Merck & Co., Inc., Rahway, NJ, USA)	Live attenuated *S.* Gallinarum (SG) 9R strain with mutated *galE* gene [94]	Reduced *Salmonella* prevalence in vaccinated flocks compared to the unvaccinated control group [95].	Two times via subcutaneous route: 6 weeks and 14–16 weeks of age [95].
Avipro® Megan Vac 1 (Elanco, Greenfield, IN, USA)	*S.* Typhimurium *cya crp* mutant [94]	Reduction in *Salmonella* colonization in ceca and reproductive tracts of vaccinated chickens [33,96]. Reduction in horizontal transfer and liver, spleen, ovary, and cecal colonization of *S.* Enteritidis [97].	Three times via drinking water: 1 day, 2 weeks, and 5 weeks of age [33,96].
Avipro® Megan Egg (Elanco, Greenfield, IN, USA)	*S.* Typhimurium *cya, crp* mutant strain χ3985 [66,98]	Reduced *Salmonella* colonization in the ceca, spleen, ovary, and bursa in vaccinated birds [66,98].	Three times via coarse spray: 2, 4, and 16 weeks of age [99].
Vaxsafe® ST, Bioproperties, (Glenorie, Australia)	Attenuated *aroA* deletion *S.* Typhimurium strain STM-1 [100]	Reduced excretion of *Salmonella* in the vaccinated group [100].	Four times: on day 1, via coarse spray; at 2 and 6 weeks of age, via drinking water; and at 12 weeks of age, via intramuscular route [101,102].
Salmovac® SE (Ceva, Libourne, France)	Attenuated *S.* Enteritidis strain 441/014 [103]	Reduced *Salmonella* colonization in ceca and invasion of internal organs [104].	Three times via drinking water: 1, 6, and 13 weeks of age [104].
Gallivac® SE (Merial, France)	*S.* Enteritidis *Ade* and *His* mutant [94]	Reduced colonization of *Salmonella* in cecum and liver [105].	Two times via drinking water: 1 and 15 days of age [105].
Poulvac® ST, Zoetis (Parsippany, NJ, USA)	*aroA* mutant *S.* Typhimurium [106]	A 50% reduction in *S.* Kentucky, *S.* Enteritidis, *S.* Heidelberg, *S.* Typhimurium, and *S. Hadar* recovery from internal organs of vaccinated birds [106].	Two times: on day 1, via coarse spray, and at 2 weeks of age, via drinking water [106].
Inactivated vaccines			
Nobilis® Salenvac TMSD animal health, Rahway, NJ, USA	Formalin killed *S.* Enteritidis and *S.* Typhimurium bacterin [107]	Reduction in *Salmonella* shedding and colonization of internal organs (liver and spleen) [70,108].	Two times via intra-muscular route: 1 day and 4 weeks of age [70,108].
Layermune® SE (Ceva Biomune, Lenexa, KS, USA)	Killed *S.* Enteritidis [109]	Reduction in *Salmonella* shedding and colonization of internal organs (liver and spleen) [109].	Two times via subcutaneous route: 5 and 9 weeks of age [109].
Corymune® 4K and 7K (CEVA Corp., Libourne, France)	Killed *S.* Enteritidis [71,109]	Reduction in *Salmonella* shedding and colonization of internal organs (liver and spleen) [109].	Two times via intramuscular route: 5 and 9 weeks of age [109].
Poulvac® SE (Zoetis, Parsippany, NJ, USA)	Formalin killed *S.* Enteritidis, Phage Types 4, 8 and 13a [110]	Reduction in *Salmonella* colonization in ceca, liver, and spleen after challenge on day 1 [72].	Two times via subcutaneous route: 12 and 20 weeks of age [72,111].
AviPro® 109 SE4 Concentrate (Elanco, Greenfield, IN, USA)	Killed *S.* Enteritidis [112]	Reduced colonization of *Salmonella* in internal organs, including reproductive tract.	Two times: first, via subcutaneous route between 12 and 16 weeks of age and booster vaccination 4 weeks later [113].
Avipro® 329 ND-IB2-SE4 Concentrate (Elanco, Greenfield, IN, USA)	Killed chicken bronchitis and Newcastle disease viruses and killed *S.* Enteritidis [114]	Reduction in *S.* Enteritidis colonization in the ceca [114].	Three times: first via subcutaneous route at 12 and 16 weeks of age or intramuscular route at 13 and 17 weeks of age, followed by vaccination with *S.* Enteritidis monovalent vaccine 4 weeks later [99].

3.3. Feed Additives

Feed additives can be classified into several main categories, such as probiotics, prebiotics, organic acids, short- and medium-chain fatty acids, essential oils, and bacteriophages, each playing a distinct role in reducing pathogen colonization. It has been reported that these additives modify intestinal microflora to enhance the overall gut health of the bird, which, in turn, helps reduce *Salmonella* colonization.

3.3.1. Probiotics

The Food and Agriculture Organization of the United Nations (FAO)/World Health Organization (WHO) defines probiotics as live microorganisms that, when administered in adequate amounts, confer a health benefit on the host [115]. Probiotics have gained significant attention as a feed additive for controlling *Salmonella* in poultry and have been extensively studied for their ability to outcompete harmful bacteria like *Salmonella* in the GIT [116–118]. In addition to competitive exclusion, these probiotics are known to produce antimicrobial compounds and modulate the immune system of the host to interfere with pathogen growth [119,120]. Probiotics have demonstrated significant potential to enhance intestinal development and microarchitecture, which contributes to improved gut health and pathogen resistance [121]. Table 2 shows the mechanism of action of some probiotics used in the poultry industry. Research indicates that probiotic supplementation can lead to increased villus height and reduced crypt depth in the intestinal epithelium, both indicators of enhanced nutrient absorption and improved intestinal integrity [122].

Table 2. Common probiotics in poultry.

Probiotic	Outcome	Reference
B. subtilis CSL2	Re-establishment of normal gut flora abundance (phylum Firmicutes and Proteobacteria and genus *Lactobacillus*) that is disrupted after *Salmonella* infection.	[123]
Poultry Star® *Enterococcus faecium*, *Pediococcus acidilactici*, *Bifidobacterium animalis*, and *Lactobacillus reuteri*	Increased the efficacy of the live attenuated vaccine (*aroA* mutant *S.* Typhimurium) and reduced the cecal colonization of *Salmonella*.	[116]
Bacillus subtilis DSM 32324, *Bacillus subtilis* DSM 32325, and *Bacillus amyloliquefaciens*	Reduction in *Salmonella* in cecal contents and establishment of normal gut flora after *Salmonella* challenge.	[117]
Bacillus amyloliquefaciens, *B. licheniformis*, and *B. pumilus*	Significant reduction in *Salmonella* in cecal contents 7 days after challenge.	[118]

3.3.2. Prebiotics

Prebiotics are defined by the International Scientific Association for Probiotics and Prebiotics (ISAPP) as "a substrate that is selectively utilized by host microorganisms conferring a health benefit" [124]. Specifically, prebiotics serve as selective nutrients for beneficial colonic bacteria, promoting their growth and metabolic activity. Although these compounds are indigestible by the host organism, they provide numerous health benefits through multiple mechanisms. For instance, mannan-oligosaccharides (MOS) and β-glucans have been identified as effective prebiotics that can inhibit the adhesion of *Salmonella* to intestinal epithelial cells, thereby reducing colonization [125]. Currently, probiotics and prebiotics are also used as a combination called synbiotics to exploit their synergistic effects in controlling *Salmonella*, and this combined approach has shown to be more effective than probiotics or prebiotics alone (Table 3). This synergistic combination is intended to enhance both the viability and functionality of probiotics in the gastrointestinal tract while promoting a balanced gut microbiome. Because synbiotics can facilitate the growth of beneficial bacteria, they may offer various health benefits to the host, such as improved digestive health, a strong immune system, and improved general well-being [126].

3.3.3. Postbiotics

Postbiotics are non-viable microbial products or metabolic byproducts from probiotic microorganisms. They have received considerable attention because of their potential ability to reduce *Salmonella* colonization in poultry. Unlike live probiotics, postbiotics consist of short-chain fatty acids, bacteriocins, enzymes, and cell wall fragments, which may offer health benefits without the associated risks of using live bacteria. These traits make them particularly appealing in poultry farming, where preserving intestinal health is crucial for preventing colonization by undesirable bacteria [127,128]. Studies have demonstrated the effectiveness of postbiotics in reducing *Salmonella* colonization in poultry. In this context, a report in Feedinfo by Expana highlighted that specific postbiotic formulations could dramatically reduce *Salmonella* colonization in poultry through modulation of the immune response in the intestines, creating an environment that inhibits the growth of pathogenic bacteria and consequently reducing the likelihood of intestinal colonization in the GIT [129]. Postbiotics such as the *Saccharomyces cerevisiae* fermentation product (SCFP) improve gut health by maintaining immune robustness and enhancing digestive efficiency. This dual action helps control *Salmonella* colonization and improve bird performance, animal welfare, and food safety, particularly benefiting antibiotic-free poultry production systems [130]. One of the advantages of postbiotics over traditional probiotics is their non-viable nature, which reduces problems with the viability and stability of living microorganisms during feed production and storage. Postbiotics also have a lower potential for transferring antibiotic resistance genes, addressing one of the major current issues in animal agriculture [131]. As the poultry industry continues to strive for sustainable and efficient methods of controlling *Salmonella* infections, postbiotics are an attractive non-antibiotic alternative that seems to resonate with consumer and regulatory demands aimed at reducing antibiotic usage.

Table 3. Efficacy of prebiotics and synbiotics in controlling *Salmonella*.

Prebiotics	Outcome	Reference
Mannan-rich yeast cell wall-derived preparation	Significant reduction in *Salmonella* recovered from ovaries and up to 1 log unit reduction in *Salmonella* in the ceca and *Salmonella*-challenged birds.	[132]
Fructo-oligosaccharides	Dose-dependent reduction in *S.* Enteritidis in the ceca up to 1.3 $\log_{10}$ in orally challenged birds. No change in *Salmonella* isolation from the internal organs (liver, gall bladder, ovary). Increase in Toll-like receptor 4 (TLR 4), interferon-γ (IFN-γ), and IgA expression indicating cell-mediated immune activation.	[133]
Synbiotics		
Bacillus subtilis and yeast cell wall-derived glucomannan	Reduction in *S.* Enteritidis counts in ceca up to 0.73 $\log_{10}$ CFU/g.	[115,116]
Enterococcus faecium, *Pediococcus acidilactici*, *Bifidobacterium animalis*, *Lactobacilus reuteri* + Fructo-oligosaccharides	Improvement in vaccine efficacy by reducing *Salmonella* counts in the cecal contents.	
BacPack® Quality Technology International, Inc., Elgin, IL, USA. Combination of a *Bacillus subtilis* strain and *Saccharomyces cerevisiae* cell wall	Reduction in cecal *S.* Enteritidis counts at 11, 15, and 19 days post-challenge.	[134]
Bacillus subtilis, *B. licheniformis* + mannooligosaccharide	Reduction in cecal *S.* Enteritidis counts in the ceca and ovaries of challenged birds.	

3.3.4. Organic Acids, Short- and Medium-Chain Fatty Acids

Organic acids, in combination with short-chain and medium-chain fatty acids (SCFAs and MCFAs), are increasingly being used within the poultry-production market as pre-harvest interventions to control *Salmonella* colonization. Such compounds have antimicrobial properties, either inhibiting or reducing the distribution of *Salmonella* in the avian gut. Similarly, organic acids, such as formic and acetic acids, in their undissociated form, can pass through bacterial cell walls and cause intracellular acidification, which results in a disturbance of metabolic functions and, consequently, the inhibition of bacterial growth [135]. SCFAs and MCFAs exhibit membrane-active properties and disrupt bacterial cell membranes, thereby enhancing permeability and leakage of critical cellular components. MCFAs have been especially potent against *Salmonella* [136]. When added to poultry feed, these acids have been associated with a significant reduction in the colonization of *Salmonella*. When combined with added ingredients, such as essential oils or probiotics, they can enhance antimicrobial activity [137]. However, it is important to consider aspects such as maximum dosage, the potential for the development of resistance, and regulatory considerations to ensure their safe and effective use in poultry production. One study demonstrated that diets supplemented with a mix of coated essential oils and organic acids in broilers (a) improved growth performance and gut health and (b) reduced *S.* Enteritidis load in challenged birds [138]. Further, research has shown that medium-chain fatty acids had a stronger antibacterial action on *Salmonella* than short-chain fatty acids, highlighting the importance of selecting the right type of fatty acids for effective control [58,136]. These examples certainly strengthen the prospect of organic acids and fatty acids as potential alternatives to antibiotics for the control of *Salmonella* in poultry.

3.3.5. Essential Oils

Essential oils (EOs) have garnered significant attention as natural antimicrobial agents in poultry production, particularly for their potential to control *Salmonella* colonization at the pre-harvest level. Derived from aromatic plants, EOs such as thyme, oregano, and lemongrass contain active compounds like thymol, carvacrol, and citral, which exhibit potent antimicrobial properties against various pathogens, including *Salmonella*. These compounds disrupt bacterial cell membranes, leading to increased permeability and leakage of essential cellular components, ultimately resulting in bacterial cell death. Additionally, EOs interfere with bacterial enzyme activity and genetic material, hindering bacterial replication and survival [139]. Incorporating EOs into poultry feed or water has been shown to reduce *Salmonella* prevalence in the GIT, thereby enhancing food safety. For instance, a study demonstrated that a combination of essential oils, including eucalyptus, thyme, and lemon, administered in drinking water at a concentration of 0.05% significantly reduced *Salmonella* contamination in the crop of the bird, leading to decreased cross-contamination during slaughter and processing [140]. Another study highlighted that dietary supplementation with a blend of essential oils and organic acids improved growth performance and intestinal health while reducing *S.* Enteritidis load in infected chickens [138]. Although EOs offer a natural alternative to synthetic antimicrobials, several factors must be considered for their effective application in poultry production. Determining optimal concentrations is crucial to ensure efficacy without any adverse effects on bird health or product quality. EOs can be volatile and may degrade during feed processing; therefore, encapsulation techniques are often employed to enhance stability. Additionally, high concentrations may impart unpleasant strong flavors or odors to meat and eggs, potentially affecting consumer acceptance. Therefore, careful formulation and dosage optimization are essential to maximize the benefits of EOs in controlling *Salmonella* in poultry.

3.3.6. Bacteriophages

Bacteriophages are viruses that specifically infect bacteria and are considered non-toxic to humans and animals. Due to their targeted action and safety, bacteriophages have been widely studied as promising alternatives to conventional antimicrobials for pathogen control in various applications [141–144]. Their host specificity allows them to effectively target pathogenic bacteria while preserving the resident microflora, making them a safer option for pathogen control [145]. Several commercial bacteriophage preparations have been developed to be used against *Salmonella* in poultry. For example, BAFASAL® (Proteon Pharmaceuticals, Lodz, Poland) is a commercial bacteriophage product designed for on-farm administration to poultry during the rearing stage. Studies have demonstrated that BAFASAL® could reduce *Salmonella* levels by up to 200-fold in treated groups compared to untreated groups, with minimal impact on the final product. Importantly, this product requires no withdrawal period for meat or eggs due to its minimal residual effects [146]. Another example is SalmoFREE, a bacteriophage mixture targeting *Salmonella* that has shown complete elimination of *Salmonella* in cloacal swabs of treated poultry. SalmoFREE also appears to confer some residual protection, as subsequent flocks exposed to the environment retained *Salmonella* resistance, likely due to remaining bacteriophage activity [147]. BioTector, produced by Cheil Jedang Corporation in Korea, is another commercial bacteriophage formulation designed to reduce *S.* Gallinarum and *Salmonella* Pullorum in poultry. This product has significantly reduced the prevalence of *Salmonella* in broilers and layers, demonstrating its effectiveness as a mitigation strategy for these pathogens in poultry operations [148]. This suggests that BioTector may also be effective against NTS.

3.4. Competitive Exclusion (CE)

The concept of competitive exclusion (CE) in poultry, introduced by E. Nurmi, is based on the idea that the natural gut flora of chickens can inhibit the growth of pathogenic bacteria [149]. Unlike probiotics, which consist of defined bacterial cultures, CE cultures contain a diverse range of microorganisms derived from the gut flora of healthy, mature birds. These CE cultures are intended to be administered to young chicks to reduce their susceptibility to *Salmonella* colonization before establishing their own stable microflora [150]. CE cultures exert pathogen-elimination effects through multiple mechanisms, including competition for attachment sites and nutrients [151], production of antimicrobial substances such as organic acids [152], and modulation of the bird's immune system [153]. In some cases, even live *Salmonella* vaccines can colonize the intestines of the vaccinated chicks, effectively excluding pathogenic bacteria through competitive interactions. It has been observed that the inhibitory effect is typically stronger within the same serovar than across different serovars of the vaccine [154]. Furthermore, combining CE cultures with live vaccines has shown additive effects in controlling *Salmonella* in vaccinated birds [155]. Several CE culture products have been developed for pathogen control in poultry [156], though only a few, such as Aviguard® (MSD Animal Health, Rahway, NJ, USA) and Broilact® (Orion Pharma Animal Health, Espoo, Finland), are commercially available today. This limited availability may stem from inconsistent results observed during various stages of product evaluation [156,157].

3.5. Genetic Approaches

Genetic approaches offer promising strategies for pre-harvest control of *Salmonella* in poultry, thereby improving the intrinsic resistance of avian species to infection. Breeding programs aim to identify and spread genetic traits associated with increased resistance against colonization by *Salmonella*. Studies have revealed that various lines of chickens exhibit varying levels of susceptibility to infection with *Salmonella*, which suggests a genetic component of resistance against the disease [158]. By selecting resistant traits, poultry producers can develop flocks that are less susceptible to *Salmonella* colonization, thereby reducing the risk of contamination in the food supply. Advancements in genomic technologies have facilitated the identification of specific genes and quantitative trait loci (QTL) associated with *Salmonella* resistance. Genome-wide association studies (GWAS) have identified specific genetic loci associated with resistance to *S. Pullorum* in chickens. Notably, a significant region on chromosome 4 has been linked to mortality resulting from *S.* Pullorum infection. Within this region, the single nucleotide polymorphism (SNP) rs314483802 accounts for 11.73% of the observed phenotypic variation in resistance. Candidate genes within this locus, such as FBXW7 and LRBA, were found to downregulate expression following infection, suggesting their roles in mediating resistance [159]. These findings might enable the development of genetic markers that can be used in marker-assisted selection (MAS) programs, allowing for the efficient breeding of resistant poultry lines. Additionally, understanding the genetic basis of immune responses in chickens can inform the design of more effective vaccines and immunotherapies, further bolstering pre-harvest *Salmonella* control measures.

However, implementing genetic approaches in poultry production presents challenges. The genetic diversity within and between poultry populations necessitates comprehensive studies to identify resistance-associated genes across different breeds and environments. Moreover, the potential trade-offs between disease resistance and other economically important traits, such as growth rate and egg production, must be carefully gauged to ensure that overall productivity is not compromised [160]. Despite these challenges, integrating genetic strategies with traditional biosecurity measures and vaccination programs holds significant potential for reducing *Salmonella* prevalence in poultry, thereby enhancing food safety and public health outcomes [161].

3.6. Antimicrobial Use

Prophylaxis use of antimicrobials has been a cornerstone in pre-harvest strategies to control *Salmonella* in poultry production for some time. Administering antibiotics to poultry flocks aims to reduce or eliminate *Salmonella* colonization in the GIT, thereby decreasing the risk of contamination during processing. Commonly used antimicrobials include tetracyclines, sulfonamides, and fluoroquinolones, which target a broad spectrum of bacterial pathogens [162]. However, the extensive use of antibiotics in poultry has raised significant concerns regarding AMR development. *Salmonella* strains resistant to multiple antibiotics have emerged, complicating treatment options for human infections and posing a public health risk. The WHO has highlighted the critical importance of addressing AMR in food-producing animals to safeguard human health [163]. In 2006, the European Union implemented a comprehensive ban on the use of antibiotics as growth promoters in animal feed, aiming to combat AMR and safeguard public health [164]. Similarly, the United States has enacted policies to restrict the use of medically important antibiotics in livestock. The US Food and Drug Administration (FDA) has transitioned over-the-counter medically important animal antimicrobial drugs to prescription status, ensuring veterinary oversight and promoting judicious use [165]. These regulatory measures have prompted the poultry industry to seek alternative strategies for *Salmonella* control, such as probiotics, prebiotics, and enhanced biosecurity practices. While antibiotics can effectively reduce *Salmonella* levels in poultry, their use must be judicious and aligned with antimicrobial stewardship principles to mitigate the risk of resistance development. Replacing prophylaxis use of antimicrobials with comprehensive management practices, such as vaccination, biosecurity, and environmental controls, is essential for sustainable *Salmonella* control in poultry production [166].

4. Established *Salmonella* Control Programs

National and international *Salmonella* control programs have been established across various countries to address the significant public health risks associated with *Salmonella* contamination in poultry. These programs, spearheaded by government agencies, regulatory bodies, and international organizations, set standards and implement strategies to reduce *Salmonella* prevalence along the poultry production chain. At the international level, agencies such as the WHO, Codex Alimentarius Commission (a joint FAO/WHO program) [167], and the World Organization for Animal Health (WOAH; formerly OIE) [166] provide guidelines and frameworks to support global poultry *Salmonella* control, recognizing that foodborne *Salmonella* infections are a worldwide concern. The EU has also implemented a harmonized *Salmonella* control strategy across its member states, mandating regular testing, biosecurity measures, and vaccination for certain poultry flocks [168,169]. In the United States, the National Poultry Improvement Plan (NPIP) established by the United States Department of Agricultural Sciences (USDA) offers a voluntary yet widely adopted program that sets standards for pathogen monitoring and control within the poultry industry [170]. In addition to the NPIP voluntary program, the FDA fully implemented the mandatory egg safety rule "Prevention of *Salmonella* Enteritidis in Shell Eggs During Production, Storage and Transportation" in 2009 for commercial layer flocks. This rule requires that the flocks with 3000 or more laying hens, whose shell eggs are not processed with treatments like pasteurization, must be tested for *S.* Enteritidis to ensure the safety of table eggs [171,172]. These national and international efforts contribute to reducing the burden of *Salmonella* infections through a unified approach that emphasizes preventive measures, consistent monitoring, and rapid response to contamination incidents. The alignment of programs across borders enhances food safety standards and supports the trade of poultry products, as compliance with recognized control measures is often a prerequisite for international market access.

4.1. Testing and Monitoring Programs

Testing and monitoring are core elements of established *Salmonella* control programs. Governments mandate routine sampling and testing for *Salmonella* at different stages of poultry production, including breeder flocks, hatcheries, and layer or broiler operations. For example, the EU's *Salmonella* Control Program requires member states to test poultry for specific serovars, such as *S.* Enteritidis and *S.* Typhimurium, which are most commonly associated with human illnesses [173]. Testing in these programs is standardized to compare results across farms, regions, and countries. In the United States, USDA's NPIP sets forth guidelines for testing poultry for *Salmonella*, supported by state and federal oversight [170,174]. Regular testing ensures early detection of *Salmonella*, allowing farms to take immediate corrective actions, such as culling, sanitation, or enhanced biosecurity, to prevent contamination spread. Testing data also contributes to national surveillance systems, informing trends and helping governments assess the effectiveness of control measures [172,175]. The FDA program requires testing the pullet environment for *S.* Enteritidis tested when pullets are 14–16 weeks old and retesting negative flocks at 40–45 weeks. If the environmental test is positive, then eggs (1000 eggs/flock) must be tested for *S.* Enteritidis within 2 weeks of the start of egg laying at 2-week intervals until four consecutive negative tests are obtained. The eggs from positive flocks cannot be sold as shell eggs and must be diverted for pasteurization or another form of treatment [176].

4.2. Vaccination Requirements

Vaccination is a preventive strategy that many *Salmonella* control programs incorporate to reduce *Salmonella* colonization in poultry. In the EU, for instance, vaccination of laying hens against *S.* Enteritidis is mandatory [177]. Vaccination policies typically target high-risk poultry populations, such as breeder and layer flocks, as these groups are crucial for preventing vertical transmission of *Salmonella* from parent to offspring. Vaccines in these programs may include live attenuated or inactivated types, each formulated to provide immunity against specific *Salmonella* serovars [62]. Governments and regulatory bodies reduce the risk of *Salmonella* transmission by mandating vaccinations, lowering bacterial loads in birds, and, ultimately, reducing contamination levels in poultry products [177,178]. Vaccination programs are carefully monitored to ensure compliance and effectiveness, with follow-up testing often used to verify immunity levels within flocks. The FDA currently supports voluntary vaccination of commercial layers against *S.* Enteritidis. The agency has noted that while laboratory studies suggest that vaccines can help reduce *S.* Enteritidis colonization in hens and eggs, the field trials have not provided consistent evidence to make vaccination mandatory. Field studies showed that vaccinated flocks often had similar *S.* Enteritidis positive rates as unvaccinated hens, indicating that vaccination alone may not be effective under field conditions. As a result, the FDA views vaccination as an optional supplementary measure to augment the effectiveness of required *S.* Enteritidis control strategies rather than a substitute for them [52].

4.3. *Biosecurity Protocols*

Biosecurity is foundational in preventing *Salmonella* from being introduced and spread within poultry farms. Established biosecurity protocols [172] require farms to control farm access, implement sanitation procedures, and restrict personnel movement between different flock areas. For example, the USDA provides biosecurity guidelines under the NPIP advising farms to use footbaths, sanitize equipment, and implement rodent control measures [179]. The FDA also recommends implementing stringent biosecurity practices to ensure no introduction or transfer of *S.* Enteritidis into or among poultry houses. These measures include protecting against cross-contamination when equipment is moved among poultry houses, preventing cross-contamination when people move between poultry houses, keeping stray poultry, wild birds, cats, and other animals out of poultry houses, and prohibiting employees from keeping birds at home [52]. In Europe, biosecurity standards are similarly rigorous, often forming a key component of each country's national *Salmonella* control plan under EU legislation. Biosecurity protocols help to control both vertical and horizontal transmission of *Salmonella*, limiting contamination risks from external sources, such as wildlife, equipment, and personnel. Regular government inspections ensure that farms adhere to biosecurity standards, addressing lapses that could increase contamination risks [26].

4.4. *Certification and Quality Assurance*

Certification programs offer poultry producers recognition for compliance with *Salmonella* control standards, boosting consumer confidence and market access. The NPIP provides a voluntary certification program in the United States that indicates pathogen control compliance, including *Salmonella* monitoring [180,181]. Certification programs support producers by validating their biosecurity and hygiene measures, often making their products more marketable domestically and internationally. In the EU, similar certification programs award "*Salmonella*-controlled" or "*Salmonella*-free" status to flocks that meet specific pathogen reduction criteria. Certified producers benefit from consumer trust; certain markets may require these certifications as part of import regulations. Certification programs reinforce adherence to control standards and incentivize producers to maintain best practices in *Salmonella* management [173,182].

4.5. *Surveillance and Reporting Systems*

Surveillance and reporting systems track *Salmonella* prevalence in poultry and human populations, providing insights into outbreak patterns, high-risk serovars, and the effectiveness of control measures. These systems integrate data from routine testing, farm inspections, and human health surveillance, offering a comprehensive view of *Salmonella* trends. The European Food Safety Authority (EFSA) consolidates *Salmonella* data from EU member states, monitoring changes in prevalence and identifying emerging risks [183]. In the United States, the Centers for Disease Control and Prevention (CDC) collaborates with the USDA and other agencies to monitor *Salmonella* trends through PulseNet and FoodNet national laboratory network systems that detect and track foodborne outbreaks [184,185]. Surveillance systems enable rapid response to outbreaks, allowing government agencies to trace contamination sources and adjust policies to address identified weaknesses in the supply chain.

4.6. *Research and Development Initiatives*

Research and development (R&D) are essential for established *Salmonella* control programs, as they help evolve and improve existing control strategies. Many programs invest in R&D to develop new vaccines, enhance biosecurity technologies, and explore antibiotic alternatives. The USDA, for instance, funds research focused on alternative pathogen control methods, which aim to reduce *Salmonella* prevalence in poultry without relying on antibiotics. Similarly, the EFSA supports research to better understand *Salmonella* epidemiology in poultry, identifying high-risk transmission points and exploring innovative control solutions. These research efforts provide evidence-based guidance for refining control programs, addressing challenges such as AMR, and adapting to evolving *Salmonella* serovar profiles. Investment in R&D strengthens *Salmonella* control programs, ensuring that they remain effective in changing agricultural and public health landscapes [183,186]

4.7. Farmer Education and Outreach Programs

Education and outreach are crucial for the successful implementation of *Salmonella* control measures. Government agencies and industry bodies conduct training programs and workshops, providing resources to inform farmers about *Salmonella* risks, biosecurity practices, and hygiene protocols. Education programs often cover areas such as handling procedures, equipment sanitation, pest control, and proper waste disposal. For example, the USDA collaborates with poultry industry associations to conduct biosecurity training for farm staff, while the EU funds awareness campaigns to help producers comply with *Salmonella* control regulations. Educating producers and farm workers on best practices enables them to implement effective control measures, reducing *Salmonella* risks at the farm level. Outreach programs promote consistency across farms of all sizes, enhancing overall program compliance and food safety outcomes [179,187].

4.8. Implementation of Alternative Pathogen Control Methods

Some control programs have begun to adopt alternative pathogen control methods to reduce reliance on antibiotics, thereby addressing AMR concerns. These alternatives include the use of probiotics, prebiotics, synbiotics, postbiotics, and organic acids, which help maintain a healthy gut microbiome in poultry and reduce *Salmonella* colonization. Programs may provide guidelines on approved products, including their usage and dosage, to ensure safe and effective application. For instance, in certain European countries, government guidelines recommend that organic acids be added to feed and water as a preventive measure against *Salmonella*. Similarly, the USDA supports research and industry adoption of probiotic use to manage gut health in poultry [188,189]. By promoting alternatives to antibiotics, *Salmonella* control programs help mitigate AMR and support sustainable poultry production.

5. Challenges and Limitations in Pre-Harvest Control Measures

Controlling NTS in poultry during the pre-harvest phase presents several challenges and limitations. One significant issue is the asymptomatic carriage and intermittent shedding of *Salmonella* in poultry, allowing the bacteria to persist undetected within flocks and complicating early identification and intervention efforts [190]. Environmental factors, such as contaminated feed, water, and litter, serve as reservoirs for *Salmonella*, facilitating its introduction and spread within poultry operations. Implementing stringent biosecurity measures is essential but can be resource-intensive and challenging to maintain consistently across diverse farming systems. Additionally, while vaccination programs can reduce *Salmonella* prevalence, their effectiveness varies depending on the serovars present and the specific vaccines used [191]. The emergence of antibiotic-resistant *Salmonella* strains further complicates control efforts, as it limits the efficacy of antimicrobial treatments and necessitates the development of alternative strategies. Moreover, the complex interactions between *Salmonella* and the poultry gut microbiome can influence colonization dynamics, making it difficult to predict and manage infection patterns effectively. These challenges underscore the need for an integrated, multifaceted approach to pre-harvest *Salmonella* control in poultry, combining biosecurity, vaccination, environmental management, and ongoing surveillance to effectively mitigate the risk of contamination. While pre-harvest interventions such as probiotics are widely studied for controlling *Salmonella* colonization in poultry, their effects could be transient. Research indicates that probiotic treatments may only reduce *Salmonella* colonization for short periods, sometimes lasting as briefly as one week [192]. Another considerable challenge with probiotic supplements is ensuring the accuracy of the bacterial composition in the final product. This uncertainty creates regulatory hurdles and raises concerns about these formulations' reliability and effectiveness [193]. Since probiotics are live bacterial cultures, incorporating them into the diets of chickens presents unique challenges. These include preventing probiotic degradation during pelleting while also ensuring shelf stability and cost-effectiveness [194].

6. Future Directions and Innovations

Advancements in pre-harvest control of NTS in poultry have focused on new ways to enhance food safety. A very promising area includes the development of bacteriophage-based interventions, using viruses that specifically target *Salmonella* bacteria and destroy them. These phage therapies may offer a natural and precise way of reducing *Salmonella* colonization in poultry flocks. Additionally, research is exploring the use of prebiotics and synbiotics to modulate the gut microbiota, thereby creating an environment less conducive to *Salmonella* colonization. Genomic technologies are also being employed to identify genetic markers associated with *Salmonella* resistance, facilitating selective breeding programs aimed at developing poultry lines with enhanced resistance to NTS. Furthermore, new developments in rapid diagnostic tools now allow for better detection and monitoring of *Salmonella* at the farm level, enabling timely interventions. These innovations may represent a complete approach to mitigating *Salmonella* risks in poultry production, when applied with traditional biosecurity measures.

In parallel, advanced vaccine preparation and delivery methodologies, such as cochleate-based delivery systems, are being utilized to enhance the immunogenicity of subunit vaccines. Cochleates are specialized structures characterized by their unique spiral morphology composed of solid lipid bilayers. Studies have demonstrated that cochleate formulations elicit enhanced systemic and mucosal immune responses, thereby improving vaccine efficacy [183]. Integrating these advanced vaccine technologies with existing control measures holds promise for more effective control of NTS in poultry.

7. Conclusions

Controlling NTS presents an important aspect of safeguarding public health and ensuring sustainability in the table egg industry and other poultry sectors. This requires an integrated approach that includes vaccination, biosecurity measures, environmental management, and rigorous *Salmonella* surveillance and monitoring, as expounded here. This will reduce the prevalence of *Salmonella* in chickens and help control AMR, ensuring the safety of poultry products intended for human consumption. The near future appears promising with the development and introduction of new strategies, such as bacteriophage-based intervention, advances in genomics and poultry breeding, and the introduction of improved vaccines and vaccine delivery techniques. These innovations are expected to create more resilient and efficient *Salmonella* control measures. Collaboration among researchers, policymakers, and industry stakeholders will mutually benefit the poultry industry, public health, and consumers worldwide by improving animal health and economic sustainability and promoting consumer confidence in poultry products.

Author Contributions: Conceptualization, S.K.; writing—original draft preparation, R.N.N. and L.K.E.; writing—review and editing, S.K.; supervision, S.K.; project administration, S.K.; funding acquisition, S.K. All authors have read and agreed to the published version of the manuscript.

Funding: This research was funded by Agriculture and Food Research Initiative Competitive Grant no. 2020-03308 from the United States Department of Agriculture (USDA) National Institute of Food and Agriculture (NIFA).

Institutional Review Board Statement: Not applicable.

Informed Consent Statement: Not applicable.

Data Availability Statement: No additional data is available.

Conflicts of Interest: The authors declare no conflicts of interest.

References

1. Andino, A.; Hanning, I. *Salmonella* Enterica: Survival, Colonization, and Virulence Differences among Serovars. *Sci. World J.* **2015**, *2015*, 520179. [CrossRef] [PubMed]
2. Hoelzer, K.; Switt, A.I.M.; Wiedmann, M. Animal Contact as a Source of Human Non-Typhoidal Salmonellosis. *Vet. Res.* **2011**, *42*, 34. [CrossRef] [PubMed]
3. Pandey, M.; Sharath, E.; Goud, K. Non-Typhoidal Salmonellosis: A Major Concern for Poultry Industry. In *Salmonella spp.—A Global Challenge*; IntechOpen: London, UK, 2021.
4. *Salmonella* (Non-Typhoidal). Available online: https://www.who.int/news-room/fact-sheets/detail/salmonella-(non-typhoidal) (accessed on 17 November 2024).
5. McWhorter, A.R.; Chousalkar, K.K. A Long-Term Efficacy Trial of a Live, Attenuated *Salmonella* Typhimurium Vaccine in Layer Hens. *Front. Microbiol.* **2018**, *9*, 1380. [CrossRef] [PubMed]
6. Samper-Cativiela, C.; Prieto, M.E.; Collado, S.; De Frutos, C.; Branscum, A.J.; Saez, J.L.; Alvarez, J. Risk Factors for *Salmonella* Detection in Commercial Layer Flocks in Spain. *Animals* **2023**, *13*, 3181. [CrossRef] [PubMed]
7. Parsons, B.N.; Humphrey, S.; Salisbury, A.M.; Mikoleit, J.; Hinton, J.C.D.; Gordon, M.A.; Wigley, P. Invasive Non-Typhoidal *Salmonella Typhimurium* ST313 Are Not Host-Restricted and Have an Invasive Phenotype in Experimentally Infected Chickens. *PLoS Negl. Trop. Dis.* **2013**, *7*, e2487. [CrossRef]
8. Shaji, S.; Selvaraj, R.K.; Shanmugasundaram, R. *Salmonella* Infection in Poultry: A Review on the Pathogen and Control Strategies. *Microorganisms* **2023**, *11*, 2814. [CrossRef]
9. Bangera, S.R.; Umakanth, S.; Chowdhury, G.; Saha, R.N.; Mukhopadhyay, A.K.; Ballal, M. Poultry: A Receptacle for Non-Typhoidal *Salmonella* and Antimicrobial Resistance. *Iran. J. Microbiol.* **2019**, *11*, 31. [CrossRef]
10. Wibisono, F.M.; Wibisono, F.J.; Helmi Effendi, M.; Plumeriastuti, H.; Hidayatullah, A.R.; Hartadi, E.B.; Sofiana, E.D. A Review of Salmonellosis on Poultry Farms: Public Health Importance. *Syst. Rev. Pharm.* **2020**, *11*, 481–486.
11. Wigley, P. Salmonella and the Chicken: Reflections on Salmonellosis and Its Control in the United Kingdom. *Poult. Sci. Manag.* **2024**, *1*, 1. [CrossRef]
12. Odoch, T.; Wasteson, Y.; L'Abée-Lund, T.; Muwonge, A.; Kankya, C.; Nyakarahuka, L.; Tegule, S.; Skjerve, E. Prevalence, Antimicrobial Susceptibility and Risk Factors Associated with Non-Typhoidal *Salmonella* on Ugandan Layer Hen Farms. *BMC Vet. Res.* **2017**, *13*, 365. [CrossRef]
13. Wang, J.; Vaddu, S.; Bhumanapalli, S.; Mishra, A.; Applegate, T.; Singh, M.; Thippareddi, H. A Systematic Review and Meta-Analysis of the Sources of *Salmonella* in Poultry Production (Pre-Harvest) and Their Relative Contributions to the Microbial Risk of Poultry Meat. *Poult. Sci.* **2023**, *102*, 102566. [CrossRef] [PubMed]
14. Denagamage, T.; Jayarao, B.; Patterson, P.; Wallner-Pendleton, E.; Kariyawasam, S. Risk Factors Associated with Salmonella in Laying Hen Farms: Systematic Review of Observational Studies. *Avian Dis.* **2015**, *59*, 291–302. [CrossRef] [PubMed]
15. Van Hoorebeke, S.; Van Immerseel, F.; Haesebrouck, F.; Ducatelle, R.; Dewulf, J. The Influence of the Housing System on *Salmonella* Infections in Laying Hens: A Review. *Zoonoses Public Health* **2011**, *58*, 304–311. [CrossRef] [PubMed]
16. Bashir, A.; Lambert, P.A.; Stedman, Y.; Hilton, A.C. Combined Effect of Temperature and Relative Humidity on the Survival of *Salmonella* Isolates on Stainless Steel Coupons. *Int. J. Environ. Res. Public Health* **2022**, *19*, 909. [CrossRef]
17. Chinivasagam, H.N.; Tran, T.; Maddock, L.; Gale, A.; Blackall, P.J. Mechanically Ventilated Broiler Sheds: A Possible Source of Aerosolized *Salmonella*, *Campylobacter*, and *Escherichia coli*. *Appl. Environ. Microbiol.* **2009**, *75*, 7417–7425. [CrossRef]
18. An HSUS Report: Human Health Implications of Intensive Poultry An HSUS Report: Human Health Implications of Intensive Poultry Production and Avian Influenza Production and Avian Influenza. Available online: https://www.wellbeingintlstudiesrepository.org/cgi/viewcontent.cgi?params=/context/hsus_reps_environment_and_human_health/article/1005/&path_info=HSUS_Public_Health_Report_on_Avian_Influenza_and_Poultry_Production.pdf (accessed on 3 December 2024).
19. Abo-Al-Ela, H.G.; El-Kassas, S.; El-Naggar, K.; Abdo, S.E.; Jahejo, A.R.; Al Wakeel, R.A. Stress and Immunity in Poultry: Light Management and Nanotechnology as Effective Immune Enhancers to Fight Stress. *Cell Stress Chaperones* **2021**, *26*, 457–472. [CrossRef]
20. Lara, L.J.; Rostagno, M.H. Impact of Heat Stress on Poultry Production. *Animals* **2013**, *3*, 356–369. [CrossRef]
21. de Toledo, T.d.S.; Roll, A.A.P.; Rutz, F.; Dallmann, H.M.; Prá, M.A.D.; Leite, F.P.L.; Roll, V.F.B. An Assessment of the Impacts of Litter Treatments on the Litter Quality and Broiler Performance: A Systematic Review and Metaanalysis. *PLoS ONE* **2020**, *15*, e0232853. [CrossRef]
22. Pal, A.; Bailey, M.A.; Talorico, A.A.; Krehling, J.T.; Macklin, K.S.; Price, S.B.; Buhr, R.J.; Bourassa, D.V. Impact of Poultry Litter *Salmonella* Levels and Moisture on Transfer of *Salmonella* through Associated in Vitro Generated Dust. *Poult. Sci.* **2021**, *100*, 101236. [CrossRef]
23. Trampel, D.W.; Holder, T.G.; Gast, R.K. Integrated Farm Management to Prevent *Salmonella* Enteritidis Contamination of Eggs. *J. Appl. Poult. Res.* **2014**, *23*, 353–365. [CrossRef]

24. Obe, T.; Boltz, T.; Kogut, M.; Ricke, S.C.; Brooks, L.A.; Macklin, K.; Peterson, A. Controlling *Salmonella*: Strategies for Feed, the Farm, and the Processing Plant. *Poult. Sci.* **2023**, *102*, 103086. [CrossRef] [PubMed]
25. Wales, A.D.; Allen, V.M.; Davies, R.H. Chemical Treatment of Animal Feed and Water for the Control of *Salmonella. Foodborne Pathog. Dis.* **2010**, *7*, 3–15. [CrossRef] [PubMed]
26. Koutsoumanis, K.; Allende, A.; Alvarez-Ordóñez, A.; Bolton, D.; Bover-Cid, S.; Chemaly, M.; De Cesare, A.; Herman, L.; Hilbert, F.; Lindqvist, R.; et al. *Salmonella* Control in Poultry Flocks and Its Public Health Impact. *EFSA J.* **2019**, *17*, e05596. [CrossRef] [PubMed]
27. McMullin, P. Infectious Diseases in Free-Range Compared to Conventional Poultry Production. *Avian Pathol.* **2022**, *51*, 424–434. [CrossRef] [PubMed]
28. European Food Safety Authority (EFSA). Report of the Task Force on Zoonoses Data Collection on the Analysis of the Baseline Study on the Prevalence of *Salmonella* in Holdings of Laying Hen Flocks of *Gallus Gallus. EFSA J.* **2007**, *5*, 97. [CrossRef]
29. Van Hoorebeke, S.; Van Immerseel, F.; Schulz, J.; Hartung, J.; Harisberger, M.; Barco, L.; Ricci, A.; Theodoropoulos, G.; Xylouri, E.; De Vylder, J.; et al. Determination of the within and between Flock Prevalence and Identification of Risk Factors for *Salmonella* Infections in Laying Hen Flocks Housed in Conventional and Alternative Systems. *Prev. Vet. Med.* **2010**, *94*, 94–100. [CrossRef]
30. Dewulf, J.; Van Hoorebeke, S.; Van Immerseel, F. 6—Epidemiology of *Salmonella* Infections in Laying Hens with Special Emphasis on the Influence of the Housing System. In *Improving the Safety and Quality of Eggs and Egg Products*; Van Immerseel, F., Nys, Y., Bain, M., Eds.; Woodhead Publishing Series in Food Science, Technology and Nutrition; Woodhead Publishing: Cambridge, UK, 2011; pp. 107–119. ISBN 978-0-85709-072-0.
31. Carrique-Mas, J.J.; Breslin, M.; Snow, L.; McLaren, I.; Sayers, A.R.; Davies, R.H. Persistence and Clearance of Different *Salmonella* Serovars in Buildings Housing Laying Hens. *Epidemiol. Infect.* **2009**, *137*, 837–846. [CrossRef]
32. Namata, H.; Méroc, E.; Aerts, M.; Faes, C.; Abrahantes, J.C.; Imberechts, H.; Mintiens, K. *Salmonella* in Belgian Laying Hens: An Identification of Risk Factors. *Prev. Vet. Med.* **2008**, *83*, 323–336. [CrossRef]
33. Dórea, F.C.; Cole, D.J.; Hofacre, C.; Zamperini, K.; Mathis, D.; Doyle, M.P.; Lee, M.D.; Maurer, J.J. Effect of *Salmonella* Vaccination of Breeder Chickens on Contamination of Broiler Chicken Carcasses in Integrated Poultry Operations. *Appl. Environ. Microbiol.* **2010**, *76*, 7820–7825. [CrossRef]
34. Selective Breeding: A Complementary Strategy to Control Avian Infectious Diseases and Zoonoses. Available online: https://www.onehealthpoultry.org/blog-posts/selective-breeding-a-complementary-strategy-to-control-avian-infectious-diseases-and-zoonoses/ (accessed on 1 December 2024).
35. Olson, E.G.; Grenda, T.; Ghosh, A.; Ricke, S.C. *Microbial Pathogen Contamination of Animal Feed. Present Knowledge in Food Safety*; Elsevier: Amsterdam, The Netherlands, 2023; pp. 378–393.
36. Feed Sanitation: Heat Treatment as a Method for the Decontamination of Poultry Feeds. Available online: https://www.thepoultrysite.com/articles/feed-sanitation-heat-treatment-as-a-method-for-the-decontamination-of-poultry-feeds?utm_source=chatgpt.com (accessed on 3 December 2024).
37. Liu, H.; Whitehouse, C.A.; Li, B. Presence and Persistence of *Salmonella* in Water: The Impact on Microbial Quality of Water and Food Safety. *Front. Public Health* **2018**, *6*, 159. [CrossRef]
38. Abreu, R.; Semedo-Lemsaddek, T.; Cunha, E.; Tavares, L.; Oliveira, M. Antimicrobial Drug Resistance in Poultry Production: Current Status and Innovative Strategies for Bacterial Control. *Microorganisms* **2023**, *11*, 953. [CrossRef] [PubMed]
39. Hedman, H.D.; Vasco, K.A.; Zhang, L. A Review of Antimicrobial Resistance in Poultry Farming within Low-resource Settings. *Animals* **2020**, *10*, 1264. [CrossRef] [PubMed]
40. Liu, F.; Luo, Y.; Xu, T.; Lin, H.; Qiu, Y.; Li, B. Current Examining Methods and Mathematical Models of Horizontal Transfer of Antibiotic Resistance Genes in the Environment. *Front. Microbiol.* **2024**, *15*, 1371388. [CrossRef] [PubMed]
41. Oladeinde, A.; Abdo, Z.; Zwirzitz, B.; Woyda, R.; Lakin, S.M.; Press, M.O.; Cox, N.A.; Thomas, J.C., IV; Looft, T.; Rothrock, M.J.; et al. Litter Commensal Bacteria Can Limit the Horizontal Gene Transfer of Antimicrobial Resistance to *Salmonella* in Chickens. *Appl. Environ. Microbiol.* **2022**, *88*, e0251721. [CrossRef]
42. Oladeinde, A.; Abdo, Z.; Press, M.O.; Cook, K.; Cox, N.A.; Zwirzitz, B.; Woyda, R.; Lakin, S.M.; Thomas, J.C.; Looft, T.; et al. Horizontal Gene Transfer Is the Main Driver of Antimicrobial Resistance in Broiler Chicks Infected with Salmonella Enterica Serovar Heidelberg. *Msystems* **2021**, *6*, e0072921. [CrossRef]
43. Biosecurity in Poultry Farming: Protecting Your Flock. Available online: https://champrix.com/articles/biosecurity-in-poultry-farming-protecting-your-flock (accessed on 14 November 2024).
44. Biosecurity on the Poultry Farm. Available online: https://www.veterinariadigital.com/en/articulos/biosecurity-on-the-poultry-farm/ (accessed on 14 November 2024).
45. Biosecurity for Poultry Producers. Available online: https://agriculture.vic.gov.au/biosecurity/animal-diseases/poultry-diseases/biosecurity-for-poultry-producers (accessed on 14 November 2024).
46. Wegener, H.C.; Hald, T.; Wong, D.L.F.; Madsen, M.; Korsgaard, H.; Bager, F.; Gerner-Smidt, P.; Mølbak, K. *Salmonella* Control Programs in Denmark. *Emerg. Infect. Dis.* **2003**, *9*, 774. [CrossRef]
47. Anderson, T.C.; Nguyen, T.A.; Adams, J.K.; Garrett, N.M.; Bopp, C.A.; Baker, J.B.; McNeil, C.; Torres, P.; Ettestad, P.J.; Erdman, M.M.; et al. Multistate Outbreak of Human *Salmonella Typhimurium* Infections Linked to Live Poultry from Agricultural Feed Stores and Mail-Order Hatcheries, United States 2013. *One Health* **2016**, *2*, 144–149. [CrossRef]

48. Ziaul Haque, A.K.M.; Akter, M.R.; Islam, S.K.S.; Alam, J.; Neogi, S.B.; Yamasaki, S.; Lutful Kabir, S.M. *Salmonella Gallinarum* in Small-Scale Commercial Layer Flocks: Occurrence, Molecular Diversity and Antibiogram. *Vet. Sci.* **2021**, *8*, 71. [CrossRef]
49. Dorea, F.C.; Berghaus, R.; Hofacre, C.; Cole, D.J. Survey of Biosecurity Protocols and Practices Adopted by Growers on Commercial Poultry Farms in Georgia, U.S.A. *Avian Dis.* **2010**, *54*, 1007–1015. [CrossRef]
50. Biosecurity Procedures in Poultry Production. Available online: https://www.woah.org/fileadmin/Home/eng/Health_standards/tahc/current/chapitre_biosecu_poul_production.pdf (accessed on 1 December 2024).
51. Meher, M.M.; Sharif, M.A.; Bayazid, A. Al Seroprevalence of *Salmonella* spp. Infection in Different Types of Poultry and Biosecurity Measures Associated with Salmonellosis. *Int. J. Agric. Environ. Food Sci.* **2022**, *6*, 557–567. [CrossRef]
52. Prevention of Salmonella Enteritidis in Shell Eggs During Production, Storage, and Transportation. Available online: https://www.federalregister.gov/documents/2009/07/09/E9-16119/prevention-of-salmonella-enteritidis-in-shell-eggs-during-production-storage-and-transportation (accessed on 3 December 2024).
53. Garber, L.; Smeltzer, M.; Fedorka-Cray, P.; Ladely, S.; Ferris, K. *Salmonella enterica* Serotype Enteritidis in Table Egg Layer House Environments and in Mice in U.S. Layer Houses and Associated Risk Factors. *Avian Dis.* **2003**, *47*, 134–142. [CrossRef] [PubMed]
54. Poirier, E.; Watier, L.; Espie, E.; Weill, F.X.; de Valk, H.; Desenclos, J.C. Evaluation of the Impact on Human Salmonellosis of Control Measures Targeted to Salmonella Enteritidis and Typhimurium in Poultry Breeding Using Time-Series Analysis and Intervention Models in France. *Epidemiol. Infect.* **2007**, *136*, 1217–1224. [CrossRef] [PubMed]
55. Jones, F.T. A Review of Practical Salmonella Control Measures in Animal Feed. *J. Appl. Poult. Res.* **2011**, *20*, 102–113. [CrossRef]
56. Muckey, M.; Huss, A.R.; Yoder, A.; Jones, C. Research Note: Evaluating the Roles of Surface Sanitation and Feed Sequencing on Mitigating Salmonella Enteritidis Contamination on Animal Food Manufacturing Equipment. *Poult. Sci.* **2020**, *99*, 3841–3845. [CrossRef] [PubMed]
57. Morita, T.; Kitazawa, H.; Iida, T.; Kamata, S. Prevention of Salmonella Cross-Contamination in an Oilmeal Manufacturing Plant. *J. Appl. Microbiol.* **2006**, *101*, 464–473. [CrossRef]
58. Van Immerseel, F.; De Zutter, L.; Houf, K.; Pasmans, F.; Haesebrouck, F.; Ducatelle, R. Strategies to Control Salmonella in the Broiler Production Chain. *Worlds Poult. Sci. J.* **2009**, *65*, 367–392. [CrossRef]
59. Raccoursier, M.; Siceloff, A.T.; Shariat, N.W. In Silico and PCR Screening for a Live Attenuated Salmonella Typhimurium Vaccine Strain. *Avian Dis.* **2024**, *68*, 18–24. [CrossRef]
60. Cogan, T.A.; Humphrey, T.J. The Rise and Fall of Salmonella Enteritidis in the UK. *J. Appl. Microbiol.* **2003**, *94*, 114–119. [CrossRef]
61. Howard, A.J.; Chousalkar, K.K.; McWhorter, A.R. In Vitro and in Vivo Efficacy of a Live Attenuated Salmonella Typhimurium Vaccine at Preventing Intestinal Colonization in Chicks. *Zoonoses Public Health* **2018**, *65*, 736–741. [CrossRef]
62. Varmuzova, K.; Faldynova, M.; Elsheimer-Matulova, M.; Sebkova, A.; Polansky, O.; Havlickova, H.; Sisak, F.; Rychlik, I. Immune Protection of Chickens Conferred by a Vaccine Consisting of Attenuated Strains of Salmonella Enteritidis, Typhimurium and Infantis. *Vet. Res.* **2016**, *47*, 1–11. [CrossRef]
63. Li, Q.; Zhu, Y.; Ren, J.; Qiao, Z.; Yin, C.; Xian, H.; Yuan, Y.; Geng, S.; Jiao, X. Evaluation of the Safety and Protection Efficacy of SpiC and NmpC or RfaL Deletion Mutants of Salmonella Enteritidis as Live Vaccine Candidates for Poultry Non-Typhoidal Salmonellosis. *Vaccines* **2019**, *7*, 202. [CrossRef] [PubMed]
64. Yin, J.; Cheng, Z.; Xu, L.; Li, Q.; Geng, S.; Pan, Z.; Jiao, X. Immunogenicity and Protective Efficacy of Salmonella Enterica Serovar Pullorum Pathogenicity Island 2 Mutant as a Live Attenuated Vaccine Candidate. *BMC Vet. Res.* **2015**, *11*, 162. [CrossRef] [PubMed]
65. Jia, S.; McWhorter, A.R.; Andrews, D.M.; Underwood, G.J.; Chousalkar, K.K. Challenges in Vaccinating Layer Hens against Salmonella Typhimurium. *Vaccines* **2020**, *8*, 696. [CrossRef] [PubMed]
66. Hassan, J.O.; Curtiss, R. Development and Evaluation of an Experimental Vaccination Program Using a Live Avirulent Salmonella Typhimurium Strain to Protect Immunized Chickens against Challenge with Homologous and Heterologous Salmonella Serotypes. *Infect. Immun.* **1994**, *62*, 5519–5527. [CrossRef]
67. Hassan, J.O.; Curtis, R.I.I.I. Efficacy of a Live Avirulent Salmonella Typhimurium Vaccine in Preventing Colonization and Invasion of Laying Hens by Salmonella Typhimurium and Salmonella Enteritidis. *Avian Dis.* **1997**, *41*, 783–791. [CrossRef]
68. van Immerseel, F.; Methner, U.; Rychlik, I.; Nagy, B.; Velge, P.; Martin, G.; Foster, N.; Ducatelle, R.; Barrow, P.A. Vaccination and Early Protection against Non-Host-Specific Salmonella Serotypes in Poultry: Exploitation of Innate Immunity and Microbial Activity. *Epidemiol. Infect.* **2005**, *133*, 959–978. [CrossRef]
69. Johnson, T.J.; Flores-Figueroa, C.; Munoz-Aguayo, J.; Pinho, G.; Miller, E. Persistence of Vaccine Origin Salmonella Typhimurium through the Poultry Production Continuum, and Development of a Rapid Typing Scheme for Their Differentiation from Wild Type Field Isolates. *Poult. Sci.* **2024**, *103*, 103707. [CrossRef]
70. Clifton-Hadley, F.A.; Breslin, M.; Venables, L.M.; Sprigings, K.A.; Cooles, S.W.; Houghton, S.; Woodward, M.J. A Laboratory Study of an Inactivated Bivalent Iron Restricted Salmonella Enterica Serovars Enteritidis and Typhimurium Challenge in Chickens. *Vet. Microbiol.* **2002**, *89*, 167–179. [CrossRef]
71. Filho, R.A.C.P.; de Paiva, J.B.; Argüello, Y.M.S.; da Silva, M.D.; Gardin, Y.; Resende, F.; Junior, A.B.; Sesti, L. Efficacy of Several Vaccination Programmes in Commercial Layer and Broiler Breeder Hens against Experimental Challenge with Salmonella Enterica Serovar Enteritidis. *Avian Pathol.* **2009**, *38*, 367–375. [CrossRef]
72. Inoue, A.Y.; Berchieri, A.J.; Bernardino, A.; Paiva, J.B.; Sterzo, E.V. Passive Immunity of Progeny from Broiler Breeders Vaccinated with Oil-Emulsion Bacterin Against Salmonella Enteritidis. *Avian Dis.* **2008**, *52*, 567–571. [CrossRef]

73. Huberman, Y.D.; Caballero-García, M.; Rojas, R.; Ascanio, S.; Olmos, L.H.; Malena, R.; Lomónaco, J.; Nievas, P.; Chero, P.; Lévano-Gracía, J.; et al. The Efficacy of a Trivalent Inactivated Salmonella Vaccine Combined with the Live S. Gallinarum 9R Vaccine in Young Layers after Experimental Infections with *S.* Enteritidis, *S.* Typhimurium, and *S.* Infantis. *Vaccines* **2022**, *10*, 1113. [CrossRef] [PubMed]
74. Barrow, P.A. Salmonella Infections: Immune and Non-Immune Protection with Vaccines. *Avian Pathol.* **2007**, *36*, 1–13. [CrossRef] [PubMed]
75. Desin, T.S.; Köster, W.; Potter, A.A. Salmonella Vaccines in Poultry: Past, Present and Future. *Expert. Rev. Vaccines* **2013**, *12*, 87–96. [CrossRef] [PubMed]
76. Liu, Q.; Liu, Q.; Yi, J.; Liang, K.; Hu, B.; Zhang, X.; Curtiss, R.; Kong, Q. Outer Membrane Vesicles from Flagellin-Deficient Salmonella Enterica Serovar Typhimurium Induce Cross-Reactive Immunity and Provide Cross-Protection against Heterologous Salmonella Challenge. *Sci. Rep.* **2016**, *6*, 34776. [CrossRef] [PubMed]
77. Meenakshi, M.; Bakshi, C.S.; Butchaiah, G.; Bansal, M.P.; Siddiqui, M.Z.; Singh, V.P. Adjuvanted Outer Membrane Protein Vaccine Protects Poultry against Infection with Salmonella enteritidis. *Vet. Res. Commun.* **1999**, *23*, 81–90. [CrossRef]
78. Li, Q.; Ren, J.; Xian, H.; Yin, C.; Yuan, Y.; Li, Y.; Ji, R.; Chu, C.; Qiao, Z.; Jiao, X. ROmpF and OMVs as Efficient Subunit Vaccines against Salmonella Enterica Serovar Enteritidis Infections in Poultry Farms. *Vaccine* **2020**, *38*, 7094–7099. [CrossRef]
79. Dolatyabi, S.; Renu, S.; Schrock, J.; Renukaradhya, G.J. Chitosan-Nanoparticle-Based Oral Salmonella Enteritidis Subunit Vaccine Elicits Cross-Protection against Salmonella Typhimurium in Broilers. *Poult. Sci.* **2024**, *103*, 103569. [CrossRef]
80. Han, Y.; Renu, S.; Patil, V.; Schrock, J.; Feliciano-Ruiz, N.; Selvaraj, R.; Renukaradhya, G.J. Immune Response to Salmonella Enteritidis Infection in Broilers Immunized Orally With Chitosan-Based Salmonella Subunit Nanoparticle Vaccine. *Front. Immunol.* **2020**, *11*, 935. [CrossRef]
81. Renu, S.; Han, Y.; Dhakal, S.; Lakshmanappa, Y.S.; Ghimire, S.; Feliciano-Ruiz, N.; Senapati, S.; Narasimhan, B.; Selvaraj, R.; Renukaradhya, G.J. Chitosan-Adjuvanted Salmonella Subunit Nanoparticle Vaccine for Poultry Delivered through Drinking Water and Feed. *Carbohydr. Polym.* **2020**, *243*, 116434. [CrossRef]
82. Han, Y.; Renu, S.; Schrock, J.; Acevedo-Villanuev, K.Y.; Lester, B.; Selvaraj, R.K.; Renukaradhya, G.J. Temporal Dynamics of Innate and Adaptive Immune Responses in Broiler Birds to Oral Delivered Chitosan Nanoparticle-Based Salmonella Subunit Antigens. *Vet. Immunol. Immunopathol.* **2020**, *228*, 110111. [CrossRef]
83. Han, Y.; Renu, S.; Patil, V.; Schrock, J.; Feliciano-Ruiz, N.; Selvaraj, R.; Renukaradhya, G.J. Mannose-Modified Chitosan-Nanoparticle-Based Salmonella Subunit Oralvaccine-Induced Immune Response and Efficacy in a Challenge Trial in Broilers. *Vaccines* **2020**, *8*, 299. [CrossRef] [PubMed]
84. Renu, S.; Markazi, A.D.; Dhakal, S.; Lakshmanappa, Y.S.; Shanmugasundaram, R.; Selvaraj, R.K.; Renukaradhya, G.J. Oral Deliverable Mucoadhesive Chitosan-Salmonella Subunit Nanovaccine for Layer Chickens. *Int. J. Nanomed.* **2020**, *15*, 761–777. [CrossRef] [PubMed]
85. Acevedo-Villanueva, K.Y.; Akerele, G.O.; Al Hakeem, W.G.; Renu, S.; Shanmugasundaram, R.; Selvaraj, R.K. A Novel Approach against Salmonella: A Review of Polymeric Nanoparticle Vaccines for Broilers and Layers. *Vaccines* **2021**, *9*, 1041. [CrossRef] [PubMed]
86. Koenen, M.E.; Boonstra-Blom, A.G.; Jeurissen, S.H.M. Immunological Differences between Layer- and Broiler-Type Chickens. *Vet. Immunol. Immunopathol.* **2002**, *89*, 47–56. [CrossRef]
87. Senevirathne, A.; Hewawaduge, C.; Park, J.Y.; Park, S.; Lee, J.H. Parenteral Immunization of Salmonella Typhimurium Ghosts with Surface-Displayed Escherichia Coli Flagellin EnhancesTLR-5 Mediated Activation of Immune Responses That Protect the Chicken against Salmonella Infection. *Microb. Pathog.* **2020**, *147*, 104252. [CrossRef]
88. Hajam, I.A.; Kim, J.H.; Lee, J.H. Incorporation of Membrane-Anchored Flagellin into Salmonella Gallinarum Bacterial Ghosts Induces Early Immune Responses and Protection against Fowl Typhoid in Young Layer Chickens. *Vet. Immunol. Immunopathol.* **2018**, *199*, 61–69. [CrossRef]
89. Senevirathne, A.; Hewawaduge, C.; Lee, J.H. Immunization of Chicken with Flagellin Adjuvanted Salmonella Enteritidis Bacterial Ghosts Confers Complete Protection against Chicken Salmonellosis. *Poult. Sci.* **2021**, *100*, 101205. [CrossRef]
90. Chaudhari, A.A.; Jawale, C.V.; Kim, S.W.; Lee, J.H. Construction of a Salmonella Gallinarum Ghost as a Novel Inactivated Vaccine Candidate and Its Protective Efficacy against Fowl Typhoid in Chickens. *Vet. Res.* **2012**, *43*, 1–11. [CrossRef]
91. Jawale, C.V.; Lee, J.H. Characterization of a Salmonella Typhimurium Ghost Carrying an Adjuvant Protein as a Vaccine Candidate for the Protection of Chickens against Virulent Challenge. *Avian Pathol.* **2014**, *43*, 506–513. [CrossRef]
92. Senevirathne, A.; Hewawaduge, C.; Lee, J.H. Salmonella Enterica Serovar Enteritidis Ghosts Displaying a Surface FliC Adjuvant Elicit a Robust Immune Response and Effective Protection against Virulent Challenge. *Vet. Microbiol.* **2020**, *243*, 108633. [CrossRef]
93. Senevirathne, A.; Hewawaduge, C.; Lee, J.H. Immunization of Chickens with Salmonella Gallinarium Ghosts Expressing Salmonella Enteritidis NFliC-FimAC and CD40LC Fusion Antigen Enhances Cell-Mediated Immune Responses and Protects against Wild-Type Challenges with Both Species. *Dev. Comp. Immunol.* **2022**, *126*, 104265. [CrossRef] [PubMed]
94. Curtiss, R. Vaccines to Control Salmonella in Poultry. *Avian Dis.* **2023**, *67*, 427–440. [CrossRef] [PubMed]
95. Feberwee, A.; de Vries, T.S.; Hartman, E.G.; de Wit, J.J.; Elbers, A.R.W.; de Jong, W.A. Vaccination against Salmonella Enteritidis in Dutch Commercial Layer Flocks with a Vaccine Based on a Live Salmonella Gallinarum 9R Strain: Evaluation of Efficacy, Safety, and Performance of Serologic Salmonella Tests. *Avian Dis.* **2001**, *45*, 83–91. [CrossRef] [PubMed]
96. AviPro Megan Vac 1. Available online: https://www.drugs.com/vet/avipro-megan-vac-1.html (accessed on 3 December 2024).

97. Holt, P.S.; Gast, R.K.; Kelly-Aehle, S. Use of a Live Attenuated Salmonella Typhimurium Vaccine to Protect Hens Against Salmonella Enteritidis Infection While Undergoing Molt. *Avian Dis.* **2003**, *47*, 656–661. [CrossRef]
98. AviPro Megan Egg. Available online: https://www.drugs.com/vet/avipro-megan-egg.html (accessed on 3 December 2024).
99. Combat Salmonella with Our Proven Poultry Vaccines. Available online: https://farmanimal.elanco.com/us/poultry/collection/salmonella-control (accessed on 3 December 2024).
100. Alderton, M.R.; Fahey, K.J.; Coloe, P.J. Humoral Responses and Salmonellosis Protection in Chickens given a Vitamin-Dependent Salmonella Typhimurium Mutant. *Avian Dis.* **1991**, *35*, 435–442. [CrossRef]
101. Bioproperties—Vaxsafe ST. Available online: http://www.bioproperties.com.au/!Pages/Vaccines/VaxsafeST.html (accessed on 13 November 2024).
102. Sharma, P.; Caraguel, C.; Sexton, M.; McWhorter, A.; Underwood, G.; Holden, K.; Chousalkar, K. Shedding of Salmonella Typhimurium in Vaccinated and Unvaccinated Hens during Early Lay in Field Conditions: A Randomised Controlled Trial. *BMC Microbiol.* **2018**, *18*, 78. [CrossRef]
103. Theuß, T.; Woitow, G.; Bulang, M.; Springer, S. Demonstration of the Efficacy of a Salmonella Enteritidis Live Vaccine for Chickens According to the Current European Pharmacopoeia Monograph. *Heliyon* **2018**, *4*, e01070. [CrossRef]
104. Salmovac®—Ceva Poultry. Available online: https://poultry.ceva.com/poultry-vaccines/salmovac/?gad_source=1&gclid=CjwKCAiAudG5BhAREiwAWMlSjFs98cpFpjgtAxQt5k2SVYcOcjd8oAlkkN8cvUf-EL-83G3_0MfSzxoCuFAQAvD_BwE (accessed on 13 November 2024).
105. Porcicultura, Salud Porcina, Bienestar, Enfermedades, Noticias Porcinas, Artículos, Fotos—El Sitio Porcino. Available online: https://www.elsitioporcino.com/ (accessed on 13 November 2024).
106. POULVAC® ST | For Animal Healthcare Professionals. Available online: https://www.zoetisus.com/products/poultry/poulvac-st (accessed on 13 November 2024).
107. Nobilis® Salenvac T—MSD Animal Health India. Available online: https://www.msd-animal-health.co.in/products/nobilis-salenvac-t/ (accessed on 13 November 2024).
108. Woodward, M.J.; Gettinby, G.; Breslin, M.F.; Corkish, J.D.; Houghton, S. The Efficacy of Salenvac, a Salmonella enterica subsp. Enterica Serotype Enteritidis Iron-Restricted Bacterin Vaccine, in Laying Chickens. *Avian Pathol.* **2002**, *31*, 383–392. [CrossRef]
109. CEVAC® Corymune 4K & 7K Benefits Beyond Protection. Available online: https://www.ceva.vn/en/Technical-Informations/Poultry/Ceva-Technical-Bulletin/AXIS-Magazine/CEVAC-R-CORYMUNE-4K-7K (accessed on 8 December 2024).
110. Poulvac® SE | For Animal Healthcare Professionals. Available online: https://www.zoetisus.com/products/poultry/poulvac-se (accessed on 13 November 2024).
111. Poulvac *E. coli* Lyophilisate for Suspension for Spray Vaccination for Chickens and Turkeys or for Use in Drinking Water for Chickens (GB). Available online: https://www.noahcompendium.co.uk/?id=-486332&template=template_printview (accessed on 3 December 2024).
112. 196-278500 | Animal and Plant Health Inspection Service. Available online: https://www.aphis.usda.gov/aphis/ourfocus/animalhealth/veterinary-biologics/product-summaries/vet-label-data/8a736d32-8d89-4f1d-b1c7-931951a9e03b (accessed on 13 November 2024).
113. Comprehensive Salmonella Prevention. Available online: https://assets-eu-01.kc-usercontent.com/43db687c-fb47-015f-c990-00802940e2bc/de6dfc75-7e1f-415e-938f-3002b53b491f/Brochure%20Salmonella%20360%C2%B0%20-%2003-2021_FINAL.pdf (accessed on 3 December 2024).
114. 196-48D711 | Animal and Plant Health Inspection Service. Available online: https://www.aphis.usda.gov/aphis/ourfocus/animalhealth/veterinary-biologics/product-summaries/vet-label-data/17f69ae2-1ae2-46a1-815f-e33a54dae527 (accessed on 13 November 2024).
115. Hill, C.; Guarner, F.; Reid, G.; Gibson, G.R.; Merenstein, D.J.; Pot, B.; Morelli, L.; Canani, R.B.; Flint, H.J.; Salminen, S.; et al. Expert Consensus Document: The International Scientific Association for Probiotics and Prebiotics Consensus Statement on the Scope and Appropriate Use of the Term Probiotic. *Nat. Rev. Gastroenterol. Hepatol.* **2014**, *11*, 506–514. [CrossRef]
116. Groves, P.J.; Williamson, S.L.; Ahaduzzaman, M.; Diamond, M.; Ngo, M.; Han, A.; Sharpe, S.M. Can a Combination of Vaccination, Probiotic and Organic Acid Treatment in Layer Hens Protect against Early Life Exposure to Salmonella Typhimurium and Challenge at Sexual Maturity? *Vaccine* **2021**, *39*, 815–824. [CrossRef]
117. Khan, S.; Chousalkar, K.K. Salmonella Typhimurium Infection Disrupts but Continuous Feeding of Bacillus Based Probiotic Restores Gut Microbiota in Infected Hens. *J. Anim. Sci. Biotechnol.* **2020**, *11*, 1–16. [CrossRef] [PubMed]
118. Price, P.T.; Gaydos, T.A.; Berghaus, R.D.; Baxter, V.; Hofacre, C.L.; Sims, M.D. Salmonella Enteritidis Reduction in Layer Ceca with a Bacillus Probiotic. *Vet. World* **2020**, *13*, 184–187. [CrossRef] [PubMed]
119. Castillo, N.A.; Perdigán, G.; De Moreno De Leblanc, A. Oral Administration of a Probiotic Lactobacillus Modulates Cytokine Production and TLR Expression Improving the Immune Response against Salmonella Enterica Serovar Typhimurium Infection in Mice. *BMC Microbiol.* **2011**, *11*, 177. [CrossRef] [PubMed]
120. Bermudez-Brito, M.; Plaza-Díaz, J.; Muñoz-Quezada, S.; Gómez-Llorente, C.; Gil, A. Probiotic Mechanisms of Action. *Ann. Nutr. Metab.* **2012**, *61*, 160–174. [CrossRef]
121. Marković, R.; Šefer, D.; Krstić, M.; Petrujkić, B. Effect of Different Growth Promoters on Broiler Performance and Gut Morphology. *Arch. Med. Vet.* **2009**, *41*, 163–169. [CrossRef]

122. Samanya, M.; Yamauchi, K. Histological Alterations of Intestinal Villi in Chickens Fed Dried Bacillus Subtilis Var. Natto. *Comp. Biochem. Physiol. A Mol. Integr. Physiol.* **2002**, *133*, 95–104. [CrossRef]
123. Oh, J.K.; Pajarillo, E.A.B.; Chae, J.P.; Kim, I.H.; Kang, D.K. Protective Effects of *Bacillus Subtilis* against *Salmonella* Infection in the Microbiome of Hy-Line Brown Layers. *Asian-Australas. J. Anim. Sci.* **2017**, *30*, 1332–1339. [CrossRef]
124. Gibson, G.R.; Hutkins, R.; Sanders, M.E.; Prescott, S.L.; Reimer, R.A.; Salminen, S.J.; Scott, K.; Stanton, C.; Swanson, K.S.; Cani, P.D.; et al. Expert Consensus Document: The International Scientific Association for Probiotics and Prebiotics (ISAPP) Consensus Statement on the Definition and Scope of Prebiotics. *Nat. Rev. Gastroenterol. Hepatol.* **2017**, *14*, 491–502. [CrossRef]
125. Froebel, L.K.; Froebel, L.E.; Duong, T. Refined Functional Carbohydrates Reduce Adhesion of Salmonella and Campylobacter to Poultry Epithelial Cells in Vitro. *Poult. Sci.* **2020**, *99*, 7027–7034. [CrossRef]
126. Pandey, K.R.; Naik, S.R.; Vakil, B.V. Probiotics, Prebiotics and Synbiotics—A Review. *J. Food Sci. Technol.* **2015**, *52*, 7577–7587. [CrossRef]
127. Waqas, M.; Nastoh, N.; Çinar, A.; Farooq, M.; Salman, M. Advantages of the Use of Postbiotics in Poultry Production: A New Concept. *Braz. J. Poult. Sci.* **2024**, *26*, eRBCA-2024. [CrossRef]
128. The Power of Postbiotics for Poultry. Available online: https://phileo-lesaffre.com/en/the-power-of-postbiotics-for-poultry/ (accessed on 17 November 2024).
129. Postbiotics Effective in Reducing Salmonella Prevalence in Poultry, Says Phileo—INDUSTRY PERSPECTIVES. Available online: https://www.feedinfo.com/perspectives/postbiotics-effective-in-reducing-salmonella-prevalence-in-poultry-says-phileo-industry-perspectives/251513 (accessed on 17 November 2024).
130. The Role of Postbiotics in Promoting Gut Health in Poultry. Available online: https://www.feedandadditive.com/the-role-of-postbiotics-in-promoting-gut-health-in-poultry/ (accessed on 17 November 2024).
131. Mishra, B.; Mishra, A.K.; Mohanta, Y.K.; Yadavalli, R.; Agrawal, D.C.; Reddy, H.P.; Gorrepati, R.; Reddy, C.N.; Mandal, S.K.; Shamim, M.Z.; et al. Postbiotics: The New Horizons of Microbial Functional Bioactive Compounds in Food Preservation and Security. *Food Prod. Process. Nutr.* **2024**, *6*, 28. [CrossRef]
132. Girgis, G.; Powell, M.; Youssef, M.; Graugnard, D.E.; King, W.D.; Dawson, K.A. Effects of a Mannan-Rich Yeast Cell Wall-Derived Preparation on Cecal Concentrations and Tissue Prevalence of Salmonella Enteritidis in Layer Chickens. *PLoS ONE* **2020**, *15*, e0232088. [CrossRef] [PubMed]
133. Velez, E.; Castillo, N.; Mesón, O.; Grau, A.; Bibas Bonet, M.E.; Perdigón, G. Study of the effect exerted by fructo-oligosaccharides from yacon (Smallanthus sonchifolius) root flour in an intestinal infection model with Salmonella Typhimurium. *Br. J. Nutr.* **2013**, *109*, 1971–1979. [CrossRef] [PubMed]
134. Araba, M.; Girgis, G.; McBride, H.; Lohrmann, T. Effect of a Bacillus Subtilis plus Yeast Cell Wall Synbiotic on Salmonella Enteritidis Colonization in Ceca of Layer Pullets. *Poultry* **2024**, *3*, 26–35. [CrossRef]
135. Gomez-Osorio, L.M.; Yepes-Medina, V.; Ballou, A.; Parini, M.; Angel, R. Short and Medium Chain Fatty Acids and Their Derivatives as a Natural Strategy in the Control of Necrotic Enteritis and Microbial Homeostasis in Broiler Chickens. *Front. Vet. Sci.* **2021**, *8*, 773372. [CrossRef]
136. Van Immerseel, F.; Russell, J.; Flythe, M.; Gantois, I.; Timbermont, L.; Pasmans, F.; Haesebrouck, F.; Ducatelle, R. The Use of Organic Acids to Combat Salmonella in Poultry: A Mechanistic Explanation of the Efficacy. *Avian Pathol.* **2006**, *35*, 182–188. [CrossRef]
137. Abd El-Ghany, W.A. Applications of Organic Acids in Poultry Production: An Updated and Comprehensive Review. *Agriculture* **2024**, *14*, 1756. [CrossRef]
138. Hu, Z.; Liu, L.; Guo, F.; Huang, J.; Qiao, J.; Bi, R.; Huang, J.; Zhang, K.; Guo, Y.; Wang, Z. Dietary Supplemental Coated Essential Oils and Organic Acids Mixture Improves Growth Performance and Gut Health along with Reduces Salmonella Load of Broiler Chickens Infected with Salmonella Enteritidis. *J. Anim. Sci. Biotechnol.* **2023**, *14*, 95. [CrossRef]
139. Böhme, K.; Barros-Velázquez, J.; Calo-Mata, P.; Aubourg, S.P. Antibacterial, Antiviral and Antifungal Activity of Essential Oils: Mechanisms and Applications. In *Antimicrobial Compounds: Current Strategies and New Alternatives*; Springer: Berlin/Heidelberg, Germany, 2014; pp. 51–81. ISBN 9783642404443.
140. Essential Oils for Chickens: The (Science-Backed) Natural Approach For Flock Health and Productivity. Available online: https://Rosehillfarm.ca/2020/07/31/Essential-Oils-for-Chickens/ (accessed on 12 November 2024).
141. Andreatti Filho, R.L.; Higgins, J.P.; Higgins, S.E.; Gaona, G.; Wolfenden, A.D.; Tellez, G.; Hargis, B.M. Ability of Bacteriophages Isolated from Different Sources to Reduce Salmonella Enterica Serovar Enteritidis in Vitro and in Vivo. *Poult. Sci.* **2007**, *86*, 1904–1909. [CrossRef]
142. Atterbury, R.J.; Van Bergen, M.A.P.; Ortiz, F.; Lovell, M.A.; Harris, J.A.; De Boer, A.; Wagenaar, J.A.; Allen, V.M.; Barrow, P.A. Bacteriophage Therapy to Reduce Salmonella Colonization of Broiler Chickens. *Appl. Environ. Microbiol.* **2007**, *73*, 4543–4549. [CrossRef] [PubMed]
143. Borie, C.; Albala, I.; Sànchez, P.; Sánchez, M.L.; Ramírez, S.; Navarro, C.; Morales, M.A.; Retamales, J.; Robeson, J. Bacteriophage Treatment Reduces Salmonella Colonization of Infected Chickens. *Avian Dis.* **2008**, *52*, 64–67. [CrossRef] [PubMed]
144. Colom, J.; Cano-Sarabia, M.; Otero, J.; Cortés, P.; Maspoch, D.; Llagostera, M. Liposome-Encapsulated Bacteriophages for Enhanced Oral Phage Therapy against *Salmonella* spp. *Appl. Environ. Microbiol.* **2015**, *81*, 4841. [CrossRef] [PubMed]
145. O'Sullivan, L.; Bolton, D.; McAuliffe, O.; Coffey, A. Bacteriophages in Food Applications: From Foe to Friend. *Annu. Rev. Food Sci. Technol.* **2019**, *10*, 151–172. [CrossRef]

146. Wójcik, E.A.; Stańczyk, M.; Wojtasik, A.; Kowalska, J.D.; Nowakowska, M.; Łukasiak, M.; Bartnicka, M.; Kazimierczak, J.; Dastych, J. Comprehensive Evaluation of the Safety and Efficacy of BAFASAL® Bacteriophage Preparation for the Reduction of Salmonella in the Food Chain. *Viruses* **2020**, *12*, 742. [CrossRef]
147. Clavijo, V.; Baquero, D.; Hernandez, S.; Farfan, J.C.; Arias, J.; Arévalo, A.; Donado-Godoy, P.; Vives-Flores, M. Phage Cocktail SalmoFREE® Reduces Salmonella on a Commercial Broiler Farm. *Poult. Sci.* **2019**, *98*, 5054. [CrossRef]
148. Żbikowska, K.; Michalczuk, M.; Dolka, B. The Use of Bacteriophages in the Poultry Industry. *Animal* **2020**, *10*, 872. [CrossRef]
149. Nurmi, E.; Nuotio, L.; Schneitz, C. The Competitive Exclusion Concept: Development and Future. *Int. J. Food Microbiol.* **1992**, *15*, 237–240. [CrossRef]
150. Gast, R.K. Serotype-Specific and Serotype-Independent Strategies for Preharvest Control of Food-Borne Salmonella in Poultry. *Avian Dis.* **2007**, *51*, 817–828. [CrossRef]
151. Schneitz, C. Competitive Exclusion in Poultry—30 Years of Research. *Food Control* **2005**, *16*, 657–667. [CrossRef]
152. Nisbet, D. Defined Competitive Exclusion Cultures in the Prevention of Enteropathogen Colonisation in Poultry and Swine. *Antonie Van Leeuwenhoek* **2002**, *81*, 481–486. [CrossRef]
153. Bar-Shira, E.; Friedman, A. Development and Adaptations of Innate Immunity in the Gastrointestinal Tract of the Newly Hatched Chick. *Dev. Comp. Immunol.* **2006**, *30*, 930–941. [CrossRef] [PubMed]
154. Methner, U.; Haase, A.; Berndt, A.; Martin, G.; Nagy, B.; Barrow, P.A. Exploitation of Intestinal Colonization-Inhibition between Salmonella Organisms for Live Vaccines in Poultry—Potential and Limitations. *Zoonoses Public Health* **2011**, *58*, 540–548. [CrossRef] [PubMed]
155. Braukmann, M.; Barrow, P.A.; Berndt, A.; Methner, U. Combination of Competitive Exclusion and Immunisation with a Live Salmonella Vaccine in Newly Hatched Chickens: Immunological and Microbiological Effects. *Res. Vet. Sci.* **2016**, *107*, 34–41. [CrossRef]
156. Cox, J.M.; Pavic, A. Advances in Enteropathogen Control in Poultry Production. *J. Appl. Microbiol.* **2010**, *108*, 745–755. [CrossRef] [PubMed]
157. Ferreira, A.J.P.; Ferreira, C.S.A.; Knobl, T.; Moreno, A.M.; Bacarro, M.R.; Chen, M.; Robach, M.; Mead, G.C. Comparison of Three Commercial Competitive-Exclusion Products for Controlling Salmonella Colonization of Broilers in Brazil. *J. Food Prot.* **2003**, *66*, 490–492. [CrossRef]
158. Calenge, F.; Kaiser, P.; Vignal, A.; Beaumont, C. Selection Evolution Open Access REVIEW BioMed Central Genetic Control of Resistance to Salmonellosis and to Salmonella Carrier-State in Fowl: A Review. *Genet. Sel. Evol.* **2010**, *42*, 1–11. [CrossRef]
159. Li, X.; Nie, C.; Liu, Y.; Chen, Y.; Lv, X.; Wang, L.; Zhang, J.; Li, K.; Jia, Y.; Ban, L.; et al. A Genome-Wide Association Study Explores the Genetic Determinism of Host Resistance to Salmonella Pullorum Infection in Chickens. *Genet. Sel. Evol.* **2019**, *51*, 1–12. [CrossRef]
160. Jie, H.; Liu, Y.P. Breeding for Disease Resistance in Poultry: Opportunities with Challenges. *Worlds Poult. Sci. J.* **2011**, *67*, 687–696. [CrossRef]
161. Gul, H.; Habib, G.; Khan, I.M.; Rahman, S.U.; Khan, N.M.; Wang, H.; Khan, N.U.; Liu, Y. Genetic Resilience in Chickens against Bacterial, Viral and Protozoal Pathogens. *Front. Vet. Sci.* **2022**, *9*, 1032983. [CrossRef]
162. Castro-Vargas, R.E.; Herrera-Sánchez, M.P.; Rodríguez-Hernández, R.; Rondón-Barragán, I.S. Antibiotic Resistance in *Salmonella* spp. Isolated from Poultry: A Global Overview. *Vet. World* **2020**, *13*, 2070–2084. [CrossRef] [PubMed]
163. Wang, Y.; Ge, H.; Wei, X.; Zhao, X. Research Progress on Antibiotic Resistance of Salmonella. *Food Qual. Saf.* **2022**, *6*, fyac035. [CrossRef]
164. Ban on Antibiotics as Growth Promoters in Animal Feed Enters into Effect. Available online: https://ec.europa.eu/commission/presscorner/detail/en/ip_05_1687 (accessed on 8 November 2024).
165. FDA Announces Transition of Over-the-Counter Medically Important Antimicrobials for Animals to Prescription Status. Available online: https://Www.Fda.Gov/Animal-Veterinary/Cvm-Updates/Fda-Announces-Transition-over-Counter-Medically-Important-Antimicrobials-Animals-Prescription-Status (accessed on 12 November 2024).
166. Prevention, Detection and Control of Salmonella in Poultry. Available online: https://food.ec.europa.eu/system/files/2016-10/ia_standards_oie_78_eu-position_annex-xvii_revised-ch-6.5-salmonella-in-poultry.pdf (accessed on 14 November 2024).
167. Guidelines for the Control of Nontyphoidal *Salmonella* spp. in Beef and Pork Meat. Available online: https://www.fao.org/fao-who-codexalimentarius/sh-proxy/pt/?lnk=1&url=https%253A%252F%252Fworkspace.fao.org%252Fsites%252Fcodex%252FStandards%252FCXG+87-2016%252FCXG_087e.pdf (accessed on 3 December 2024).
168. Control of Salmonella. Available online: https://food.ec.europa.eu/food-safety/biological-safety/food-borne-diseases-zoonoses/control-salmonella_en#:~:text=Special%20guarantees%20for%20Salmonella%20in%20certain%20EU%20countries,-When%20the%20prevalence&text=Such%20guarantees%20include%20extended%20monitoring,(table%20eggs)%20and%20Norway (accessed on 14 November 2024).
169. Montoro-Dasi, L.; Lorenzo-Rebenaque, L.; Marco-Fuertes, A.; Vega, S.; Marin, C. Holistic Strategies to Control Salmonella Infantis: An Emerging Challenge in the European Broiler Sector. *Microorganisms* **2023**, *11*, 1765. [CrossRef] [PubMed]
170. NVAP Reference Guide: National Poultry Improvement Plan. Available online: https://www.aphis.usda.gov/nvap/reference-guide/poultry/npip (accessed on 14 November 2024).

171. Guidance for Industry: Questions and Answers Regarding the Final Rule, Prevention of Salmonella Enteritidis in Shell Eggs During Production, Storage, and Transportation. Available online: https://www.fda.gov/regulatory-information/search-fda-guidance-documents/guidance-industry-questions-and-answers-regarding-final-rule-prevention-salmonella-enteritidis-shell (accessed on 3 December 2024).
172. Egg Safety Final Rule. Available online: https://www.fda.gov/food/egg-guidance-regulation-and-other-information/egg-safety-final-rule (accessed on 3 December 2024).
173. Regulation (EC) No 2160/2003 of the European Parliament and of the Council of 17 November 2003 on the Control of Salmonella and Other Specified Food-Borne Zoonotic Agents. Available online: https://eur-lex.europa.eu/legal-content/en/ALL/?uri=CELEX:32003R2160 (accessed on 14 November 2024).
174. NPIP Salmonella Update 2024-History of Salmonella and the NPIP. Available online: https://www.poultryimprovement.org/documents/2024-GCC-Day-1-Waltman-NPIP-Salmonella-Update.pdf (accessed on 3 December 2024).
175. National Poultry Improvement Plan (NPIP). Available online: https://www.poultryimprovement.org/ (accessed on 14 November 2024).
176. Food and Drug Administration-CFR—Code of Federal Regulations Title 21. Available online: https://www.accessdata.fda.gov/scripts/cdrh/cfdocs/cfCFR/CFRSearch.cfm?fr=118.6&utm_source=chatgpt.com (accessed on 19 November 2024).
177. New Solutions for Better Salmonella Control in Europe. Available online: https://www.poultryworld.net/health-nutrition/health/new-solutions-for-better-salmonella-control-in-europe/#:~:text=Ceva%20Sant%C3%A9%20Animale%20registered%20a,Salmonella%20Enteritidis%20and%20Salmonella%20Typhimurium (accessed on 14 November 2024).
178. Ceva Introduces Its New Solutions for Better *Salmonella* Control in Europe. Available online: https://www.thepoultrysite.com/news/2023/10/ceva-introduces-its-new-solutions-for-better-salmonella-control-in-europe (accessed on 14 November 2024).
179. How to Protect Your Bird Backyard Biosecurity. Available online: www.oregon.gov/oda/shared/Documents/Publications/AnimalHealth/ProtectBirdsPoster.pdf (accessed on 14 November 2024).
180. Why You Should Only Buy Baby Chicks from NPIP-Certified Hatcheries. Available online: https://www.freedomrangerhatchery.com/blog/understanding-npip-certification-standards-for-poultry/#:~:text=By%20obtaining%20an%20NPIP-certified,of%20the%20birds%20they%20produce (accessed on 14 November 2024).
181. Salmonella, Mycoplasma and Avian Influenza Monitoring in Parent Breeder Flocks. Available online: https://www.hyline.com/ViewFile?id=a8f0f50a-81c9-48a7-99e5-e4a84eaf40ba (accessed on 14 November 2024).
182. *Salmonella* Control: Comparing US and EU Regulations for Poultry. Available online: https://xtalks.com/salmonella-control-comparing-us-and-eu-regulations-for-poultry-3750/#:~:text=The%20EU%20takes%20a%20different,products%20reach%20foodservice%20or%20retail (accessed on 14 November 2024).
183. Salmonella. Available online: https://www.efsa.europa.eu/en/topics/topic/salmonella#:~:text=Salmonella%20is%20a%20bacterium%20that,indirectly%20between%20animals%20and%20humans (accessed on 14 November 2024).
184. CDC PulseNet Outbreak Detection. Available online: https://www.cdc.gov/pulsenet/hcp/about/outbreak-detection.html#:~:text=PulseNet%20rapidly%20identifies%20potential%20outbreaks,assist%20epidemiologists%20in%20investigating%20outbreaks (accessed on 14 November 2024).
185. CDC Foodborne Diseases Active Surveillance Network. Available online: https://www.cdc.gov/foodnet/reports/preliminary-data.html (accessed on 14 November 2024).
186. Rizzi, V.; Boelaert, F.; Mäkelä, P.; Takkinen, J.; Ammon, A. Decrease in Human Salmonellosis in the European Union and Reduction of the *Salmonella* Prevalence in Poultry Populations. Available online: https://www.google.com.tw/url?sa=t&source=web&rct=j&opi=89978449&url=https://www.efsa.europa.eu/sites/default/files/assets/corporate101111-ax3.pdf&ved=2ahUKEwi-x8rQhZCKAxXx97sIHcpyA8cQFnoECBkQAQ&usg=AOvVaw3jaYGbI9hyWgvoAMds5Q1S (accessed on 3 December 2024).
187. Checklist for Biosecurity Training. Available online: https://www.aphis.usda.gov/sites/default/files/2023-11/fsc-birds-checklist-biosecurity-training.pdf (accessed on 15 November 2024).
188. Alternatives to Antibiotics (ATA) Challenges and Solutions in Animal Production. Available online: https://www.ars.usda.gov/alternativestoantibiotics/ResearchCenter/ATAindex.html (accessed on 14 November 2024).
189. Kumar, M.; Sarma, D.K.; Shubham, S.; Kumawat, M.; Verma, V.; Nina, P.B.; JP, D.; Kumar, S.; Singh, B.; Tiwari, R.R. Futuristic Non-Antibiotic Therapies to Combat Antibiotic Resistance: A Review. *Front. Microbiol.* **2021**, *12*, 609459. [CrossRef]
190. Ayana, G.U.; Kamutambuko, R. Probiotics in Disease Management for Sustainable Poultry Production. *Adv. Gut Microbiome Res.* **2024**, *2024*, 4326438. [CrossRef]
191. Antunes, P.; Mourão, J.; Campos, J.; Peixe, L. Salmonellosis: The Role of Poultry Meat. *Clin. Microbiol. Infect.* **2016**, *22*, 110–121. [CrossRef]
192. Halder, N.; Sunder, J.; De, A.K.; Bhattacharya, D.; Joardar, S.N. Probiotics in Poultry: A Comprehensive Review. *J. Basic. Appl. Zool.* **2024**, *85*, 23. [CrossRef]

193. Probiotic Intervention to Prevent *Salmonella* Infection in Poultry. Available online: https://today.uconn.edu/2021/06/probiotic-intervention-to-prevent-salmonella-infection-in-poultry/ (accessed on 14 November 2024).
194. Hong, H.A.; Le, H.D.; Cutting, S.M. The Use of Bacterial Spore Formers as Probiotics. *FEMS Microbiol. Rev.* **2005**, *29*, 813–835. [CrossRef]

animals MDPI

Article

Risk Factors for *Salmonella* Detection in Commercial Layer Flocks in Spain

Clara Samper-Cativiela [1,2], Maria Esther Prieto [3], Soledad Collado [3], Cristina De Frutos [3], Adam J. Branscum [4], Jose Luis Saez [3] and Julio Alvarez [1,2,*]

1 VISAVET Health Surveillance Centre, Universidad Complutense, 28040 Madrid, Spain
2 Department of Animal Health, Faculty of Veterinary Medicine, Universidad Complutense, 28040 Madrid, Spain
3 Ministerio de Agricultura, Pesca y Alimentación, 28010 Madrid, Spain
4 Biostatistics Program, School of Biological and Population Health Sciences, Oregon State University, Corvallis, OR 97331, USA
* Correspondence: jalvarez@visavet.ucm.es

Simple Summary: Foodborne salmonellosis remains one of the top zoonotic diseases affecting public health worldwide, and its incidence has remained stable in the last years in the European Union (EU) triggering questions on the usefulness of currently available measures to prevent its occurrence. A main focus of *Salmonella* national control programs is monitoring the presence of the bacteria in animal reservoirs, especially in poultry, and for this reason, thousands of samples are collected every year in poultry farms in EU countries, but the importance of certain factors in the probability of detecting *Salmonella* remains poorly understood. A thorough analysis conducted on data collected in all laying hen flocks sampled in Spain in 2015–2020 revealed that even though the presence of *Salmonella* was rare (<3.5% of positive sampling events), when samples were collected in certain months (fall–winter) and housing systems (caged flocks) and by competent authorities (as opposed to food business operators), the probability of detecting *Salmonella* increased significantly. These results demonstrate that the sensitivity of the sampling strategy may be influenced by how and when samples are collected and that certain flocks may be at an increased risk of infection.

Abstract: Trends in *Salmonella* human infections are assumed to be related to the distribution of the pathogen in the animal reservoir/food products, and cases in humans are most often linked to poultry and poultry products (eggs, meat). Therefore, ongoing *Salmonella* national control programs (NCPs) in European Union Member States have the objective of monitoring and reducing its prevalence in commercial poultry flocks. Results from NCPs have shown certain factors (housing systems, season of sampling and if sampling is conducted by food business operators (FBOps) or competent authorities (CAs), among others) can influence detection rates, but associations are often not consistent. Here, we analyzed data from the Spanish NCP on 7216 laying hen flocks subjected to 36,193 sampling events over a six-year period to characterize its performance and identify variables influencing detection rates. Overall, 1205 sampling events were positive for *Salmonella* spp. (any serovar) and 132 for *S.* Enteritidis-*S.* Typhimurium/monophasic. Bayesian multivariable models adjusting for multiple covariates concluded that sampling events later in the year, in caged flocks with older animals and conducted by CAs had increased odds of positivity for *Salmonella* spp., revealing aspects linked with a differential estimation of *Salmonella* levels in laying hen flocks.

Keywords: *Salmonella*; poultry; laying hen; national control program; epidemiology; risk factor; Bayesian hierarchical model

1. Introduction

Despite significant efforts to prevent and control non-typhoidal *Salmonella enterica* subsp. *enterica* (NTS) infection, it remains one of the top causes of foodborne diseases in low-, medium- and high-income countries, with over 75,000 cases reported annually in Europe between 2016 and 2019 [1], although this likely represents a severe underestimation [2]. Even though the number of confirmed salmonellosis cases in the European Union (EU) decreased between 2008 and 2012, case counts have stabilized in recent years (2016–2020) [1].

Transmission of non-typhoidal salmonellosis is usually due to ingestion of contaminated food, with pork and, especially in the case of outbreaks, poultry products (in particular eggs and egg-based products) often identified as the main vehicles of infection [3–5]. Even though multiple NTS serovars can cause human infection, most cases are due to a relatively low number of them, with *S.* Enteritidis and *S.* Typhimurium being identified as the two top disease-causing serovars in Europe and worldwide [1]. Because of the perceived importance of poultry as a source of NTS, numerous countries have implemented control programs to monitor the presence of *Salmonella* in production flocks and prevent the transmission of NTS through the food chain. In Europe, the establishment of national control programs (NCPs) for *Salmonella* serovars of "public health significance" in all relevant stages of production, processing and distribution in poultry is stipulated in Regulation (EC) 2160/2003. In the case of laying hens of *Gallus gallus*, responsible for an estimated 5.7–22.2% of all human salmonellosis cases in the EU in 2016 [6], the goal of NCPs is to maintain the percentage of positive flocks of adult laying hens at $\leq$2% or to ensure an annual percent reduction in prevalence of at least 10% (if prevalence in the previous year was below 10%), $\geq$20% (if prevalence in the previous year was between 10% and 20%), $\geq$30% (if prevalence in the previous year was between 20% and 40%) or $\geq$40% (if prevalence in the previous year was at or above 40%) according to Commission Regulation 517/2011. Commission Regulation 517/2011 also lists the serovars with public health significance in the case of laying hen flocks (*S.* Enteritidis and *S.* Typhimurium including its monophasic variant with the antigenic formula 1,4,[5],12:i:-, referred from here on as target serovars).

Control of *Salmonella* infection in laying hen flocks is complicated by its complex dynamics in infected flocks, which are influenced by multiple factors including management and host factors and the multiple possible sources of infection, such as entry of new (infected) animals or contaminated eggs, domestic and wild reservoirs, insects and contaminated equipment, feed or water [7]. The use of antimicrobials to control *Salmonella* infection in poultry is not allowed in the EU, and therefore, adequate biosecurity measures and vaccination are the main tools to prevent the establishment of NTS infections in layer flocks [8]. Once NTS is present in a farm, infected animals may act as asymptomatic carriers, perpetuating the infection in the flock and further contaminating the environment and the food products produced [9].

According to Commission Regulation 517/2011, detection of the presence of NTS in layer flocks by NCPs is based on periodical sampling events conducted by the food business operator (FBOp) and the competent authorities (CAs) at different frequencies (see Material and Methods). In addition, the CA must also carry out additional sampling in other situations: in flocks present in a farm where a flock has tested positive for a target serovar or in new flocks housed in the same building where a previously positive flock was present; in the case of suspicion due to, for example, foodborne outbreak investigation; and in any other situation where the CA considers it appropriate. Sampling protocols vary depending on the housing system (see Materials and Methods). The different sampling strategies, along with factors such as the housing system (e.g., caged vs. cage-free flocks), flock size and flock age, have been shown to influence the probability of positive test results [6,10,11], which could be attributable to both an increased prevalence of infection in certain flocks/periods and variability in the performance of different surveillance protocols [12]. The sampler (FBOp or CA) has also been shown to influence results from the EU *Salmonella* NCPs in poultry housing systems in which separate reporting is mandatory (broilers and fattening and breeding turkeys), with samples collected by CAs having a significantly higher likelihood of a test positive result, prompting the need to conduct further investigations to define the reasons behind such discrepancies [1]. However, depending on the country, separate data for FBOps and CAs is also available for other poultry categories, such as laying hens in Spain. Currently, it is unclear whether this is also the case in laying hens since there is no requirement for separate reporting of CA/FBOp data, an issue worth investigating as recently recommended by EFSA and ECDC [1].

In this context, we conducted an analysis of the data from NCPs performed in laying hen flocks in Spain from 2015 to 2020 to (1) characterize the sampled population and the frequency of detection of *Salmonella* spp. and of target serovars and (2) identify factors associated with increased probability of *Salmonella* detection related to the flock and the sampling strategy while accounting for other potentially influencing factors.

2. Materials and Methods

2.1. Study Population

This study was performed using data from the NCP for *Salmonella* in Spain carried out between 2015 and 2020 according to the European and national legislation (Commission Regulation 517/2011, Spanish Royal Decrees 328/2003 and 637/2021 and the annually approved NCP for *Salmonella*). Briefly, these NCPs are based on periodical sampling events conducted by the FBOp (according to the EU guidelines, adult flocks of laying hens are sampled at least every 15 weeks, with the first sampling in the production stage taking place at a flock age of 24 ± 2 weeks, but the NCP for Salmonella in Spain starts with 1-day-old birds, with sampling events already at the rearing stage) or the CA (that will sample at least one adult flock per farm comprising at least 1000 birds, preferably at the end of the production phase, in addition to other scenarios—see below "reasons for sampling"). Samples collected may include pooled feces (in birds housed in cages), boot swabs (in barn or free-range houses) and dust samples (collected either directly or through the use of fabric swabs). Samples are then transported to authorized laboratories where they will be processed for the isolation of *Salmonella* according to ISO6579-1:2003 and ISO6579-1:2017, the method recommended by the European Union Reference Laboratory for *Salmonella* in fecal and environmental samples, as described in Commission Regulation 517/2011. If a positive result is found (i.e., bacterial growth is obtained), at least the serotyping for target serovars must be carried out to confirm/discard the presence of *S.* Enteritidis and *S.* Typhimurium (including its monophasic variant).

The database analyzed here contained information on all samples collected from laying hen flocks for the detection of *Salmonella* spp. using bacteriology and included samples collected by CA and FBOp. The database included information on several variables at the different hierarchical levels:

- Farm level: location of the farm (municipality);
- Flock level: size of the flock (number of birds) and housing type;
- Sampling level: date of sampling (day), reason for sampling (see below), sampler (CA or FBOp), number of samples analyzed independently and age of the flock at sampling;
- Sample level: sample type (i.e., specimen) and laboratory results from the sample (isolation of *Salmonella* and, when positive, if a target serotype—Enteritidis or Typhimurium—was identified).

2.2. Data Preparation

Samples of a given type (e.g., boot swabs, feces and dust) collected on the same day were grouped in sampling events, which represented the unit of analysis. The hierarchical structure of the data therefore had three levels: farms, flocks (with farms typically housing multiple flocks over the study period) and sampling events (with flocks typically being sampled multiple times over the study period). In this study, flocks were defined as a group of birds entering the same day in a farm and housed in the same barn so that they share the same air space. Prior to data analysis, variables collected at the different levels were categorized/recategorized as follows:

Farm-level variables: Since very few farms were sampled per municipality or province, location was considered at the autonomous region (from here on, "region") level. Regions with less than 50 farms were collapsed into a single category ("Region 8").

Flock-level variables: Size of flock was categorized into quartiles by using the empirical 25th, 50th and 75th percentiles as cutoffs. Housing type was analyzed using four categories (birds housed in cages, floor, free-ranging or organic systems) and as a dichotomous variable (cage systems vs. other systems).

Sampling-level variable: Date of sampling was categorized by year as well as bimester (Jan–Feb, Mar–Apr, May–Jun, Jul–Aug, Sep–Oct and Nov–Dec). There were 13 options for reason for sampling, many of which were associated with a previous isolation of *Salmonella* (confirmation of a positive result, sampling events conducted in laying flocks (of age 24 ± 2 weeks) housed in premises in which the previous flock was positive, control of all other flocks after a positive flock has been detected in a farm, epidemiological investigations due to connections with a positive flock/foodborne outbreak) or other reasons (e.g., tests of antimicrobial usage). Given that the objective of the study was to identify variables associated with *Salmonella* detection in routine sampling events conducted in the frame of the NCP (i.e., in the absence of any suspicion) and that all the previously mentioned categories accounted for a very small proportion of all observations in the initial database (<2.5%), they were excluded from further analysis leaving only environmental (i.e., sampling conducted to verify the efficacy of cleaning and disinfection practices conducted by the FBOp) and periodical sampling as possible values. In the analyses, these were combined with the sampler (CA/FBOp) variable so that three categories remained: FBOp_periodical, FBOp_environmental and CA_periodical. Environmental sampling events conducted by the CA were excluded since these were typically performed after cleaning and disinfection of *Salmonella*-positive flocks, while environmental samplings conducted by FBOp were maintained since they were not typically linked with positive flocks but conducted usually after cleaning and disinfection once a barn had been emptied before the new flock arrived irrespective of the status of the emptied flock. Thus, this category was maintained separately in the analysis to detect possible variations in detection rates in this context.

For age, due to the detection of abnormally low and high recorded values and the lack of variation in the age of a single flock at different FBOp sampling events in some cases in the database, a quality check was conducted. This consisted of comparing the expected age of the birds (based on the difference between the sampling date and the date of entry of the animals assuming they entered at 17 weeks of age) and the recorded age at the time of first CA sampling or, for flocks in which no CA sampling had been conducted, in the first FBOp sampling. Only observations for which the expected and the recorded date differed by no more than 6 weeks (an arbitrary threshold selected to exclude only records where a significant deviation was suspected) were maintained in the database, and the recorded age in the initial CA/FBOp sampling was used as a reference to calculate the age of the flock at subsequent sampling events by adding the difference (days) between the first sampling and each subsequent sampling.

Sample type, once sampling events due to non-routine reasons were excluded, could be one of boot swabs, feces, dust and fabric swabs, and each sampling could involve the collection of one or more samples. Finally, the laboratory results were categorized into two dichotomous variables, each based on whether one or more of the samples collected in the frame of a given sampling yielded a positive result: isolation of (any serovar of) *Salmonella* (yes/no) and isolation of a target serovar, i.e., Typhimurium—including its monophasic variant—or Enteritidis (yes/no).

Variables available for the study and their nature are summarized in Table S1.

2.3. Data Analysis

Two models with different response variables were considered for data analysis: one based on isolation of *Salmonella* (any serovar) and the other based on isolation of target serovar (only Enteritidis or Typhimurium including its monophasic variant). In order to assess the relationship between the available covariates at different hierarchical levels and the response variables of interest, a Bayesian mixed-effects logistic regression model was fitted that accounted for the hierarchical structure of the data and within-farm and within-flock correlations. The laboratory result (positive/negative) obtained in sampling "*i*" conducted in flock "*j*" housed in farm "*k*" was assumed to follow a Bernoulli distribution:

$$y_{i,j,k} \sim \text{Bernoulli}\left(p_{i,j,k}\right),$$

with $p_{i,j,k}$ being the probability of obtaining a positive laboratory result (isolation of *Salmonella* spp. or isolation of a target serovar). This probability was modeled as a function of the variables available at the sampling level and, in order to account for the dependence between flocks housed in the same farm and of sampling events conducted in the same flock, we included random effects for farm and flock (nested in farm):

$$\text{logit}\left(p_{i,j,k}\right) = U_{j,k} + X_{i,j,k}\beta$$

where $X_{i,j,k}$ denotes the vector of covariates (e.g., date of sampling) for sampling i in flock j from farm k, with β being the corresponding vector of regression coefficients. The random effect for flock-level ($U_{j,k}$) was modeled as

$$U_{j,k} \sim Normal\left(\mu\_flock_{j,k}, \sigma_{\text{flock}}\right),$$

with mean that depends on a vector of flock-level covariates ($Z_{j,k}$, e.g., housing type), with corresponding coefficient vector γ and a farm-specific term α_k:

$$\mu\_flock_{j,k} = \alpha_k + Z_{j,k}\gamma$$

Finally, the farm-level random effect was modeled as a function of the only covariate available at that level, namely the farm location (V_k):

$$\alpha_k \sim Normal(\mu\_farm_k, \sigma_{\text{farm}}),$$

where $\mu\_farm_k = \delta_0 + \delta_1 V_k$.

The variable number of analyses was excluded from the multivariable analysis due to its correlation with the sampler (93.6% of samples conducted by CA involving more than one sample vs. 1.1% in the case of those conducted by FBOp).

Independent diffuse Gaussian prior distributions with mean 0 and variance 10 restricted to the (−10, 10) range were assumed for the regression coefficients β, γ and δ, while uniform distributions U(0.01, 10) were used for the variance components σ_{flock} and σ_{farm}. The association between a given covariate and the probability of a positive test result for *Salmonella* was evaluated using the posterior medians and 95% posterior probability intervals (PPIs) for odds ratios. Variables were included in the final model depending on their association (based on the 95% PPI) and the deviance information criteria (DIC), such that the model with the lowest DIC was preferred [13]. The predictive ability of the final model was assessed via a receiver operating curve (ROC) using the model-based predicted probability of the model as the continuous response and the observed binary outcome as the classification variable.

Models were fitted in OpenBUGS 3.2.3 [14] through the R2OpenBUGS package [15] in R 4.0.2 [16]. Markov Chain Monte Carlo (MCMC) runs were performed using three chains with different randomly selected initial values, and chains were thinned to avoid autocorrelation by selecting one in every 10 consecutive samples. Convergence was assessed visually by examining the mixing of the chains and more formally using the Gelman–Rubin statistic [17,18]. Models were run for 2500 iterations after discarding the first 500 burn-in samples. Model checking was performed through posterior predictive simulations in which the observed data were compared with the replicated datasets generated from the posterior predictive distribution of the final model across different levels of the predictors considered.

3. Results

3.1. Descriptive Results

Upon exclusion of non-routine sampling events and flocks in which the age recorded was not considered reliable, the database analyzed contained initially 36,193 sampling events conducted between January 2015 and December 2020 in 7216 flocks housed in 1153 farms. Cataluña, Castilla y Leon, Andalucia and Castilla la Mancha were the autonomous regions with the highest number of sampling events (55.9% of all sampling events), flocks (53.0% of all flocks) and farms (53.3% of all farms), although the relative contribution of each region depended on the level of the hierarchy in the database considered (Figure 1).

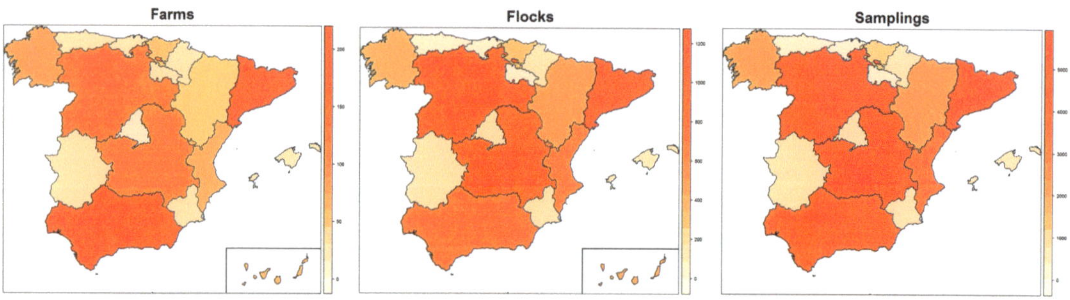

Figure 1. Number of farms, flocks and samplings analyzed in the frame of the national control program for Salmonella in laying hens between 2015 and 2020 in Spain per autonomous region.

Farms included in the study housed on average 6.3 flocks over the study period (median = 4, IQR = 2–8), and flocks were sampled on average five times while present in the farm (median = 5, IQR = 3–7). The average size of the flocks was 26,316 birds (median = 10,000, IQR = 2500–32,070), and these were most commonly in the cage system (3833/7216, 53.1%%) followed by floor (20.3%), free-ranging (16.5%) and organic flocks (10.1%) (Table 1). However, these proportions were not constant over time, with organic and free-ranging systems becoming more common in the last years of the study period (Figure 2). Caged flocks were much larger than other categories (median size = 29,949 birds vs. 5100 for floor system flocks, 5253 for free-ranging and 1700 for organic).

Most of the sampling events were performed by the FBOp (32,794/36,193, 90.6%) and were most commonly periodical sampling (86.8% of all FBOp sampling events), with the remaining being environmental sampling events. The remaining 3399 sampling events were conducted by the CA and belonged to the periodical sampling category (Table 1). Sampling events conducted by the FBOp were generally based on the analysis of one pool made up of two individual samples (99.6% and 94.4% of all FBOp periodical and environmental sampling events, respectively), with the remaining sampling events considering the result of between two and nine independent bacteriological analyses. In contrast, sampling events performed by the CA typically (93.6% of all CA sampling events) considered the results of two independent bacteriological analyses (of a pool made up of two samples as in those conducted by the FBOp plus an additional sample analyzed separately). The annual number of sampling events conducted ranged between 5112 in 2015 and 7158 in 2020, with a clear increasing trend over the study period (Table 1, Figure S1). In each year, sampling events were most often conducted in the last two bimesters (between September and December). The average age of the sampled flocks was 62.8 weeks (median = 61, IQR = 39–83). Finally, samples collected consisted most often of fecal samples (used in 20,065 sampling events, 55.4%) followed by boot swabs (11,720, 32.4%), both of which were used in the vast majority of the periodical sampling events conducted by both the FBOp and the CA (all but 96 of the 31,880 periodical sampling events included in the database) (Table 1). Dust was the sample of choice in 3542 sampling events (all but 90 classified as environmental), and fabric swab samples were used in 866 sampling events, of which 860 were environmental (Table 1).

Table 1. Characteristics of farms, flocks and sampling events included in the *Salmonella* national control program in Spain 2015–2020 and proportion of those testing positive for *Salmonella* (any serovar and target serovars—Enteritidis, Typhimurium and its monophasic variant).

Level	Variable	Category	N (%)	*Salmonella* spp. Positive (%)	Target Serovar Positive (%)
Farm (n = 1153)	Region	Region 0	206 (17.9)	84 (40.8)	16 (7.8)
		Region 1	173 (15.0)	86 (49.7)	28 (16.2)
		Region 2	53 (4.6)	24 (45.3)	6 (11.3)
		Region 3	112 (9.7)	54 (48.2)	7 (6.3)
		Region 4	124 (10.8)	28 (22.6)	4 (3.2)
		Region 5	75 (6.5)	45 (60.0)	14 (18.7)
		Region 6	97 (8.4)	14 (14.4)	3 (3.1)
		Region 7	84 (7.3)	33 (39.3)	12 (9.1)
		Region 8	164 (14.2)	56 (34.1)	15 (9.1)
		Region 9	65 (5.6)	13 (20.0)	3 (4.6)
Flock (n = 7216)	Size	Q1 (<2500)	1810 (25.1)	170 (9.4)	36 (2.0)
		Q2 (2500–10,000)	1817 (25.2)	172 (9.5)	35 (1.9)
		Q3 (10,000–32,055)	1784 (24.7)	291 (16.3)	36 (2.0)
		Q4 (>32,055)	1805 (25.0)	205 (11.4)	16 (0.9)
	Housing type	Caged	3833 (53.1)	586 (15.3)	62 (1.6)
		Floor	1465 (20.3)	86 (5.9)	19 (1.3)
		Free-ranging	1188 (16.5)	86 (7.2)	29 (2.4)
		Organic	730 (10.1)	80 (11.0)	13 (1.8)
Sampling (n = 36,193)	Sampler/reason sampling	FBOp—periodical	28,481 (78.7)	730 (2.6)	23 (0.08)
		FBOp—env	4313 (11.9)	45 (1.0)	6 (0.14)
		CA—periodical	3399 (9.4)	430 (12.7)	103 (3.0)
	Year	2015	5112 (14.1)	148 (2.9)	8 (0.16)
		2016	5436 (15.0)	198 (3.6)	23 (0.42)
		2017	5756 (15.9)	199 (3.5)	21 (0.36)
		2018	6175 (17.1)	215 (3.5)	26 (0.42)
		2019	6556 (18.1)	224 (3.4)	29 (0.44)
		2020	7158 (19.8)	221 (3.1)	25 (0.35)
	Bimester	Jan–Feb	5796 (16.0)	148 (2.6)	11 (0.19)
		Mar–Apr	5792 (16.0)	151 (2.6)	13 (0.22)
		May–Jun	5910 (16.3)	159 (2.7)	18 (0.30)
		Jul–Aug	5903 (16.3)	170 (2.9)	19 (0.32)
		Sep–Oct	6512 (18.0)	305 (4.7)	32 (0.49)
		Nov–Dec	6280 (17.4)	272 (4.3)	39 (0.62)
	Age (weeks)	Q1 (<39)	8912 (24.6)	216 (2.4)	21 (0.24)
		Q2 (39–61)	8947 (24.7)	238 (2.7)	16 (0.18)
		Q3 (61–83)	9223 (25.5)	309 (3.4)	31 (0.34)
		Q4 (>83)	9111 (25.2)	442 (4.9)	64 (0.70)
	Number of analyses	1	32,645 (90.2)	801 (2.5)	40 (0.12)
		2–10	3548 (9.8)	404 (11.4)	92 (2.6)
	Sample type	Feces	20,065 (55.4)	898 (4.5)	64 (0.32)
		Boot swabs	11,720 (32.4)	249 (2.1)	59 (0.50)
		Dust	3542 (9.8)	48 (1.4)	7 (0.20)
		Fabric swab	866 (2.4)	10 (1.2)	2 (0.23)

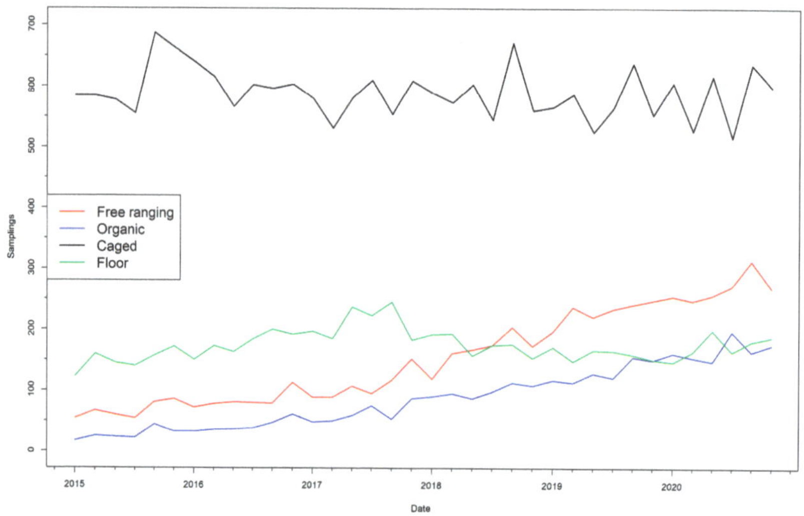

Figure 2. Number of sampling events per bimester conducted in flocks in the frame of the national control program for *Salmonella* in Spain between 2015 and 2020 depending on the housing system.

3.2. *Salmonella spp. Detection*

Salmonella spp. was detected in 1205 of the 36,193 sampling events analyzed throughout the whole study period (3.33% positivity). Detection rates over time depending on the sampler are shown in Figure S2. Among these cases, *S.* Enteritidis or Typhimurium (including its monophasic variant) were identified in 132 sampling events (Enteritidis in 81 sampling events, Typhimurium in 37 and the monophasic variant in 14), representing 0.36% of all sampling events and 11.0% of the *Salmonella* positive cases. Among isolates belonging to non-target serovars, in 382 sampling events (31.9%), no further serotyping was carried out once the presence of target serovars was discarded, and for those that were fully serotyped, the most frequent types were Infantis (retrieved in 136 sampling events, 16.5% of the 823 sampling events in which the *Salmonella* recovered were fully serotyped), Ohio (110 sampling events, 13.4%) and Corvallis (50 sampling events, 6.1%) (Table S2). More than one serovar was retrieved in 20 sampling events.

Out of the 1153 farms and 7216 flocks sampled at least once during the study period, 437 farms (37.9%) and 838 flocks (11.6%) were positive at least once. Among the locations with positive sampling events for target serovars, these numbers dropped to 108 farms (9.4%) and 123 flocks (1.7%). At the farm level, the percentage of positive farms varied widely depending on the region, with some regions having 40–60% of their farms testing positive for *Salmonella* spp. at least once while in others, the proportion of positive farms remained below 25%. When only sampling events positive to target serovars are considered, the proportion of positive farms per region dropped to 3.1–18.7% but was highly correlated with the proportion of positive samples to *Salmonella* (any serovar) (Pearson's rho = 0.83) (Figure S3). When considering the positivity at the flock level, similar regions had higher/lower proportions of positive flocks with again a high correlation between results for *Salmonella* and for target serovars only (Pearson's rho = 0.86) (Figure S3).

A clear trend in positive sampling events as a function of flock size was not observed, although lower proportions of positive sampling events for *Salmonella* spp. were found in smaller flocks (9.4% in flocks with <10,000 birds compared to 16.3% in flocks with 10,000–32,055 birds and 11.4% in those with >32,055 birds), while the proportion of positive sampling events for target serovars was consistently around 2.0% for flocks with <32,055 birds and around 0.9% for those above that size (Table 1). Regarding the **housing system**, a larger proportion of positive sampling events for *Salmonella* spp. was found in flocks housed in caged systems (586/3833, 15.3%). For target serovars, the highest proportion of positive sampling events was observed for free-ranging flocks (29/1188, 2.4%) (Table 1).

Finally, at the sampling level, differences in the proportions of positive sampling events both for *Salmonella* spp. and only for specific serovars were observed for multiple variables. In terms of the **sampler**, sampling events conducted by the CA were more often positive than those conducted by FBOps when considering periodical sampling events (12.7% vs. 2.6% for *Salmonella*, 3.0% vs. 0.08% for target serovars), while FBOp environmental sampling events yielded the lowest proportion of positive results for *Salmonella* spp. (1.0%) but not for target serovars (0.14%). Regarding the **sampling date**, the proportion of positive sampling events every year ranged between 2.9 and 3.6% for all *Salmonella* and between 0.16 and 0.44% for target serovars, with more positive sampling events detected in both cases in the last two bimesters (4.3–4.7% and 0.49–0.62 for *Salmonella* spp. and target serovars) than in the other four (<3% and <0.33%) (Table 1). Concerning the **age of the flock**, a higher prevalence of sampling events tested positive in older (>83 weeks) birds (4.9% positive for *Salmonella* spp. and 0.70% for target serovars) than in younger birds (Table 1). Similarly, and in terms of the **number of analyses**, sampling events based on more than one independent bacteriological analysis were more often positive for all *Salmonella* serovars (11.3%) and for target serovars (2.59%) than those involving a single laboratory analysis (2.5% and 0.12%, respectively, for *Salmonella* spp. and only target serovars). Finally, for the **sample type**, sampling events based on feces were more commonly positive (4.5%) for all serovars compared to those based on boot swabs (2.1%), dust (1.4%) or fabric swabs (1.2%), but differences were reduced when considering target serovars (0.2 to 0.50% positive rate regardless of the sample type) (Table 1).

3.3. Model Results

For the response variable of testing positive for any type of *Salmonella* spp. (yes/no), the final Bayesian mixed-effects logistic regression model included all variables considered except the size of the flock and the number of analyses (farm level: region; flock level: housing type as a dichotomous variable; sampling level: sampler reason for sampling, year, bimester, age of the flock at the sampling, sample type). Regression coefficients for the flock-level variables (flock size in quartiles and housing type) could not be estimated when the model included both, due in part to multicollinearity, and housing type was preferred and thus kept in the model. Furthermore, housing type was included in the model as a dichotomous variable (caged vs. other systems) due to the improved DIC over the model containing all four categories and because similar values were obtained for the coefficients of the organic, floor and free-ranging categories when these were considered separately. Finally, when including sample type in the model the dust/fabric swabs categories were merged since both had similar percentages of positivity (Table 1) and were used in environmental sampling events.

Farm location (region) was associated with an increased likelihood of finding positive results in sampling events (Table 2). Specifically, compared to the reference region (Region 0) three regions (Region 4, Region 6 and Region 9) had odds ratios indicating a lower risk (upper limit of the 95% PPI below 0.65), while the 95% PPI of the coefficients associated with Regions 2, 3, 7 and 8 barely included 1 (Table 2). In contrast, Region 5 to some extent and particularly Region 1 had an increased risk compared with the reference category, with part or all of their 95% PPI above 1, respectively (Table 1). This was reflected in the estimates of the farm-level random effect α, which was largely influenced by the region where the farm was located (Figure S4). At the flock level and regarding **housing type**, odds ratios greater than 1 were found in caged flocks (median OR = 2.07, 95% PPI 1.52–2.85) compared with the reference category (organic/free-ranging/floor systems) (Table 2), with flock-level random effects influenced by the region where the flock was located though with higher variability compared with the farm-level random effect (Figure S5).

Table 2. Estimates of the farm, flock and sampling-level covariates on the probability of obtaining a positive result for *Salmonella* in 36,193 sampling events conducted in the frame of the national control program in Spain in 2015–2020.

Level (n)	Variable	Category	Median β, γ, δ	Rhat	Median OR (95% PPI)
Farm (n = 1153)	Region	Region 0	Ref		
		Region 1	0.369	1.001	1.45 (0.97–2.13)
		Region 2	−0.539	1.001	0.58 (0.33–1.03)
		Region 3	−0.328	1.006	0.72 (0.45–1.1)
		Region 4	−1.590	1.004	0.2 (0.13–0.33)
		Region 5	0.132	1.001	1.14 (0.71–1.81)
		Region 6	−1.340	1.002	0.26 (0.15–0.46)
		Region 7	−0.419	1.003	0.66 (0.38–1.14)
		Region 8	−0.317	1.003	0.73 (0.48–1.11)
		Region 9	−1.085	1.006	0.34 (0.18–0.63)
Flock (n = 7216)	Housing type	Others	Ref		
		Caged	0.729	1.005	2.07 (1.52–2.85)
Sampling (n = 36,193)	Sampler/reason sampling	FBOp—periodical	Ref		
		FBOp—env	−1.142	1.001	0.32 (0.15–0.74)
		CA—periodical	1.893	1.001	6.64 (5.63–7.75)
	Year	2015	−0.400	1.005	0.67 (0.52–0.87)
		2016	−0.097	1.003	0.91 (0.71–1.15)
		2017	−0.045	1.002	0.96 (0.75–1.21)
		2018	0.081	1.002	1.08 (0.87–1.36)
		2019	0.066	1.001	1.07 (0.86–1.33)
		2020	Ref		
	Bimester	Jan–Feb	Ref		
		Mar–Apr	−0.180	1.001	0.84 (0.65–1.07)
		May–Jun	−0.195	1.002	0.82 (0.64–1.06)
		Jul–Aug	0.010	1.001	1.01 (0.79–1.29)
		Sep–Oct	0.531	1.002	1.70 (1.36–2.12)
		Nov–Dec	0.218	1.001	1.24 (0.99–1.56)
	Age (weeks)	Q1 (<39)	Ref		
		Q2 (39–61)	−0.039	1.002	0.96 (0.78–1.18)
		Q3 (61–83)	0.090	1.001	1.09 (0.90–1.33)
		Q4 (>83)	0.492	1.002	1.63 (1.35–2.00)
	Sample type	Boot swabs	Ref		
		Feces	−0.116	1.004	0.89 (0.65–1.23)
		Dust/fabric swab	−0.350	1.001	0.7 (0.31–1.51)

Finally, at the sampling level, a strong effect of the variables **sampler, bimester and age of the flock** was observed, so that higher odds of obtaining a positive result in sampling events conducted by CAs (OR = 6.64, 95% PPI 5.63–7.75), between the months of September and December (95% PPI 1.36–2.12 for Sep–Oct and 0.99–1.56 for Nov–Dec) and in flocks with older (>83 weeks) birds (OR = 1.63, 95% PPI 1.35–2.00) were observed compared with the reference categories for each variable, while lower odds were observed for sampling events in 2015 compared with 2020 (Table 2, Figure 3). The model had a good predictive ability as demonstrated by the ROC curve (AUC = 94.2, 95%CI 93.7–94.8) and the precision–recall curve (Figure S6)

Model convergence was achieved as demonstrated in part by Gelman–Rubin statistics that were below 1.01 (Table 2) and two farm-level random effects. The model fitted the data well as demonstrated by the predictive checks plots, with the sum and standard deviation of the observed total number of positive samples falling well within the predicted values both for the total dataset (Figure S7) and for the different categories in all variables considered.

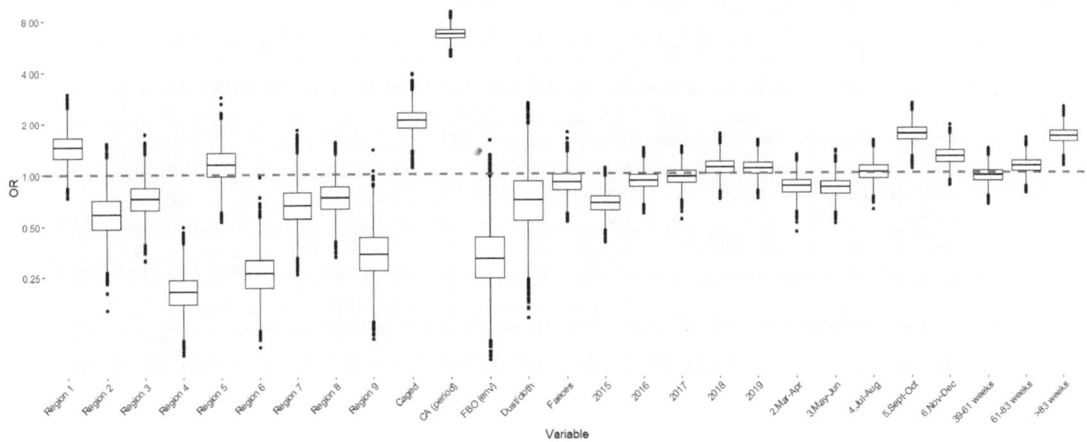

Figure 3. Estimates of the exponentiated coefficients (ORs) of the variables included in the final multivariable model using detection of *Salmonella* (any serovar) as the outcome and fitted with the data from *Salmonella* national control programs in Spain between 2015 and 2020 (*Y*-axis in the logarithmic scale).

For the response variable of testing positive for target serovars, we were unable to achieve model convergence. This could be partially because of the very low number of positive observations (i.e., sampling events in which target serovars were detected) in the database (0.4% of all sampling events). However, when the predictions generated by the model fitted with the detection of *Salmonella* (any serovar) as the outcome of interest were used to predict the positivity to target serovars, good sensitivity and specificity were still found as demonstrated by the ROC and precision–recall plots (Figure S8).

4. Discussion

The importance of poultry as a major source of foodborne salmonellosis has triggered in the past the implementation of control programs in multiple countries aiming at minimizing *Salmonella* infection in poultry production in order to prevent the contamination of meat and eggs [19]. In the EU, the application of these programs in laying hens, which currently aim at a reduction target of $\leq$2% positive flocks to serovars of public health importance, led to a steep reduction in flock prevalence of *Salmonella* spp. and target serovars between 2008 and 2014 (down to 2.1% and 0.9%, respectively), but since then, flock prevalence has increased slightly and stabilized around 3.3% (*Salmonella* spp.) and 1.2% (target serovars) [1]. In this context, and considering the ongoing changes experienced by the industry in the last decades (with past changes banning use of battery cages in favor of enriched cages and the current trend following the "End of cage age" initiative) [20], it is of paramount importance to identify factors associated with increased risk of *Salmonella* infection in poultry production, including the characteristics of the surveillance activities themselves (e.g., which samples are collected, by who, etc.) in order to optimize control programs. Here, we performed an exhaustive analysis of the data generated in the Spanish NCP to describe the characteristics of farms/flocks/sampling events positive for *Salmonella* spp. and for target serovars and also to identify factors associated with an increased probability of detection.

In this study, several factors at the farm, flock and sampling level were strongly associated with the risk of detecting *Salmonella* spp. Results obtained for the only variable at the farm level, i.e., the **location of the farm** (autonomous region), suggested that farms located in certain regions could experience baseline odds of positivity up to four times larger than others (Table 2). This geographic variation, which can be also observed at the

European level [1], could be related to farm-level characteristics not considered in the model (that is, different from flock size or housing type) but also to environmental conditions associated with certain regions such as more extreme temperatures, which could lead to increased stress and in turn increased risk of spread and persistence of *Salmonella* [21].

Among the flock-level covariables, **housing type** was strongly associated with an increased risk of detecting *Salmonella* spp., with flocks in cages experiencing a larger risk compared with other categories. Multiple observational studies to assess the effect of the housing system on the prevalence of *Salmonella* in laying hens have been conducted in the past with conflicting results, which could be due to differences in methodologies and sample sizes considered [22]. Similar results to those described here were found in a large EU baseline study in laying hens [23] and other smaller observational studies in several European countries [11,24–26], with hens raised in cages experiencing significantly higher odds of contamination with *Salmonella*, although these studies considered mostly laying hens in conventional (battery) cage systems, which were phased out in 2012. This increased risk in caged systems could be attributed to difficulties in effectively cleaning cages and a higher infectious pressure derived from higher animal densities leading to increased shedding, at least for some serovars [27–29]. In contrast, other studies have indicated no differences or even higher odds of *Salmonella* contamination in laying hens and/or eggs from hens housed in non-caged vs. caged systems [30–32], which could be due to an increased risk of *Salmonella* infection from environmental sources [33]. Our results demonstrate that, at the national level, currently the risk of *Salmonella* spp. detection in caged flocks is approximately twice as high (95% PPI 1.52–2.85) compared with non-caged systems; however, it was not possible to further discriminate between the risk among the different varieties included in the non-caged category (organic, free-range and barn housing systems).

The proportion of *Salmonella* spp.—positive sampling events was higher for flocks > 10,000 birds (though not when considering target serovars, Table 1), in agreement with previous work indicating that larger flocks (>20,000 or 30,000 birds) were subjected to increased risk of *Salmonella* detection [25,34]. However, **flock size** was not included in the multivariable model due to a lack of convergence when both this variable and housing type were considered at the flock level likely due to their high correlation, and housing type was preferred.

Vaccination of flocks with both *S.* Enteritidis and non-*S.* Enteritidis vaccines has been previously linked with a decreased risk of contamination of the farm compared with unvaccinated flocks [10]. In Spain, there are several *S.* Enteritidis and/or Typhimurium-based attenuated and inactivated vaccines that can be used in poultry, and in fact, vaccination is compulsory in pullets (rearing phase) of laying hens except in farms with adequate biosecurity and subjected to appropriate *Salmonella* controls (with negative results for target serovars) according to Commission Regulation 1177/2006. Unfortunately, no specific information regarding the vaccination status of the flocks included in the study was available, and therefore, the effect of this practice in our results could not be assessed here. Nevertheless, vaccination coverage in layer flocks in Spain is close to 100% (Ministerio de Agricultura, Pesca y Alimentación, personal communication), and therefore, it should not have had a major impact on the differential risks observed in certain cases.

Regarding variables at the sampling level, the analysis of the **sampler effect** (FBOp or CA) confirmed what had been already described in several EU Member States for broiler and breeder and fattening turkey farms [1]: a substantially increased odds of obtaining a positive result when a periodical sampling was conducted by CAs compared with FBOps (OR = 6.64, 95% PPI 5.63–7.75) was found once other known risk factors were accounted for in the model. This difference was even larger when compared with FBOp environmental sampling events (Table 2), which is not unexpected since these represent a different epidemiological scenario (barns recently emptied and subjected to cleaning and disinfection). This reflected the difference observed in the proportion of *Salmonella* spp. sampling events depending on the sampler (almost five times higher for CAs than for FBOps considering periodical sampling events, Table 1), which was nevertheless lower than what was found in the *Salmonella* NCPs in Spain in 2020 for other poultry species (2.1% CA vs. 0.07 FBOp in broiler flocks and 3.9% CA vs. 0% in FBOp in fattening turkeys, while no positive breeder flocks were detected by either) [1]. This difference could be at least in part due to the usual practice of analyzing independently more than one sample in sampling events conducted by the CA (93.6% considering more than one independent analysis) compared with those conducted by the FBOp (0.4% and 5.6% for periodical and environmental sampling events, respectively) since testing more samples can help to maximize the probability of detecting infected holdings as demonstrated in the EU baseline study, in which only one or two positive samples were detected in 38% of around 1540 *Salmonella* spp. positive flocks [23]. Sampling events conducted by the CA were more often focused on older flocks (e.g., 35% of all CA periodical sampling events were conducted in flocks > 83 weeks compared with 18% in FBOp periodical sampling events), but this should not have influenced the estimates provided here since the age of the flock was also included in the model. The same laboratory procedures for *Salmonella* detection are used regardless of the sampler (CA or FBOp) and regardless of the laboratory where this process is conducted (with some laboratories processing both samples collected by CAs and FBOps), and thus, it is unlikely that the sensitivity of the culture method is affected by this. However, the sensitivity of sampling events conducted by FBOps could be influenced by the more heterogeneous expertise of the personnel that can be involved in sample collection in FBOp sampling events compared with those conducted by CAs, which could in part also explain our results.

A strong effect of the **age** of the animals on the odds of positivity was also observed, with older (>83 weeks) hens having 1.63 (95% PPI 1.35–2.00) higher odds of testing positive for *Salmonella* spp. compared with birds < 39 weeks (Table 2). Previous studies have found similar results pointing at a higher risk of infection in older animals [24,35], which could be due to an increased bacterial load in the environment (accumulation of *Salmonella*), an extended opportunity to be exposed to the bacteria and/or increased susceptibility as animals age (coupled with a decrease in the protection conferred by vaccination administered as pullets), although this effect has been also linked with moulting (a practice not allowed in the EU) in some cases [36]. However, this effect was not found in some cases [7], highlighting the difficulty in comparing studies based on different populations (also regarding age ranges considered and ways of considering these in the analysis).

Both **the year and the time of the year** in which the sampling was carried out were also associated with the results. The year effect was mostly due to the lower risk of positive results for *Salmonella* spp. in sampling events performed in the first year of the study period (2015) compared with the reference (2020), while the odds of positivity were very similar for the rest of the study period (Table 2). In the case of season, however, a strong effect was observed for the latter months of each year (November–December and particularly September–October). A seasonal effect on the probability of foodborne salmonellosis and *Salmonella* infection in layer farms has been described in the past [6,37,38], though typically linked to summer months, which could be attributed in part to the thermal stress caused by higher temperatures that would lead to increased shedding, even though this would only apply to flocks in which the temperature inside the barn cannot be perfectly regulated (e.g., systems providing outdoor access). Nevertheless, this seasonal effect has been also connected to other causes linked with specific periods of the year that would affect the management or distribution of living vectors such as rodents or insects [7,8]. Indeed, sampling in winter was associated with increased detection of *Salmonella* in a study including five European countries, which was attributed to lower air quality in winter and a higher proportion of animals staying inside (for systems with outdoor sections) due to adverse climatic conditions [26]. Further studies are needed to better understand which changes could lead to this increased risk of detection in fall/early winter in Spain, which were also observed when comparing the raw proportion of positive sampling events for target serovars (Table 1).

The **type of sample** analyzed has been also previously found to influence *Salmonella* recovery rates in the laboratory, with dust yielding a higher proportion of positive samples compared with fecal samples [23,24,34], which could be due to the increased survivability of *Salmonella* in dust compared with other *Enterobacteriaceae* [12]. This is in contrast with our results, since dust (and fabric swab)-based sampling events had slightly lower odds of retrieving *Salmonella* spp. while boot swabs and fecal samples had a more similar performance (Table 2, Figure 3). This result is, however, not unexpected considering that 97.8% of the dust and fabric swab samples were collected in the frame of environmental sampling events conducted by FBOps, which yielded the lower proportion of positive sampling events for *Salmonella* spp. (Table 2). In any case, this result should be interpreted carefully since dust and fabric swabs represented only ~12% of the sampling-level dataset, and furthermore, this information was already partially accounted for in the sampler/reason for sampling (since almost all of them were identified as FBOp environmental sampling events, mostly performed for the verification of the cleaning and disinfection process). Moreover, an effect of the housing system in the performance of dust-based sampling has been described, with lower sensitivity in sampling events based on dust samples compared with feces in non-cage flocks [39], further highlighting the complex interaction between several covariables in the probability of *Salmonella* detection.

Only a small proportion (11.0%) of all positive sampling events detected in 2015–2020 was due to the presence of target serovars, in agreement with previous data from NCPs on breeding flocks in which target serovars (also including Hadar, Infantis and Virchow in this case) were detected in around 10% of all positive flocks in Spain in 2014–2016 [6]. Nevertheless, several of the non-target serovars retrieved (such as Infantis, Agona, Kentucky and Newport, found in 136, 29, 20 and 18 sampling events) were among the top 20 most frequent serovars in confirmed human cases described in Europe in 2020 [1], and in fact, serovars other than Enteritidis and Typhimurium/monophasic have been previously associated with foodborne *Salmonella* outbreaks linked with eggs, albeit at a much lower frequency compared with Enteritidis [40]. This, linked with the ability of the model fitted using *Salmonella* spp. detection as the outcome variable to predict the risk of detection of target serovars, suggests that results found here may have public health significance. Nevertheless, there may not be a clear linkage between the detection of several serovars that may enter the flock through very diverse routes such as environmental contamination and human cases, and therefore, the significance of our findings must be interpreted carefully. Still, multiple previous studies aiming at understanding the epidemiology of *Salmonella* in laying hens have also considered other serovars as the target variable often leading to the identification of shared risk factors with those increasing the risk of infection by target serovars [10], further supporting the approach followed here to identify variables to be considered for the optimization of surveillance and monitoring activities.

Several limitations must be considered in the interpretation of the results obtained in this study: two-way interactions were not included in the model to avoid increasing its complexity derived from the hierarchical structure of the dataset. In addition, information on the specific farm locations was not available for the analysis, and thus, spatial autocorrelation could not be formally considered in the analyses. Positive spatial autocorrelation in the probability of detection of *Salmonella* in broiler farms located within 20 km was previously demonstrated in a study in Brazil focusing on a single vertically integrated company located in a specific region of the country [41], and in fact, the strong association of the region with *Salmonella* detection found in our study points at the possible existence of other spatially structured risk factors that were not included in the analysis and that should be further investigated. Temporal autocorrelation in the detection probability was not formally considered in the model either, though the lack of independence between the observations was accounted for through the use of random effects at different levels in the model. These farm- and flock-level random effects contributed to explaining the sampling-level risk, further suggesting that factors other than those explicitly considered here were also associated with an increased probability of positivity. Other studies have demonstrated that a previous *Salmonella* infection was linked to an increased risk of detection of *Salmonella* [7,34], which could help explain farm (and flock)-specific higher baseline risks. In any case, the inclusion of random effects to account for the lack of independence between observations in models evaluating the probability of *Salmonella* detection in poultry has been widely used before [34,42–44], supporting the validity of the approach followed here, which was intended to allow for a global analysis of the country-level data.

5. Conclusions

This study represents the first comprehensive description and analysis of a multi-year database including all laying hen flocks in Spain subjected to routine sampling in the frame of the *Salmonella* NCPs.

Our results demonstrate that, even though *Salmonella* detection was rare (especially involving the target serovars Enteritidis and Typhimurium/monophasic), there were certain farm-level (location), flock-level (housing system) and sampling-level variables (age of animals, sample type, sampler entity, year and bimester) that had a strong influence on the probability of detecting *Salmonella* (with an increased risk in flocks housed in caged flocks with older animals and in sampling events conducted by the CA in the last months of the year). Some of these factors had been previously identified in other countries/regions of the world, sometimes considering different production systems. Therefore, our findings can serve as a first step in the identification of factors associated with a differential probability of detection of *Salmonella* in the frame of the NCP in Spain that can be useful both to maximize the sensitivity of sampling strategies and to implement management measures destined to minimize the risk of infection on flocks presenting certain risk factors.

Supplementary Materials: The following supporting information can be downloaded at https://www.mdpi.com/article/10.3390/ani13203181/s1, Figure S1: Number of samplings conducted in each bimester, Figure S2: Monthly percentage of positive sampling events depending on the sampler and type over the study period, Figure S3: Correlation between proportion of farms/flocks with one or more positive samplings for *Salmonella* spp. or for target serovars at the autonomous region level, Figure S4: Median estimates for the farm-level random effects (α, n = 1154) depending on the region where the farm was located, Figure S5: Median estimates for the flock-level random effects (U, n = 7219) depending on the region where the flock was located, Figure S6: ROC and precision–recall curves based on the probability of a positive result in a sampling for *Salmonella* (any serovar), Figure S7: Predictive checking plots showing the total number of positive samplings observed (1205) and the predicted by the final model in 6000 iterations, Figure S8: ROC and precision–recall curves based on the probability of a positive result in a sampling for *Salmonella* (target serovars), Table S1: Variables available and transformations considered in the analysis, Table S2: Number of positive sampling events per serovar.

Author Contributions: Conceptualization, C.S.-C., J.L.S. and J.A.; data curation, C.S.-C., M.E.P., S.C. and J.A.; formal analysis, C.S.-C. and J.A.; funding acquisition, J.L.S. and J.A.; investigation, C.S.-C., M.E.P., S.C., C.D.F. and J.A.; methodology, C.S.-C., A.J.B. and J.A.; project administration, J.L.S., S.C. and J.A.; resources, M.E.P., S.C., C.D.F. and J.L.S.; supervision, J.L.S., S.C. and J.A.; validation, A.J.B. and J.A.; visualization: C.S.-C. and J.A.; writing—original draft preparation, C.S.-C. and J.A.; writing—review and editing, all authors. All authors have read and agreed to the published version of the manuscript.

Funding: This research was funded by the European Union's Horizon 2020 Research and Innovation programme under grant agreement No 773830: One Health European Joint Programme and is a contribution to the Joint Research Project ADONIS and to project PID2021-125136OB-I00 (AMR-EPI-PLAS, Ministerio de Ciencia e Innovación).

Institutional Review Board Statement: Ethical review and approval were waived for this study since animals were not subjected to additional procedures other than those contemplated in the Spanish and European legislation.

Data Availability Statement: Data presented in this study are available upon reasonable request from the authors.

Conflicts of Interest: M.E. Prieto, S. Collado, C. de Frutos and J.L. Saez work for the Spanish Ministry of Agriculture, Food and Fisheries responsible for the *Salmonella* national control programs. The funders had no role in the design of the study; in the collection, analyses or interpretation of data; in the writing of the manuscript; or in the decision to publish the results.

References

1. EFSA; ECDC. The European Union One Health 2020 Zoonoses Report. *EFSA J.* **2021**, *19*, 324. [CrossRef]
2. Havelaar, A.H.; Ivarsson, S.; Lofdahl, M.; Nauta, M.J. Estimating the true incidence of campylobacteriosis and salmonellosis in the European Union, 2009. *Epidemiol. Infect.* **2013**, *141*, 293–302. [CrossRef] [PubMed]
3. Guillier, L.; Thebault, A.; Fravalo, P.; Mughini-Gras, L.; da Silva, N.J.; David, J.; Kooh, P.; Cadavez, V.; Gonzales-Barron, U. Risk factors for sporadic salmonellosis: A systematic review and meta-analysis. *Microb. Risk Anal.* **2021**, *17*, 11. [CrossRef]
4. Pijnacker, R.; Dallman, T.J.; Tijsma, A.S.L.; Hawkins, G.; Larkin, L.; Kotila, S.M.; Amore, G.; Amato, E.; Suzuki, P.M.; Denayer, S.; et al. An international outbreak of Salmonella enterica serotype Enteritidis linked to eggs from Poland: A microbiological and epidemiological study. *Lancet Infect. Dis.* **2019**, *19*, 778–786. [CrossRef] [PubMed]
5. Pires, S.M.; Vieira, A.R.; Hald, T.; Cole, D. Source attribution of human salmonellosis: An overview of methods and estimates. *Foodborne Pathog. Dis.* **2014**, *11*, 667–676. [CrossRef] [PubMed]
6. EFSA. Scientific opinion on the Salmonella control in poultry flocks and its public health impact. *EFSA J.* **2019**, *17*, 155.
7. Schulz, J.; Van Hoorebeke, S.; Hald, B.; Hartung, J.; Van Immerseel, F.; Radtke, I.; Kabell, S.; Dewulf, J. The dynamics of Salmonella occurrence in commercial laying hen flocks throughout a laying period. *Avian Pathol.* **2011**, *40*, 243–248. [CrossRef]
8. Carrique-Mas, J.J.; Davies, R.H. Salmonella Enteritidis in commercial layer flocks in Europe: Legislative background, on-farm sampling and main challenges. *Braz. J. Poult. Sci.* **2008**, *10*, 10. [CrossRef]
9. Menanteau, P.; Kempf, F.; Trotereau, J.; Virlogeux-Payant, I.; Gitton, E.; Dalifard, J.; Gabriel, I.; Rychlik, I.; Velge, P. Role of systemic infection, cross contaminations and super-shedders in Salmonella carrier state in chicken. *Environ. Microbiol.* **2018**, *20*, 3246–3260. [CrossRef]
10. Denagamage, T.; Jayarao, B.; Patterson, P.; Wallner-Pendleton, E.; Kariyawasam, S. Risk Factors Associated With Salmonella in Laying Hen Farms: Systematic Review of Observational Studies. *Avian Dis.* **2015**, *59*, 291–302. [CrossRef]
11. Mahe, A.; Bougeard, S.; Huneau-Salaun, A.; Le Bouquin, S.; Petetin, I.; Rouxel, S.; Lalande, F.; Beloeil, P.A.; Rose, N. Bayesian estimation of flock-level sensitivity of detection of Salmonella spp., Enteritidis and Typhimurium according to the sampling procedure in French laying-hen houses. *Prev. Vet. Med.* **2008**, *84*, 11–26. [CrossRef] [PubMed]
12. Carrique-Mas, J.J.; Davies, R.H. Sampling and bacteriological detection of Salmonella in poultry and poultry premises: A review. *Rev. Sci. Tech.* **2008**, *27*, 665–677. [CrossRef] [PubMed]
13. Spiegelhalter, D.; Best, N.; Carlin, B.P.; Van der Linde, A. Bayesian measures of model complexity and fit. *J. R. Stat. Soc. B* **2002**, *64*, 583–639. [CrossRef]
14. Lunn, D.; Spiegelhalter, D.; Thomas, A.; Best, N. The BUGS project: Evolution, critique and future directions. *Stat. Med.* **2009**, *28*, 3049–3067. [CrossRef]
15. Sturtz, S.; Ligges, U.; Gelman, A. R2WinBUGS: A Package for Running WinBUGS from R. *J. Stat. Softw.* **2005**, *12*, 1–16. [CrossRef]
16. R Core Team Team. R: A Language and Environment for Statistical Computing. R Foundation for Statistical Computing: Vienna, Austria. 2020. Available online: http://www.R-project.org/ (accessed on 1 September 2023).
17. Brooks, S.P.; Gelman, A. General methods for monitoring convergence of iterative simulations. *J. Comput. Graph. Stat.* **1998**, *7*, 434–455.
18. Gelman, A.; Rubin, D.B. Inference from iterative simulation using multiple sequences. *Stat. Sci.* **1992**, *7*, 457–511. [CrossRef]
19. Messens, W.; Vivas-Alegre, L.; Bashir, S.; Amore, G.; Romero-Barrios, P.; Hugas, M. Estimating the public health impact of setting targets at the European level for the reduction of zoonotic Salmonella in certain poultry populations. *Int. J. Environ. Res. Public Health* **2013**, *10*, 4836–4850. [CrossRef]
20. European Commission (Directorate-General for Health and Food Safety). Communication from the Commission on the European Citizen's Initiative (Eci) "End the Cage Age". 2021. Available online: https://ec.europa.eu/transparency/documents-register/detail?ref=C(2021)4747&lang=en (accessed on 6 December 2022).
21. Burkholder, K.M.; Thompson, K.L.; Einstein, M.E.; Applegate, T.J.; Patterson, J.A. Influence of stressors on normal intestinal microbiota, intestinal morphology, and susceptibility to Salmonella enteritidis colonization in broilers. *Poult. Sci.* **2008**, *87*, 1734–1741. [CrossRef]
22. Van Hoorebeke, S.; Van Immerseel, F.; Haesebrouck, F.; Ducatelle, R.; Dewulf, J. The influence of the housing system on Salmonella infections in laying hens: A review. *Zoonoses Public Health* **2011**, *58*, 304–311. [CrossRef]
23. EFSA. Report of the task force on zoonoses data collection on the analysis of the baseline study on the prevalence of Salmonella in holdings of laying hen flocks of Gallus gallus. *EFSA J.* **2007**, *97*, 85.
24. Namata, H.; Meroc, E.; Aerts, M.; Faes, C.; Abrahantes, J.C.; Imberechts, H.; Mintiens, K. Salmonella in Belgian laying hens: An identification of risk factors. *Prev. Vet. Med.* **2008**, *83*, 323–336. [CrossRef] [PubMed]
25. Snow, L.C.; Davies, R.H.; Christiansen, K.H.; Carrique-Mas, J.J.; Cook, A.J.; Evans, S.J. Investigation of risk factors for Salmonella on commercial egg-laying farms in Great Britain, 2004–2005. *Vet. Rec.* **2010**, *166*, 579–586. [CrossRef] [PubMed]

26. Van Hoorebeke, S.; Van Immerseel, F.; Schulz, J.; Hartung, J.; Harisberger, M.; Barco, L.; Ricci, A.; Theodoropoulos, G.; Xylouri, E.; De Vylder, J.; et al. Determination of the within and between flock prevalence and identification of risk factors for Salmonella infections in laying hen flocks housed in conventional and alternative systems. *Prev. Vet. Med.* **2010**, *94*, 94–100. [CrossRef] [PubMed]
27. Davies, R.; Breslin, M. Observations on Salmonella contamination of eggs from infected commercial laying flocks where vaccination for Salmonella enterica serovar Enteritidis had been used. *Avian. Pathol.* **2004**, *33*, 133–144. [CrossRef] [PubMed]
28. Gast, R.K.; Guraya, R.; Jones, D.R.; Anderson, K.E.; Karcher, D.M. Frequency and Duration of Fecal Shedding of Salmonella Enteritidis by Experimentally Infected Laying Hens Housed in Enriched Colony Cages at Different Stocking Densities. *Front. Vet. Sci.* **2017**, *4*, 47. [CrossRef]
29. Gast, R.K.; Guraya, R.; Jones, D.R.; Guard, J.; Anderson, K.E.; Karcher, D.M. Frequency and Duration of Fecal Shedding of Salmonella Serovars Heidelberg and Typhimurium by Experimentally Infected Laying Hens Housed in Enriched Colony Cages at Different Stocking Densities. *Avian. Dis.* **2017**, *61*, 366–371. [CrossRef]
30. Holt, P.S. Centennial Review: A revisiting of hen welfare and egg safety consequences of mandatory outdoor access for organic egg production. *Poult. Sci.* **2021**, *100*, 101436. [CrossRef]
31. Jones, D.R.; Guard, J.; Gast, R.K.; Buhr, R.J.; Fedorka-Cray, P.J.; Abdo, Z.; Plumblee, J.R.; Bourassa, D.V.; Cox, N.A.; Rigsby, L.L.; et al. Influence of commercial laying hen housing systems on the incidence and identification of Salmonella and Campylobacter. *Poult. Sci.* **2016**, *95*, 1116–1124. [CrossRef]
32. Van Hoorebeke, S.; Van Immerseel, F.; De Vylder, J.; Ducatelle, R.; Haesebrouck, F.; Pasmans, F.; de Kruif, A.; Dewulf, J. The age of production system and previous Salmonella infections on-farm are risk factors for low-level Salmonella infections in laying hen flocks. *Poult. Sci.* **2010**, *89*, 1315–1319. [CrossRef]
33. EFSA. Opinion of the Scientific Panel on Animal Health and Welfare (AHAW) on a request from the Commission related to the welfare aspects of various systems of keeping laying hens. *EFSA J.* **2005**, *197*, 23.
34. Huneau-Salaun, A.; Marianne, C.; Sophie, L.B.; Francoise, L.; Isabelle, P.; Sandra, R.; Virginie, M.; Philippe, F.; Nicolas, R. Risk factors for Salmonella enterica subsp. enterica contamination in 519 French laying hen flocks at the end of the laying period. *Prev. Vet. Med.* **2009**, *89*, 51–58. [CrossRef] [PubMed]
35. Wales, A.; Breslin, M.; Carter, B.; Sayers, R.; Davies, R. A longitudinal study of environmental Salmonella contamination in caged and free-range layer flocks. *Avian Pathol.* **2007**, *36*, 187–197. [CrossRef] [PubMed]
36. Garber, L.; Smeltzer, M.; Fedorka-Cray, P.; Ladely, S.; Ferris, K. Salmonella enterica serotype enteritidis in table egg layer house environments and in mice in U.S. layer houses and associated risk factors. *Avian Dis.* **2003**, *47*, 134–142. [CrossRef]
37. Gole, V.C.; Woodhouse, R.; Caraguel, C.; Moyle, T.; Rault, J.L.; Sexton, M.; Chousalkar, K. Dynamics of Salmonella Shedding and Welfare of Hens in Free-Range Egg Production Systems. *Appl. Environ. Microbiol.* **2017**, *83*, e03313-16. [CrossRef]
38. Powell, M.R.; Crim, S.M.; Hoekstra, R.M.; Williams, M.S.; Gu, W. Temporal patterns in principal Salmonella serotypes in the USA; 1996–2014. *Epidemiol. Infect.* **2018**, *146*, 437–441. [CrossRef]
39. Arnold, M.E.; Martelli, F.; McLaren, I.; Davies, R.H. Estimation of the sensitivity of environmental sampling for detection of Salmonella in commercial layer flocks post-introduction of national control programmes. *Epidemiol. Infect.* **2014**, *142*, 1061–1069. [CrossRef]
40. Jackson, B.R.; Griffin, P.M.; Cole, D.; Walsh, K.A.; Chai, S.J. Outbreak-associated Salmonella enterica serotypes and food Commodities, United States, 1998–2008. *Emerg. Infect. Dis.* **2013**, *19*, 1239–1244. [CrossRef]
41. Machado Junior, P.C.; Chung, C.; Hagerman, A. Modeling Salmonella Spread in Broiler Production: Identifying Determinants and Control Strategies. *Front. Vet. Sci.* **2020**, *7*, 564. [CrossRef]
42. Le Bouquin, S.; Allain, V.; Rouxel, S.; Petetin, I.; Picherot, M.; Michel, V.; Chemaly, M. Prevalence and risk factors for *Salmonella* spp. contamination in French broiler-chicken flocks at the end of the rearing period. *Prev. Vet. Med.* **2010**, *97*, 245–251. [CrossRef]
43. Namata, H.; Welby, S.; Aerts, M.; Faes, C.; Abrahantes, J.C.; Imberechts, H.; Vermeersch, K.; Hooyberghs, J.; Meroc, E.; Mintiens, K. Identification of risk factors for the prevalence and persistence of Salmonella in Belgian broiler chicken flocks. *Prev. Vet. Med.* **2009**, *90*, 211–222. [CrossRef] [PubMed]
44. Volkova, V.V.; Bailey, R.H.; Rybolt, M.L.; Dazo-Galarneau, K.; Hubbard, S.A.; Magee, D.; Byrd, J.A.; Wills, R.W. Inter-relationships of Salmonella status of flock and grow-out environment at sequential segments in broiler production and processing. *Zoonoses Public Health* **2010**, *57*, 463–475. [CrossRef] [PubMed]

animals

MDPI

Article

Limited Emergence of *Salmonella enterica* Serovar Infantis Variants with Reduced Phage Susceptibility in PhagoVet-Treated Broilers

Sandra Sevilla-Navarro [1,†], Jennifer Otero [2,†], Júlia López-Pérez [2], Jan Torres-Boncompte [1], Tiago Prucha [3], Maarten De Gussem [3], Daniela Silva [4], Julia Burgan [4], Pablo Catalá-Gregori [1], Pilar Cortés [2,*] and Montserrat Llagostera [2]

[1] Centro de Calidad Avícola y Alimentación Animal de la Comunidad Valenciana (CECAV), 12539 Alquerias NP, Castellón, Spain; s.sevilla@cecav.org (S.S.-N.); j.torres@cecav.es (J.T.-B.); p.catala@cecav.org (P.C.-G.)

[2] Molecular Microbiology Group, Departament de Genètica i de Microbiologia, Universitat Autònoma de Barcelona, 08193 Cerdanyola del Vallès, Barcelona, Spain; jennifer.otero@uab.cat (J.O.); julia.lopez@uab.cat (J.L.-P.); montserrat.llagostera@uab.cat (M.L.)

[3] VETWORKS BVBA, Knokstraat 38, 9880 Poeke, Belgium; tiago.prucha@vetworks.eu (T.P.); maarten.degussem@vetworks.eu (M.D.G.)

[4] ALS Life Sciences Portugal, Zona Industrial de Tondela ZIM II lote 6, 3460-070 Tondela, Portugal; daniela.silva@alsglobal.com (D.S.); juliaburgan@gmail.com (J.B.)

* Correspondence: mariapilar.cortes@uab.cat

† These authors contributed equally to this work.

Simple Summary: One of the main concerns associated with phage applications is the potential emergence of bacterial variants exhibiting reduced susceptibility to bacteriophages, which may jeopardize the success of such applications, as occurs with antibiotics. In this work, we studied the emergence of reduced-phage-susceptibility variants in broiler trials challenged with *Salmonella enterica* serovar Infantis and treated with the PhagoVet product. We characterized the bacteriophages composing the PhagoVet product at both a microscopic and genomic level, which displayed a broad host range of infection against 271 strains representing 18 *Salmonella* serovars. Our results indicate that the emergence of reduced-phage-susceptibility variants is unlikely to compromise the efficacy of oral PhagoVet against *S.* Infantis.

Abstract: *Salmonella enterica* serovar Infantis (*S.* Infantis) poses a growing issue in the poultry sector, with phage-based products emerging as a safe and effective control measure. This study investigated the emergence of reduced-phage-susceptibility variants (RPSV) of *S.* Infantis in PhagoVet-treated broilers, given that RPSV could undermine phage treatment efficacy. The bacteriophages in the PhagoVet product were characterized using transmission electron microscopy (TEM), genome sequencing, and infection profiling. Furthermore, two broiler trials were conducted: a challenge group (T1) and a challenge-and-treated group (T2). The *S.* Infantis infective dose was set at 10^4 and 10^6 colony-forming units (CFUs) per animal, with PhagoVet administration at 10^6 and 10^8 plaque-forming units (PFUs) per animal, in Trials 1 and 2, respectively. The results revealed that the four PhagoVet bacteriophages belonged to different genera. PhagoVet evidenced broad-spectrum efficacy against 271 strains representing 18 *Salmonella* serovars. In Trial 1, PhagoVet reduced bacterial counts in feces to nearly undetectable levels by day 42, with no RPSV detected. However, in Trial 2, three and five RPSVs were detected in feces and ceca, respectively. Consequently, PhagoVet demonstrated efficacy against *S.* Infantis in broilers, and the potential impact of RPSV is deemed unlikely to compromise its efficacy.

Keywords: *Salmonella* Infantis; broilers; bacteriophage; PhagoVet; reduced phage susceptibility

1. Introduction

S. Infantis currently ranks among the top 10 serovars associated with human infections, standing as the fourth leading cause of salmonellosis cases in the EU [1]. Recent data from the European Food Safety Authority (EFSA) indicated that 95.6% of the *Salmonella* isolates in broiler flocks belong to the Infantis serovar, demonstrating a close association with poultry production [2]. In recent years, the increasing incidence of *S.* Infantis infections in both humans and animals has been further complicated by the dissemination of multidrug-resistant (MDR) clones across several countries. In fact, these MDR strains have been linked to prolonged illness, extended hospitalizations, and increased mortality rates, thereby posing considerable public health implications [3]. Alarming levels of AMR (45.3%) have been reported in *S.* Infantis strains isolated from broilers, particularly against sulfonamides, tetracyclines, ciprofloxacin, and cefotaxime, antibiotics classified by the World Health Organization (WHO) as of "critical importance & highest priority" for human medicine due to limited alternative treatment options [1,4–7].

S. Infantis exhibits distinct genetic characteristics, most of them encoding the pESI-like mega-plasmids, that enhance its epidemiological fitness, particularly in terms of easy acquisition and transmission of antimicrobial resistance (AMR), resistance to heavy metals, possession of mobile virulence genes, and proficiency in biofilm formation [3,6–8]. These attributes have established *S.* Infantis as a widely distributed serovar with persistent infections in animal production, particularly in the poultry sector.

Since 2007, the poultry sector has implemented stringent cleaning and disinfection protocols, biosecurity measures, and prophylactic interventions. While these measures have been demonstrated to be effective against *Salmonella* Enteritidis and Typhimurium, challenges have been encountered in the case of *S.* Infantis due to the unavailability of authorized live vaccines and the bacterium's high tolerance and adaptation to current chemical solutions. This has resulted in the failure of the cleaning and disinfection processes [9]. Therefore, finding effective tools for the prevention and control of *S.* Infantis, such as the use of bacteriophages, is imperative.

Bacteriophages, or phages, stand out as one of the safest options for the prevention, treatment, and eradication of bacterial pathogens including MDR ones. Unlike antibiotics, their specificity limits side-effects such as damage to the physiological microbiota [10]. Phages are ubiquitous in environments where bacteria proliferate, coevolving with bacteria and contributing to the regulation of their population, thereby maintaining equilibrium in ecosystems [11]. However, one of the main concerns associated with phage applications is the potential swift emergence of phage-bacterial variants that exhibit resistance or reduced susceptibility to phages, which may jeopardize the success of such applications, as occurs with antibiotics [12,13]. Within this framework, the PhagoVet consortium, established in 2018 through funding from H2020-FTI call is dedicated to registering a bacteriophage-based product (PhagoVet) for *Salmonella* control in poultry farming. In this study, we investigated the emergence of *S.* Infantis variants resistant to or with reduced susceptibility to PhagoVet in broilers because of the increasing impact of *S.* Infantis in poultry production. To our knowledge, the emergence of these variants had not been previously studied in this bacterium.

2. Materials and Methods

2.1. Bacterial Strains

Salmonella enterica serovar Typhimurium LB5000 (SGSC181; University of Calgary, Calgary, Canada) and *Salmonella enterica* serovar Enteritidis LK5 (SGSC3820; University of Calgary, Calgary, Canada) strains were used to propagate and quantify the bacteriophages. A chromosomal spontaneous mutant resistant to rifampicin (Rif^R) was obtained from the *S.* Infantis 1724105 strain and was employed for challenging *Gallus gallus* in farm trials. The *S.* Infantis 1724105 strain was obtained from a broiler farm as part of Salmonella self-controls following Regulation (EC) 2160/2003 (from Centro de Calidad Avícola y Alimentación Animal de la Comunidad Valenciana, CECAV, Castellón, Spain). All *Salmonella* strains were cultured in Luria–Bertani (LB) broth, agar plates, or XLD agar (Xylose-Lysine-Deoxycholate Agar; Becton Dickinson, Heidelberg, Germany) media, supplemented with rifampicin (75 µg/mL) when required. In all cases, plates were incubated for 18 h at 37 °C.

2.2. PhagoVet Product

PhagoVet is a bacteriophage-based product developed by a European consortium integrated by ALS (Tondela, Portugal), Vetworks (Poeke, Belgium), CECAV (Alquerias, Castellón, Spain) and UAB (Barcelona, Spain). It consists of four virulent bacteriophages (UAB_1, UAB_60, UAB_69, and UAB_Phi78), selected from our *Salmonella* phage library, with production for this study conducted by Jafral (Ljubljana, Slovenia). The PhagoVet product was prepared by mixing the lysates of the four bacteriophages to obtain a titer of 1×10^{10} PFUs/mL. Phage titration was performed by plating ten-fold serial dilutions onto LB plates using the double agar method and the appropriate bacterial host [14].

2.3. Host Range Determination of the PhagoVet Product

The lysis ability of the cocktail was tested against 271 *Salmonella* strains of the serovars Agona, Anatum, Derby, Enteritidis, Hadar, Heidelberg, Infantis, Kentucky, Mbandaka, Mikawasima, monophasic Typhimurium, Newport, Ohio, Saintpaul, Senftenberg, Stanley, Typhimurium, and Virchow. The methodology used for this study was the spot test onto bacterial lawns, as reported [14].

2.4. Bacteriophage Characterization and Genome Sequencing

Transmission electron microscopy (TEM) was employed to determine the bacteriophage morphologies as previously described [15]. For genome sequencing, high-titer lysates (10^{11}–10^{12} PFUs/mL) were obtained by ultracentrifugation at 51,000× *g* for 2 h, and DNA was purified using the phenol-chloroform method [16]. Sequencing and preliminary analysis of the sequences was performed by STAB VIDA (Caparica, Portugal) on the Illumina MiSeq platform, using 300 bp paired-end sequencing reads and an average sequencing depth of 100×. The analysis of the generated sequence raw data was carried out using CLC Genomics Workbench 12.0. (Qiagen, Redwood City, CA, USA). The trimmed sequence reads were used to perform a de novo assembly approach using an algorithm based on de Bruijn graphs [17] and a preliminary annotation was performed using the pipeline from RAST server version 2.0 (Rapid Annotation using Subsystem Technology) (http://rast.nmpdr.org/; accessed on 25 June 2019) [18]. Different analyses of the phage genomes were performed using Geneious 2020.0.5. (Biomatters, Auckland, New Zealand.). Firstly, BLAST was performed, and the closest hits were searched. From this analysis, the phage genus that they belonged to and the model phage of the specific genus were searched on the ICTV web page (https://talk.ictvonline.org/; accessed on 2 February 2020). ProgressiveMAUVE [19] was used for genome comparisons at the nucleotide level with their respective model phages, and the genomes were zeroed using those phages as references. When required, a manual search to identify open-reading frames (ORF) was conducted using BlastX. Functional predictions were conducted using BLASTp programs [20], HMMscan (https://www.ebi.ac.uk/Tools/hmmer/search/hmmscan, accessed on 5 May 2020) and eggNOG [21].

Furthermore, in silico analyses of bacteriophage genomes were carried out using the Virulence Factor Database (VFDB, http://www.mgc.ac.cn/VFs/; accessed on 21 July 2021) [22] to identify virulence-associated genes, and ResFinder [23] and the CARD database [24] were employed to detect antibiotic resistance genes. The VIRIDIC program facilitated the taxonomic classification [25].

2.5. Isolation of S. Infantis Variants with Reduced Phage Susceptibility

The *S.* Infantis 1724105 Rif^R strain used in animal trials is sensitive to two of the PhagoVet phages (UAB_60 and UAB_69). Therefore, this study focused on determining the emergence of bacterial variants resistant or with reduced susceptibility to these two phages in two animal trials using broilers. In both trials, the minimum number of animals per group ensuring independent replicates and enough data for conducting appropriate statistical analysis were used. Furthermore, the *Salmonella*-free status on the arrival of the animals was corroborated in cloacal samples from 25% of the animals. After each trial, poultry farms were emptied, washed, and disinfected for the next trial round. All trials adhered to Regulations (EC) 1831/2002 and 429/2008, according to the additive use, animal categories involved, and following advice on the adequate statistical power. All experimental procedures involving the handling of experimental animals were approved by the Ethical Review Panel of the Directorate-General for Agriculture, Fisheries and Livestock from the Valencian Community, by the code 2021/VSC/PEA/0003, according to Spanish regulations (Real Decreto 53/2013) [26]. In the following paragraphs, the experimental procedure of each trial is detailed, and the design is summarized in Table S1.

Trial 1. A total of 288 male one-day broilers were purchased from a local commercial source and located in two independent rooms separated by walls within the same barn to avoid cross-contamination with phages and *Salmonella*. Two different groups were assessed: T1 (positive control challenged with *Salmonella*) and T2 (challenged with *Salmonella* and treated with a minimum PhagoVet dose). Each group had 12 replicates with 12 animals per replicate (n = 144 animals/group). On arrival and after randomization to treatments, broilers of both groups received water and were fed ad libitum from day 1 to the end of the trial. After 24 h of rearing, 20% of the birds in both groups were orally challenged with *S.* Infantis 1724105 Rif^R at a concentration on 10^4 CFUs/animal. PhagoVet product was administered through individual drinkers via drinking water once a week at a dose of 10^6 PFUs/animal.

The isolation of *Salmonella* variants with reduced susceptibility to UAB_60 and/or UAB_69 phages involved the collection of feces with boot swabs from T1 and T2 groups on days 3, 21, and 42 of bacterial infection. From each experimental group, a pool of feces was prepared. For this, each sample was diluted 1:10 in buffered peptone water (BPW) followed by homogenization. Subsequently, 1 mL of each individual sample was added to a flask and thoroughly mixed. To isolate *Salmonella* colonies from each pool, 10-fold serial dilutions in 0.9% NaCl buffer were prepared and plated on XLD agar plates supplemented with rifampicin (75 µg/mL). After overnight incubation at 37 °C, the *Salmonella* concentration was calculated. Afterward, a maximum of 200 colonies for each time and group were randomly selected and isolated on LB plates supplemented with rifampicin (75 µg/mL). To ensure the absence of contaminating bacteriophages, each isolate was streaked on green plates three times [27]. Finally, the colonies were streaked on LB plates supplemented with rifampicin (75 µg/mL) and incubated at 37 °C for 20 h. The susceptibility of *Salmonella* isolates to UAB_60 and UAB_69 bacteriophages was determined as previously described [28]. In all assays, the *S.* Infantis Rif^R parental strain was included as a control. In those cases where a minimal number of colonies grew on the counting plates, all were individually isolated, and their sensitivity to phages was subsequently determined.

Trial 2. The design of this trial closely mirrored Trial 1, with the presented following changes. Thus, following 24 h of rearing, 50% of birds within groups T1 and T2 (n = 144 animals/group) were orally challenged with *S.* Infantis 1724105 Rif^R at a concentration of 10^6 CFUs/animal. The PhagoVet product was administered at a dosage of 10^8 PFUs/animal via the drinking water on three occasions during the first week (upon the broiler's arrival, 24 h post-infection with *S.* Infantis, and 24 h after the second PhagoVet administration). Thereafter, the product was administered weekly through individual drinkers during the second and third weeks. In addition, the animals of the T2 group underwent a 2 h period of water fasting upon arrival to the farm (i.e., before the first product application) to guarantee optimal PhagoVet consumption at the proper dose. To identify the *Salmonella* variants, the procedure was like in Trial 1 with the following modifications. Two distinct pools, one comprising 12 boot swabs (feces) and the other consisting of 10 ceca, were made on days 7, 14, and 21 of infection from the T1 and T2 groups. The *Salmonella* concentration was determined following the procedure outlined in Trial 1, and a maximum of 200 colonies per time point and type of samples were isolated to search for the desired variants.

For both trials, *Salmonella* enumeration from cecum samples (24 cecum samples from each experimental group in both T1 and T2 trial) was performed at the end of the trials by the miniaturized most probable number technique previously described (ISO/TS 6579-2:2012) [29]. Furthermore, zootechnical parameters, such as body weight (BW), mortality and feed rate conversion (FRC), were assessed.

2.6. Statistical Analysis

Each trial described above is a completely randomized design, with pen as the experimental unit for statistical purposes. Results of the ISO/TS 6579-2:2012 [29] were treated by one-way ANOVA using the General Linear Model (GLM) function in SPSS Statistics Software (IBM, v.27, IBM Corp: Armonk, NY, USA). Differences due to phage treatment in the performance parameters during the study were evaluated using a GLM. All parameters have been reported as group least squares mean. Standard error of the mean, difference of the mean and 95% confidence intervals have also been reported. Significant differences have been declared at $p \leq 0.05$.

3. Results

3.1. Characteristics of PhagoVet Product

The PhagoVet product is composed of the UAB_1, UAB_60, UAB_69, and UAB_Phi78 bacteriophages. As the UAB_Phi78 bacteriophage had been previously characterized [15,30], we proceeded to study the other three phages at both microscopic and genomic levels. As depicted in Figure 1, the UAB_1 bacteriophage features an icosahedral head (92.7 $\pm$ 2.7 nm) and a contractile tail (108.2 $\pm$ 2.1 nm). Similarly, UAB_60 exhibited an identical morphology, with a head measuring 112.0 $\pm$ 6.3 nm and a tail of 115.9 $\pm$ 2.7 nm, while UAB_69 possessed a head of 74.7 $\pm$ 2.1 nm and a tail of 113.9 $\pm$ 4.2 nm. The genomes of UAB_1, UAB_60, and UAB_69 bacteriophages were sequenced, and their complete genomes were deposited in the Genbank database under accession numbers OL656106, OL656107, and OL656108, respectively. Genomic analysis of UAB_1, UAB_60, and UAB_69 revealed their affiliation with the *Justusliebigvirus*, *Tequatrovirus*, and *Felixounavirus* genera, respectively. UAB_Phi78 belonged to the *Zindervirus* genus, as previously reported [30]. The genomes of UAB_1 and UAB_69 bacteriophages exhibited short direct terminal repeats (DTR) (Figure S1), while the genome of UAB_60 lacked DTR. Furthermore, the in silico analyses of the genomes revealed no similarities to known virulence-associated genes or antibiotic resistance genes. In addition, no genes encoding potential immunoreactive food allergens or genes suggesting factors associated with lysogeny were identified.

On other hand, it is noteworthy that the PhagoVet product demonstrated a broad host range against 271 strains encompassing 18 *Salmonella* serovars, as shown in Table 1.

Table 1. Percentage of infection of the PhagoVet product of *Salmonella* strains of different serovars.

Serotypes	Number of Strains per Serotype	PhagoVet (%) [a]
Monophasic Typhimurium	20	100
Agona	22	100
Anatum	12	100
Derby	10	100
Enteritidis	24	100
Hadar	19	100
Heidelberg	4	100
Infantis	53	96
Kentucky	15	93

Table 1. *Cont.*

Serotypes	Number of Strains per Serotype	PhagoVet (%) [a]
Mbandaka	12	100
Mikawasima	7	86
Newport	5	100
Ohio	12	92
Saintpaul	1	100
Senftemberg	11	36.4
Stanley	2	100
Typhimurium	24	92
Virchow	18	72
Total	271	93

[a] The values are the percentage (%) of the strains of each serotype infected by the PhagoVet product.

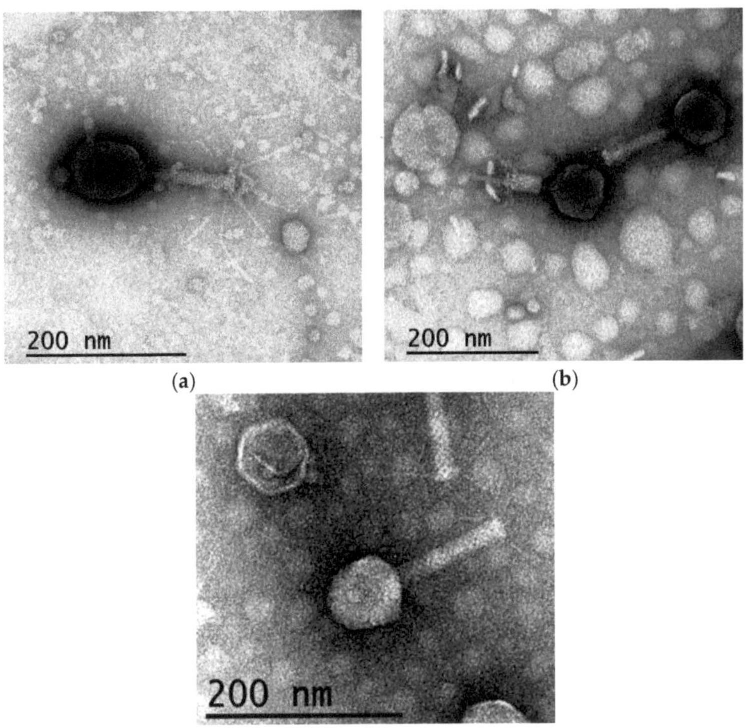

(a) (b) (c)

Figure 1. Electron micrographs of bacteriophages UAB_1 (**a**), UAB_60 (**b**), and UAB_69 (**c**). Scale bars are detailed in the images.

3.2. *Identification of S. Infantis Variants with Reduced PhagoVet-Susceptibility*

In both trials, the animals remained generally healthy throughout the study, with no observed abnormal clinical signs.

Data from Trial 1 showed that the overall mortality rate was 4.86%, with no significant differences between treatment groups T1 and T2 ($p > 0.05$). Similarly, there were no significant differences observed in body weight (T1: 2418.9 vs. T2: 2354.4; $p > 0.05$) or feed conversion ratio (FCR) (T1: 2.00 vs. T2: 1.53; $p > 0.05$) at 42 d of the study. In this trial, *S.* Infantis was administered at a low infective dose (10^4 CFUs/animal), resulting in a *Salmonella* concentration in the feces of the challenge group of approximately 5.7 $\log_{10}$ CFUs/g, maintained at 6.1 $\log_{10}$ CFUs/g until at least 21 d and reaching 4.6 $\log_{10}$ CFUs/g by the end of the study (42 d) (Table 2). Furthermore, treatment with PhagoVet (10^6 PFUs/mL) led to a reduction in the *Salmonella* concentration by approximately 1 $\log_{10}$ CFUs/g at 21 d, reaching nearly undetectable levels by the end of the study (T2 group, 42 d). However, the *Salmonella* concentration in the ceca, as determined by the MPN, was <1 $\log_{10}$ CFUs/g at the end of the study in both groups (Table S2). To assess the presence of variants with reduced susceptibility to UAB_60 and UAB_69 bacteriophages, a total of 600 and 413 colonies were isolated from the feces of T1 and T2 groups, respectively (Table 2). After susceptibility testing, all of them were found to be sensitive to both phages.

Table 2. *S.* Infantis Rif^R concentration and number of colonies isolated from feces over time in Trial 1.

Time (Day)	T1		T2	
	Concentration ($\log_{10}$ CFUs/g)	No. Isolates Tested	Concentration ($\log_{10}$ CFUs/g)	No. Isolates Tested
3	5.7	200	5.6	200
21	6.1	200	5.2	200
42	4.6	200	nc	13 [a]

T1, group of animals challenged with *Salmonella*. T2, group of animals challenged with *Salmonella* and treated with PhagoVet. nc, not calculated, the number of colonies per plate was lower than 15. [a], all colonies that grew on count plates were tested.

In Trial 2, the overall mortality rate was 3.1%, with no significant differences between treatment groups T1 and T2 ($p > 0.05$). Likewise, there were no significant differences observed in body weight (T1: 2598.72 vs. T2: 2611.43; $p > 0.05$) or feed conversion ratio (T1: 2.19 vs. T2: 2.00; $p > 0.05$) at 42 d of the study. In this trial, the *Salmonella* infective dose and the PhagoVet dose were 10^6 CFUs/animal and 10^8 PFUs/animal, respectively, and the PhagoVet administration schedule was modified. Samples for *Salmonella* counting in both feces and ceca were taken at 7, 14 and 21 d. A total of 600 colonies from the feces of T1 and T2 groups, respectively, and 457 and 564 colonies from the ceca of T1 and T2 groups, respectively, were isolated for studying phage susceptibility. Results revealed that the bacterial concentration in feces was in general lower than those observed in trial 1 for both T1 and T2 groups (Table 3). However, higher *Salmonella* counts were detected in ceca at day 7 for both groups (Table 3), decreasing to very low values at 21 d (Table 3), regardless of phage treatment. The reduction in the cecal *Salmonella* population was corroborated by the most probable number technique at the end of the study (day 42). As observed in Trial 1, the concentration of *Salmonella* in the ceca was below 1 $\log_{10}$ CFUs/g in both T1 and T2 groups (Table S2). In this trial, variants with reduced susceptibility to phages were identified (Table 4). Specifically, in the feces, one of the 200 clones isolated at 7 d from the T1 group exhibited reduced susceptibility to UAB_69 phage. The same was observed for clones isolated at 14 d and 21 d from the T2 group. However, all these variants remained sensitive to the UAB_60 phage. Among cecum isolates, only five at 7 d from the T1 group demonstrated reduced susceptibility to both UAB_60 and UAB_69 phages.

Table 3. *S.* Infantis Rif^R concentration in both feces and broiler ceca over time in Trial 2.

Time (Day)	*Salmonella* Concentration ($\log_{10}$ CFUs/g)			
	Feces		Ceca	
	T1	T2	T1	T2
7	4.5	3.7	6.7	6.7
14	4.9	5.3	3.7	3.1
21	4.4	4.8	nc	nc

T1, group of animals challenged with *Salmonella*. T2, group of animals challenged with *Salmonella* and treated with PhagoVet. nc, not calculated because the number of colonies per plate was lower than 15.

Table 4. *S.* Infantis RifR variants isolated in Trial 2 from feces and broiler ceca with reduced susceptibility to UAB_60 and UAB_69 bacteriophages.

Time (Day)	T1				T2			
	No. Tested Isolates	No. Variants (%)			No. Tested Isolates	No. Variants (%)		
		UAB_60	UAB_69	UAB_60 and UAB_69		UAB_60	UAB_69	UAB_60 and UAB_69
Feces								
7	200	0	1 (0.5)	0	200	0	0	0
14	200	0	0	0	200	0	1 (0.5)	0
21	200	0	0	0	200	0	1 (0.5)	0
Ceca								0
7	200	0	0	5 (2.5)	200	0	0	0
14	200	0	0	0	200	0	0	0
21	57 [a]	0	0	0	164 [a]	0	0	0

T1 group, animals infected with *Salmonella*. T2 group, animals infected with *Salmonella* and treated with PhagoVet. [a], all colonies that grew on count plates were tested.

4. Discussion

There is significant concern about the potential emergence of resistant variants or those with reduced susceptibility to phages, which may compromise their application in phage therapy and other uses. Aware of this problem, along with the increasing incidence of *S.* Infantis infections in both humans and animals and understanding the advantages of applying phages in avian production for controlling *S.* Infantis, we studied the impact of the emergence of this type of bacterial variants in oral phage therapy in broiler production administering the PhagoVet product, a cocktail composed of four bacteriophages.

Microscopic characterization and genomic analysis of the UAB_1, UAB_60, UAB_69 and UAB_Phi78 bacteriophages showed that all of them belonged to the *Caudoviricetes* class, but to different genera within this class (Figure 1) [30]. The genomes of UAB_1 and UAB_69 bacteriophages exhibited short direct terminal repeats (DTR) (Figure S1), like the UAB_Phi78 genome [30]. In contrast, bacteriophage UAB_60 did not have DTR. This phage belonged to the *Tequatrovirus* genus, which includes T4-like bacteriophages, whose packaging mechanism results in terminase cleavage at random sites, leading to genome termini with permutations [31]. This explains why we did not find terminal ends of the genome. On the other hand, the wide host range of the PhagoVet product (Table 1) and the absence of negative genomic determinants in the phage genomes support that the PhagoVet product can be considered safe and suitable for application in animal production. In this respect, the PhagoVet product efficiently reduced *S.* Infantis in feces around 1 $\log_{10}$ at 21 d (Trial 1) and achieved almost undetectable values at the end of the experiment (reduction > 4 $\log_{10}$) (Table 2).

The experimental conditions of infection of the broilers with *S.* Infantis allowed this bacterium to persist in feces until the end of Trial 1 (Table 2), reaching concentrations in the intestinal tract comparable to those reported by other authors during similar experimental periods [32,33]. However, its concentration remained below 1 $\log_{10}$ CFUs/g in ceca using the NMP method at the end of both trials (Table S2). This could be attributed to the fact that, in our case, only a small percentage of the animals was orally challenged with *Salmonella* or also to the characteristics of the *Salmonella* strain used in these works. We ruled out the absence of the pESI-like mega-plasmid or the virulence genes encoded within it as a contributing factor, as genome sequencing confirmed their presence in the *S.* Infantis 1724105 strain. Nevertheless, it must be noted that other authors encountered similar problems of *Salmonella* colonization of the gastrointestinal tract, even administering *Salmonella* by oral gavage [34]. Furthermore, it has been reported that changes in the gastrointestinal microbiota, which can reduce available resources or produce certain metabolic products, could adversely impact the growth and survival of *Salmonella* strains used in challenge experiments, particularly over a two-week period of experimentation [35–37]. Despite the difficulties in demonstrating phage therapy efficacy in cecal content, our results regarding *Salmonella* reduction in feces are comparable to those obtained in other studies, albeit with other *Salmonella* serovars, performed in similar experimental periods [32,34], and, to our knowledge, this study is the first to use oral phage therapy in broilers infected with *S.* Infantis.

Regarding the emergence of variants with reduced susceptibility to phages, it must be highlighted that none of these variants were detected in Trial 1. In Trial 2, in which the *Salmonella* infective dose was highest, one of them was detected in feces at 7 d from the untreated group, and one at 14 and 21 d from the treated group (Table 4). From the ceca, only five variants were isolated at 7 d from the untreated group. We speculate that these five variants could be clonal, and that variants found could either be present in the bacterial cultures used for animal infection or have arisen spontaneously during the early days of the infection when the *Salmonella* concentration in the intestine was highest. In any case, these variants failed to colonize the digestive tract of the broilers because the *Salmonella* concentration in the ceca was extremely low at 42 d (Table S2). It must be noted that the total number of variants was similar in both untreated and treated groups, suggesting that there was no effect of the phage treatment on the emergence of these variants. This finding aligns with a previous study conducted by us with broilers and *S.* Typhimurium [28].

Previous works have reported dissimilar results concerning the detection of phage resistance. A review on the development of bacteriophage resistance during bacteriophage therapy revealed that phage-resistant variants of different bacterial species emerged in up to 80% of studies targeting the intestinal tract (out of 11 studies) on different animal species and in 50% of studies (out of 6) using sepsis models on mice [38]. Interestingly, the intestinal tract seemed more susceptible to the emergence of phage-resistance, and although in some circumstances it has been associated with the alteration of known virulence factors, such as O-antigen or LPS [38], it seems more likely to be linked to target bacteria acquiring genes encoding mechanisms to interfere with the phage multiplicative cycle through horizontal transfer from the abundant intestinal microbiota [28]. Despite the increasing number of studies on the use of bacteriophages in animal production, few have been carried out on *Gallus gallus* and *Salmonella*, specifically exploring the emergence of bacterial variants with reduced susceptibility [33,34,36,39]. One study demonstrated that phages reduced the cecal colonization of *S. enterica* serovars Enteritidis and Typhimurium in broilers, at least within 4 days of treatment [39]. The authors isolated bacteriophage-insensitive mutants able to colonize chicken ceca within 24 to 48 h of phage treatment, but these mutants were not maintained for extended periods in ceca. Hurley et al. [36] performed a trial for 30 days on *S.* Typhimurium-infected broilers without a clear reduction in *Salmonella* levels in feces. They found phage-resistant mutants at 15 and 29 days in animals irrespective of phage administration. More recently, two studies conducted with *S.* Typhimurium and *S.* Enteritidis did not find resistant variants in feces and cloacal swabs at 35 days [34] and in caeca at 42 days of trials [33]. In fact, it seems that bacterial resistance to phages often entails a fitness cost [38,40], although this may not consistently result in reduced infectivity, at least in the intestinal tract [38].

5. Conclusions

This is a pioneering study applying phage therapy against *S.* Infantis under conditions that closely mirror those encountered in broiler production. The absence of phage-resistant variants or those with reduced susceptibility following the administration of the PhagoVet product highlights its potential effectiveness in reducing or eliminating *S.* Infantis on poultry farms.

The product led to a significant reduction in *S.* Infantis concentrations in feces, demonstrating its potential as a control measure. Furthermore, even in broilers with low levels of intestinal colonization by *S.* Infantis, the PhagoVet product did not give rise to the emergence of resistant bacterial variants, suggesting that the risk of compromising the efficacy of this treatment is minimal.

Supplementary Materials: The following supporting information can be downloaded at https://www.mdpi.com/article/10.3390/ani14162352/s1: Figure S1: Sequences of the short direct terminal repeats identified in the genomes of UAB_1 and UAB_69 bacteriophages; Table S1: Details of the animal trials; Table S2: *S.* Infantis Rif^R counts in broilers ceca (log_{10} CFUs/g) by the end of study in Trials 1 and 2, determined by the NMP method.

Author Contributions: Conceptualization, S.S.-N., P.C. and M.L.; software, J.L.-P.; validation, S.S.-N., P.C. and M.L.; formal analysis, S.S.-N.; investigation, S.S.-N., J.O., J.L.-P., J.T.-B. and J.B.; writing—original draft preparation, S.S.-N., P.C. and M.L.; writing—review and editing, all authors; visualization, S.S.-N., P.C. and M.L.; supervision, S.S.-N., P.C.-G., M.D.G., D.S., P.C. and M.L. All authors have read and agreed to the published version of the manuscript.

Funding: This research was funded by the European Union's Horizon 2020 research and innovation programme under grant agreement No 820523 and from AGAUR-Generalitat de Catalunya (2021 SGR 00646) and the Ministerio de Ciencia e Innovación (Spain) (PID2020-117708GB-I00). Furthermore, J. L-P received a predoctoral fellowship from the Ministerio de Educación y Formación Profesional de España.

Institutional Review Board Statement: This study was conducted according to the guidelines of the Declaration of Helsinki and by the Ethical Review Panel of the Directorate-General for Agriculture, Fisheries and Livestock from the Valencian Community (protocol code 2021/VSC/PEA/0003, date of approval 3 February 2021), according to Spanish regulations (Real Decreto 53/2013).

Informed Consent Statement: Not applicable.

Data Availability Statement: Data are available upon request.

Acknowledgments: We thank Susana Escribano, Joan Ruiz, and Josep García Llorens, for their excellent technical assistance.

Conflicts of Interest: The authors declare no conflicts of interest. The funders had no role in the design of the study; in the collection, analyses, or interpretation of data; in the writing of the manuscript; or in the decision to publish the results.

References

1. European Food Safety Authority; European Centre for Disease Prevention and Control. The European Union One Health 2021 Zoonoses Report. *EFSA J.* **2022**, *20*, e07666. [CrossRef]
2. European Food Safety Authority (EFSA); European Centre for Disease Prevention and Control (ECDC). The European Union One Health 2022 Zoonoses Report. *EFSA J.* **2023**, *21*, e8442. [CrossRef]
3. Proietti, P.C.; Stefanetti, V.; Musa, L.; Zicavo, A.; Dionisi, A.M.; Bellucci, S.; Mensa, A.; Menchetti, L.; Branciari, R.; Ortenzi, R.; et al. Genetic profiles and antimicrobial resistance patterns of *Salmonella* Infantis strains isolated in Italy in the food chain of broiler meat production. *Antibiotics* **2020**, *9*, 814. [CrossRef]
4. Gymoese, P.; Kiil, K.; Torpdahl, M.; Østerlund, M.T.; Sørensen, G.; Olsen, J.E.; Nielsen, E.M.; Litrup, E. WGS based study of the population structure of *Salmonella enterica* serovar Infantis. *BMC Genom.* **2019**, *20*, 870. [CrossRef] [PubMed]
5. Jovčić, B.; Novović, K.; Filipić, B.; Velhner, M.; Todorović, D.; Matović, K.; Rašić, Z.; Nikolić, S.; Kiškarolj, F.; Kojić, M. Genomic characteristics of colistin-resistant *Salmonella enterica* subsp. *enterica* serovar Infantis from poultry farms in the republic of Serbia. *Antibiotics* **2020**, *9*, 886. [CrossRef]
6. Lapierre, L.; Cornejo, J.; Zavala, S.; Galarce, N.; Sánchez, F.; Benavides, M.B.; Guzmán, M.; Sáenz, L. Phenotypic and genotypic characterization of virulence factors and susceptibility to antibiotics in *Salmonella* Infantis strains isolated from chicken meat: First Findings in Chile. *Animals* **2020**, *10*, 1049. [CrossRef]
7. Mughini-Gras, L.; van Hoek, A.H.A.M.; Cuperus, T.; Dam-Deisz, C.; van Overbeek, W.; Van den Beld, M.; Wit, B.; Rapallini, M.; Wullings, B.; Franz, E.; et al. Prevalence, risk factors and genetic traits of *Salmonella* Infantis in Dutch broiler flocks. *Vet. Microbiol.* **2021**, *258*, 109120. [CrossRef] [PubMed]
8. Alba, P.; Leekitcharoenphon, P.; Carfora, V.; Amoruso, R.; Cordaro, G.; Di Matteo, P.; Ianzano, A.; Iurescia, M.; Diaconu, E.L.; ENGAGE-EURL-AR Network Study Group; et al. Molecular epidemiology of *Salmonella* Infantis in Europe: Insights into the success of the bacterial host and its parasitic pESI-like megaplasmid. *Microb. Genom.* **2020**, *6*, e000365. [CrossRef]
9. Pardo-Esté, C.; Lorca, D.; Castro-Severyn, J.; Krüger, G.; Alvarez-Thon, L.; Zepeda, P.; Sulbaran-Bracho, Y.; Hidalgo, A.; Tello, M.; Molina, F.; et al. Genetic characterization of *Salmonella* Infantis with multiple drug resistance profiles isolated from poultry-farm in Chile. *Microorganisms* **2021**, *9*, 2370. [CrossRef]
10. Clavijo, V.; Flórez, M.J.V. The gastrointestinal microbiome and its association with the control of pathogens in broiler chicken production: A review. *Poult. Sci.* **2018**, *97*, 1006–1021. [CrossRef]

11. Thanki, A.M.; Hooton, S.; Gigante, A.M.; Atterbury, R.J.; Clokie, M.R.J. Potential Roles for bacteriophages in reducing *Salmonella* from poultry and swine. In *Salmonella spp. a Global Challenge*; Lamas, A., Regal, P., Franco, C.M., Eds.; IntechOpen: London, UK, 2021; ISBN 978-1-83969-018-1.
12. Mangalea, M.R.; Duerkop, B.A. Fitness Trade-offs resulting from bacteriophage resistance potentiate synergistic antibacterial strategies. *Infect. Immun.* **2020**, *88*, e00926-19. [CrossRef] [PubMed]
13. Egido, J.E.; Costa, A.R.; Aparicio-Maldonado, C.; Haas, P.J.; Brouns, S.J.J. Mechanisms and clinical importance of bacteriophage resistance. *FEMS Microbiol. Rev.* **2022**, *46*, fuab048. [CrossRef] [PubMed]
14. Kutter, E. Phage host range and efficiency of plating. *Methods Mol. Biol.* **2009**, *501*, 141–149. [CrossRef] [PubMed]
15. Bardina, C.; Spricigo, D.A.; Cortés, P.; Llagostera, M. Significance of the bacteriophage treatment schedule in reducing *Salmonella* colonization of poultry. *Appl. Environ. Microbiol.* **2012**, *78*, 6600–6607. [CrossRef] [PubMed]
16. Sambrook, J.; Russell, D.W. *Molecular Cloning: A Laboratory Manual*, 3rd ed.; Cold Spring Harbor Laboratory Press: New York, NY, USA, 2001; Volume 1.
17. Compeau, P.E.; Pevzner, P.A.; Tesler, G. How to apply de Bruijn graphs to genome assembly. *Nat. Biotechnol.* **2011**, *29*, 987–991. [CrossRef] [PubMed]
18. Overbeek, R.; Olson, R.; Pusch, G.D.; Olsen, G.J.; Davis, J.J.; Disz, T.; Edwards, R.A.; Gerdes, S.; Parrello, B.; Shukla, M.; et al. The SEED and the Rapid Annotation of microbial genomes using Subsystems Technology (RAST). *Nucleic Acids Res.* **2014**, *42*, D206–D214. [CrossRef]
19. Darling, A.E.; Mau, B.; Perna, N.T. ProgressiveMauve: Multiple genome alignment with gene gain, loss and rearrangement. *PLoS ONE* **2010**, *5*, e11147. [CrossRef]
20. Atschul, S.F.; Gish, W.; Miller, W.; Myers, E.W.; Lipman, D.J. Basic local alignment search tool. *J. Mol. Biol.* **1990**, *215*, 403–410. [CrossRef] [PubMed]
21. Huerta-Cepas, J.; Szklarczyk, D.; Heller, D.; Hernández-Plaza, A.; Forslund, S.K.; Cook, H.; Mende, D.R.; Letunic, I.; Rattei, T.; Jensen, L.J.; et al. eggNOG 5.0: A hierarchical, functionally and phylogenetically annotated orthology resource based on 5090 organisms and 2502 viruses. *Nucleic Acids Res.* **2019**, *47*, D309–D314. [CrossRef]
22. Liu, B.; Zheng, D.; Jin, Q.; Chen, L.; Yang, J. VFDB 2019: A comparative pathogenomic platform with an interactive web interface. *Nucleic Acids Res.* **2019**, *47*, D687–D692. [CrossRef] [PubMed]
23. Bortolaia, V.; Kaas, R.S.; Ruppe, E.; Roberts, M.C.; Schwarz, S.; Cattoir, V.; Philippon, A.; Allesoe, R.L.; Rebelo, A.R.; Florensa, A.F.; et al. ResFinder 4.0 for predictions of phenotypes from genotypes. *J. Antimicrob. Chemother.* **2020**, *75*, 3491–3500. [CrossRef] [PubMed]
24. Alcock, B.P.; Raphenya, A.R.; Lau, T.T.Y.; Tsang, K.K.; Bouchard, M.; Edalatmand, A.; Huynh, W.; Nguyen, A.; Cheng, A.A.; Liu, S.; et al. CARD 2020: Antibiotic resistome surveillance with the comprehensive antibiotic resistance database. *Nucleic Acids Res.* **2020**, *48*, D517–D525. [CrossRef]
25. Moraru, C.; Varsani, A.; Kropinski, A.M. VIRIDIC-A novel tool to calculate the intergenomic similarities of prokaryote-infecting viruses. *Viruses* **2020**, *12*, 1268. [CrossRef]
26. Real Decreto 53/2013, De 1 de Febrero, Por El Que Se Establecen Las Normas Básicas Aplicables Para La Protección de Los Animales Utilizados en Experimentación y Otros Fines Científicos, Incluyendo La Docencia. BOE-A-2013-1337. Available online: https://www.boe.es/eli/es/rd/2013/02/01/53 (accessed on 11 July 2024).
27. Chan, R.K.; Botstein, D.; Watanabe, T.; Ogata, Y. Specialized transduction of tetracycline resistance by phage P22 in *Salmonella* Typhimurium. II. Properties of a high-frequency-transducing lysate. *Virology* **1972**, *50*, 883–898. [CrossRef]
28. López-Pérez, J.; Otero, J.; Sánchez-Osuna, M.; Erill, I.; Cortés, P.; Llagostera, M. Impact of mutagenesis and lateral gene transfer processes in bacterial susceptibility to phage in food biocontrol and phage therapy. *Front. Cell. Infect. Microbiol.* **2023**, *13*, 1266685. [CrossRef]
29. *ISO-ISO/TS 6579-2:2012*; Microbiology of Food and Animal Feed-Horizontal Method for the Detection, Enumeration and Serotyping of Salmonella—Part 2: Enumeration by a Miniaturized Most Probable Number Technique. ISO: Geneva, Switzerland, 2012. Available online: https://www.iso.org/standard/56713.html (accessed on 11 July 2024).
30. Bardina, C.; Colom, J.; Spricigo, D.A.; Otero, J.; Sánchez-Osuna, M.; Cortés, P.; Llagostera, M. Genomics of three new bacteriophages useful in the biocontrol of *Salmonella*. *Front. Microbiol.* **2016**, *7*, 545. [CrossRef]
31. Streisinger, G.; Emrich, J.; Stahl, M.M. Chromosome structure in phage T4, iii. Terminal redundancy and length determination. *Proc. Natl. Acad. Sci. USA* **1967**, *57*, 292–295. [CrossRef]
32. Lim, T.H.; Kim, M.S.; Lee, D.H.; Lee, Y.N.; Park, J.K.; Youn, H.N.; Lee, H.J.; Yang, S.Y.; Cho, Y.W.; Lee, J.B.; et al. Use of bacteriophage for biological control of *Salmonella* Enteritidis infection in chicken. *Res. Vet. Sci.* **2012**, *93*, 1173–1178. [CrossRef] [PubMed]
33. Thanki, A.M.; Hooton, S.; Whenham, N.; Salter, M.G.; Bedford, M.R.; O'Neill, H.V.M.; Clokie, M.R.J. A bacteriophage cocktail delivered in feed significantly reduced *Salmonella* colonization in challenged broiler chickens. *Emerg. Microbes Infect.* **2023**, *12*, 2217947. [CrossRef]
34. Kosznik-Kwaśnicka, K.; Podlacha, M.; Grabowski, Ł.; Stasiłojć, M.; Nowak-Zaleska, A.; Ciemińska, K.; Cyske, Z.; Dydecka, A.; Gaffke, L.; Mantej, J.; et al. Biological aspects of phage therapy versus antibiotics against *Salmonella enterica* serovar Typhimurium infection of chickens. *Front. Cell. Infect. Microbiol.* **2022**, *12*, 941867. [CrossRef] [PubMed]

35. Hume, M.E.; Nisbet, D.J.; Scanlan, C.M.; Corrier, D.E.; De Loach, J.R. Fermentation of radiolabelled substrates by batch cultures of caecal microflora maintained in a continuous-flow culture. *J. Appl. Bacteriol.* **1995**, *78*, 677–683. [CrossRef] [PubMed]
36. Hurley, A.; Maurer, J.J.; Lee, M.D. Using bacteriophages to modulate *Salmonella* colonization of the chicken's gastrointestinal tract: Lessons learned from in silico and in vivo modeling. *Avian Dis.* **2008**, *52*, 599–607. [CrossRef] [PubMed]
37. Bjerrum, L.; Engberg, R.M.; Pedersen, K. Infection models for *Salmonella* Typhimurium DT110 in day-old and 14-day-old broiler chickens kept in isolators. *Avian Dis.* **2003**, *47*, 1474–1480. [CrossRef] [PubMed]
38. Oechslin, F. Resistance development to bacteriophages occurring during bacteriophage therapy. *Viruses* **2018**, *10*, 351. [CrossRef]
39. Atterbury, R.J.; Van Bergen, M.A.; Ortiz, F.; Lovell, M.A.; Harris, J.A.; De Boer, A.; Wagenaar, J.A.; Allen, V.M.; Barrow, P.A. Bacteriophage therapy to reduce *Salmonella* colonization of broiler chickens. *Appl. Environ. Microbiol.* **2007**, *73*, 4543–4549. [CrossRef]
40. Gao, D.; Ji, H.; Wang, L.; Li, X.; Hu, D.; Zhao, J.; Wang, S.; Tao, P.; Li, X.; Qian, P. Fitness Trade-Offs in phage cocktail-resistant *Salmonella enterica* Serovar Enteritidis results in increased antibiotic susceptibility and reduced virulence. *Microbiol. Spectr.* **2022**, *10*, e0291422. [CrossRef]

animals

MDPI

Article

Post-Antibiotic and Post-Antibiotic Sub-Minimum Inhibitory Concentration Effects of Carvacrol against *Salmonella* Typhimurium

Eva Boyer [1], Ángela Galán-Relaño [1,2,*], Antonio Romero-Salmoral [1,2], Paula Barraza [1], Lidia Gómez-Gascón [1,2], Carmen Tarradas [1,2], Inmaculada Luque [1,2], Fabiana Carolina de Aguiar [3] and Belén Huerta Lorenzo [1,2]

[1] Animal Health Department, Veterinary Faculty, University of Cordoba, 14014 Cordoba, Spain; ebobuvet@gmail.com (E.B.); antromsal98@gmail.com (A.R.-S.); paula2000bm@gmail.com (P.B.); v32gogal@uco.es (L.G.-G.); sa1taigc@uco.es (C.T.); sa1lumoi@uco.es (I.L.); sa2hulob@uco.es (B.H.L.)

[2] Zoonotic and Emerging Diseases (ENZOEM), University of Cordoba, 14014 Cordoba, Spain

[3] Seara Foods, Av. Paludo, 155-Industrial, Seara 89770-000, SC, Brazil; fabianac.aguiar@yahoo.com.br

* Correspondence: agalanr12@gmail.com

Simple Summary: Carvacrol, an essential oil compound, has been shown to exhibit antimicrobial activity against pathogens. The aim of this study was to evaluate the post-antibiotic effects (PAE) and post-antibiotic sub-minimum inhibitory concentration (PA-SME) effects of carvacrol on *Salmonella* Typhimurium at different concentrations (1×, 2×, 4× MIC) and inoculum sizes (10^6 and 10^8 CFU/mL). The minimum inhibitory concentration (MIC) of carvacrol was established at 0.6 mg/mL, with a required exposure time of 10 min for bacterial inhibition. The results demonstrated that carvacrol exhibited no PAE at the MIC. However, at the 2× MIC, the PAE was observed to be 2 h for the standard inoculum and 1 h for the high-density inoculum. At 4× MIC, the ECP exceeded 43.5 h for both inocula. Further exposure to sub-MIC carvacrol (0.15 mg/mL) after the post-antibiotic phase resulted in an extended ECP of over 43.5 h. These findings suggest that elevated carvacrol sub-MICs can markedly prolong PA-SME, potentially enabling reduced dosing frequencies, minimised adverse effects and enhanced efficacy in the treatment of infected animals and disinfection of agri-food facilities.

Abstract: Carvacrol is a compound present in essential oils with proven antimicrobial activity against numerous pathogens. We firstly determine the post-antibiotic effect (PAE) of carvacrol (1×, 2×, 4× MIC) and post-antibiotic sub-minimum inhibitory concentration (MIC) effect (1× + 0.25× MIC and 2× + 0.25× MIC) for two concentrations of *Salmonella* Typhimurium ATCC14028 (10^6 and 10^8 CFU/mL). Prior to testing, the minimum concentration and exposure time to achieve the bacterial inhibition (MIC 0.6 mg/mL and 10 min) were determined by broth microdilution and time–kill curve methods, respectively. At the MIC, carvacrol did not generate any PAE. At twice the MIC, the PAE was 2 h with the standard inoculum (10^6 CFU/mL) and 1 h with the high-density inoculum (10^8 CFU/mL). At 4× MIC concentrations, the PAE was higher in both cases > 43.5 h. Continuous exposure of post-antibiotic phase bacteria (1× and 2× MIC) to carvacrol at 0.25× MIC (0.15 mg/mL) resulted in an increase in PAE (PA-SME) above 43.5 h with both inocula. These results suggest that the PA-SME of carvacrol for *S.* Typhimurium can be significantly prolonged by increasing the sub-MICs, which would allow dose spacing, reduce adverse effects and improve its efficacy in the treatment of infected animals and as a disinfectant in agri-food facilities.

Keywords: minimum inhibitory concentration; post-antibiotic effect; post-antibiotic sub-minimum inhibitory concentration effect; bacterial growth suppression; natural antimicrobials

1. Introduction

Essential oils and their active ingredients, biosynthesised by aromatic plants, are chemical substances authorised for use as disinfectants and food additives (Generally Recognised As Safe, GRAS) [1]. Several studies have demonstrated their antimicrobial potential against a wide range of microorganisms of interest in animal health, as *Salmonella* Typhimurium, which is the second most reported food borne pathogen at the European Union [2]. The antibacterial activity of essential oils has garnered increasing interest in recent years and shown to be effective even against multidrug-resistant isolates [3,4]. The mechanism of action of essential oils is complex, as several components are involved, but the characteristics of these oils can vary depending on the origin; for this reason, there is a trend to analyse the main active principles and their combined effects [5,6].

Carvacrol ($C_{10}H_{14}O$) is a liquid phenolic monoterpenoid, 2-methyl-5-(1-methylethyl) phenol, present in the essential oil (EO) of oregano (*Origanum vulgare*), thyme (*Thymus vulgaris*), pepperwort (*Lepidium flavum*), wild bergamot (*Citrus aurantium var. bergamia Loisel*), and other plants [7]. Carvacrol shows minimum inhibitory concentration (MIC) values, understood as the lowest concentration of product capable of inhibiting visible growth of the inoculum, of 194 mg/L, with an interquartile range of 26–780 mg/L against *Salmonella* spp. [8]. However, these parameters, commonly used to determine the in vitro susceptibility of microorganisms, do not provide information on the most appropriate dosage schedule [9].

The PAE refers to the ability of an antimicrobial to inhibit the growth of a bacterium once its serum concentration falls below the MIC. Its most important application is the ability to space out doses, reducing side effects and costs of therapy [10], and it can be determined by the pathogen and the infective dose, the antimicrobial agent, the duration of exposition and the drug concentration [11]. The success of the treatment will also depend on the host immune response and the effect of sub-inhibitory concentrations of the drug prior to its elimination. The PAE is often prolonged if after a short exposure to supra-inhibitory concentrations, the bacteria are re-exposed to sub-MIC concentrations, a process referred to as the PA-SME [9]. Both parameters can be affected by antimicrobial concentration and infecting bacterial dose [12,13].

There are reports on the PAE of conventional antimicrobials against *Salmonella* spp.; however, data on EOs or their active ingredients are very limited. In the present study, we determined, for the first time, the PAE and PA-SME of different concentrations of carvacrol against two different doses of the reference strain *Salmonella* Typhimurium ATCC14028: a standard inoculum ($\approx 10^6$ CFU/mL), equivalent to the minimum infective dose in sub-clinical carrier animals, and a high-density inoculum ($\approx 10^8$ CFU/mL), equivalent to the infective dose associated with clinical pictures [14].

2. Materials and Methods

2.1. Bacterial Strain and Tested Product

The *Salmonella* Typhimurium strain (ATCC14028) was obtained from the American Type Culture Collection (ATCC, Manassas, VA, USA), cultured according to the manufacturer's instructions, and stored at −80 °C in Brain Heart Infusion (BHI) (Oxoid Ltd., Wade Road, Basingstoke, Hampshire, RG24 8PW, UK) supplemented with 20% glycerol. This strain was isolated from the pooled heart and liver tissue of four-week-old chickens and is considered a reference for bioinformatics, enteric disease research, food testing, infectious disease research and pharmaceutical testing (https://www.atcc.org/products/14028, accessed on 1 May 2024).

This strain was tested against different concentrations of carvacrol from Sigma-Aldrich Laboratories (St. Louis, MO, USA) (Natural, 99%, FG).

2.2. Determination of MIC

It is crucial to determine the MIC in order to conduct PAE and PA-SME assays. Following the broth microdilution method established by CLSI guidelines [15] for in vitro susceptibility testing with bacteria of animal origin, equal volumes of carvacrol and bacterial inoculum were plated in 96-well microtiter plates.

From a 16–18 h pure culture, 2–3 colonies were resuspended in sterile saline to an optical density of 0.08–0.1 (measured on an ELISA reader at λ 595 nm), corresponding to a concentration of 1.5×10^8 CFU/mL. From this solution, 100 μL were mixed in 9.9 mL of Müeller-Hinton broth (MHB, Oxoid Ltd., Wade Road, Basingstoke, Hampshire, RG24 8PW, UK) to achieve an inoculum of 1.5×10^6 CFU/mL (final test concentration 5×10^5 CFU/mL).

Carvacrol dilutions were obtained by diluting the compound in MHB with dimethyl sulfoxide (0.1%) in a range from 1.3 to 0.15 mg/mL. To avoid potential pipetting errors, the carvacrol dilutions were prepared in 50 mL tubes and dispensed into the corresponding wells of the 96-well plate. Subsequently, 100 μL of the inoculum and 100 μL of the corresponding dilution of carvacrol were dispensed into each well. To ensure accuracy and reliability, all assays included a positive control (100 μL of broth without antimicrobial and 100 μL of bacterial inoculum) and a negative control (200 μL of broth without antimicrobial and without microorganism). Additionally, quality controls were incorporated, including enrofloxacin and *E. coli* ATCC 25922.

The final bacterial inoculum concentration was determined by mixing 10 μL from the positive control well with 9.9 mL of MHB and then plated 100 μL on three Tryptic Soy Agar plates (TSA, Oxoid Ltd., Wade Road, Basingstoke, Hampshire, RG24 8PW, UK). All plates were incubated at 35 ± 2 °C in aerobiosis for 18 h. The assay was performed in triplicate on different days, and the mean of the results was taken as the final value.

2.3. Time–Kill Curve Assay

To determine the minimum exposure time to carvacrol in the PAE and PAE-SME assays, the 24-hours' time–kill curve study was performed according to the methodology described in document M26-A (CLSI, 2023) [15], using now two bacterial inocula with different concentrations ($\approx 10^6$ and $\approx 10^8$ CFU/mL).

Each inoculum was exposed to 0× (growth control), 1×, 2× and 4× carvacrol MIC during 24 h in a water shaking bath (100 rpm) at 35 ± 2 °C. Bacterial growth was monitored by plating serial dilutions onto Mueller–Hinton (MH) plates at times 0, 10 min, and 1, 2, 4, 6, 8 and 24 h. The detection limit of the test was 10 CFU/mL and only plates with 30 to 300 colonies were counted [16]. Each experiment was conducted in triplicate and the average value taken as result.

The concentration of viable bacteria ($\log_{10}$ CFU/mL) per exposure time at the different test concentrations was represented on a semi-log scale using the Microsoft Excel 18.0 software (Microsoft Office 365, Redmond, WA, USA). The antibacterial activity (*E*) was estimated as the difference between the viable bacterial count ($\log_{10}$ CFU/mL) at the end ($n_{t\text{-}24}$) and at the start ($n_{t\text{-}0}$) of the test. Following the criteria of Sidhu et al. [17], three theoretical cut-off points were established: (a) bacteriostatic effect: $E = 0$; there are no changes in value of $n_{t\text{-}0}$; (b) bactericidal effect: $E = -3$; there is a reduction of $\geq 3 \log_{10}$ (99.90%) with respect to the $\log_{10}$ of $n_{t\text{-}0}$ and (c) virtual eradication of bacteria: $E = -4$; there is a reduction of $\geq 4 \log_{10}$ (99.99%) with respect to $\log_{10}$ of $n_{t\text{-}0}$.

The relationship between the carvacrol concentrations and the respective *E* values observed at the end of the assay (24 h) was determined, and the efficiency of each concentration per time was estimated as the reduction percentage of the viable bacteria count with respect to the initial inoculum [16].

2.4. PAE and PA-SME Assays

The test was conducted in accordance with the methodology described by Odenholt-Tornqvist et al. [9] and Nedbalcova et al. [12], with adjustments made to the concentrations and exposure time to ensure optimal efficacy according to published results of lethality tests and toxicity profile of carvacrol [18–20].

The two bacterial cultures of the strain in logarithmic growth phase were again prepared to a concentration of $\approx 10^6$ CFU/mL (standard-density inoculum) and $\approx 10^8$ CFU/mL (high-density inoculum). For each inoculum, two sets (A and B) of 4 tubes were prepared for exposure for 10 min to different concentrations of carvacrol (1×, 2× and 4× MIC) in a water bath at 35 ± 2 °C with agitation (100 rpm). Unexposed bacteria were evaluated in parallel as growth control (carvacrol 0×).

After the initial incubation, the carvacrol was removed by sedimentation of the cells through centrifugation at 3500× *g* for 10 min. The cells were then washed with fresh cation adjusted MHB (without carvacrol) that had been pre-warmed to 35 °C. The process was repeated twice, after which the cells were resuspended in fresh broth (without carvacrol) at 35 °C and incubated at 37 °C with shaking for up to 24 h.

Viable counts were determined by serial dilution and plating on duplicate on MH plates before exposure (time −0.16 h), immediately after the removal of carvacrol (time 0) and after 1, 2, 3, 4, 5, 6, 8, 24, 30 and 48 h of incubation. The control organisms received an identical treatment.

To study the PA-SME of carvacrol, after removal of the product, the second set of tubes (B) was continuously exposed at a sub-MIC concentration of carvacrol (0.25× MIC). One tube with PA-phase bacteria to which no drug was added served as the control. All samples and controls were incubated in a water bath at 35 ± 2 °C with agitation (100 rpm).

The viable counts ($\log_{10}$ CFU/mL) of PAE were plotted vs. time with the software Microsoft Excel (Microsoft Office 365, USA). The PAE was defined as PAE = T − C, where T is the time required for the viable count in the test culture to increase 1 $\log_{10}$ over the count observed at time zero, and C represents the corresponding time for the carvacrol-free control. The PA-SME was defined according to the formula PA-SME = $T_{PA} - C$, where T_{PA} is the time for the cultures previously exposed to carvacrol, which thereafter had been exposed to sub-MIC, to increase 1 $\log_{10}$ unit above the counts observed immediately after washing, and C is the corresponding time for the unexposed control [21]. The PAE and PA-SME were measured in three independent experiments and the average value taken as result. The Student's *t* test was used to determine the significant differences ($p < 0.05$) between carvacrol tested concentrations and bacterial inoculum densities [21].

3. Results

3.1. MIC Assay Results

The mean MIC value of *S.* Typhimurium ATCC 14028 against carvacrol was 0.6 mg/mL.

3.2. Time–Kill Curve Results

Carvacrol showed concentration-dependent antimicrobial activity against the tested strain (Figures 1 and 2). All the carvacrol concentrations (1×, 2× and 4× MIC: 0.6, 1.2 and 2.4 mg/mL, respectively) achieved 100% elimination (virtual eradication) of both inocula within ten minutes of exposure (Table 1). Efficacy was maintained in all cases until the end of the assay.

Table 1. Percentage reduction in the number of viable bacteria with respect to the initial inoculum obtained for each concentration of carvacrol and assay time [16].

Post-Exposition Time	Carvacrol Efficacy with Inoculum 10^6		
	1× CMI (0.6 mg/mL)	**2× CMI (1.2 mg/mL)**	**4× CMI (2.4 mg/mL)**
10 min	100%	100%	100%
1 h	100%	100%	100%
2 h	100%	100%	100%
4 h	100%	100%	100%
8 h	100%	100%	100%
24 h	100%	100%	100%

Table 1. *Cont.*

Post-Exposition Time	Carvacrol Efficacy with Inoculum 10^8		
	1× CMI (0.6 mg/mL)	2× CMI (1.2 mg/mL)	4× CMI (2.4 mg/mL)
10 min	100%	100%	100%
1 h	100%	100%	100%
2 h	100%	100%	100%
4 h	100%	100%	100%
8 h	100%	100%	100%
24 h	100%	100%	100%

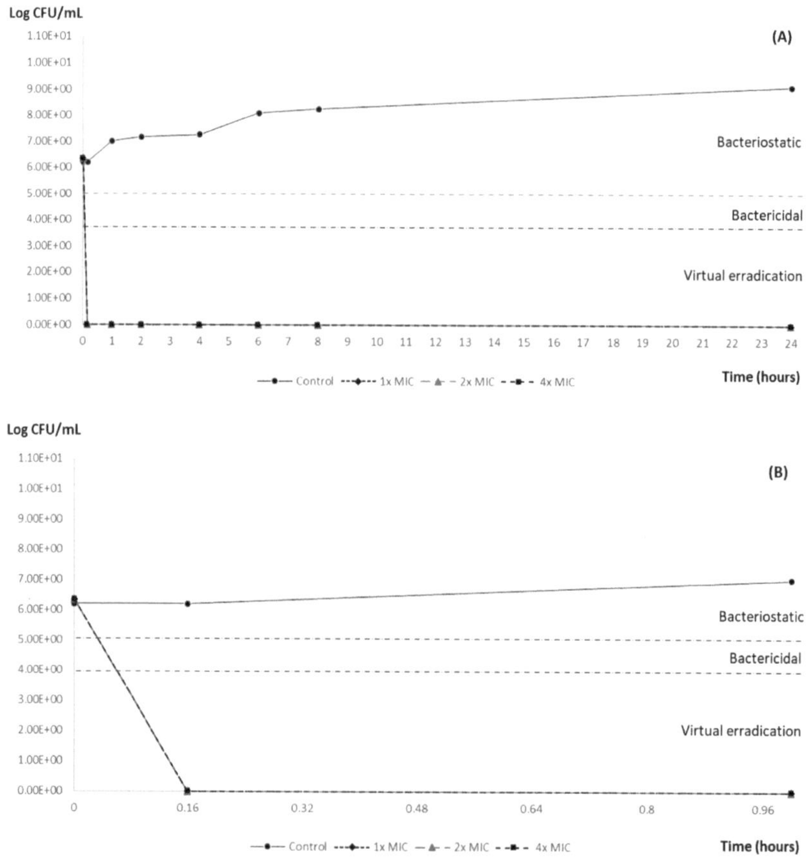

Figure 1. (**A**) Lethality curve of different concentrations of carvacrol against a 10^6 CFU/mL inoculum of *Salmonella* Typhimurium ATCC 14028. (**B**) Enlargement of the results obtained in the first hour of the test. The horizontal dotted lines represent the theoretical cut-off points to evaluate the efficacy of the antimicrobial (reduction with respect to the initial inoculum): bacteriostatic effect (reduction $\geq 2 \log_{10}$), bactericidal (reduction $\geq 3 \log_{10}$) and virtual eradication of bacteria (reduction $\geq 4 \log_{10}$). MIC 600 μg/mL.

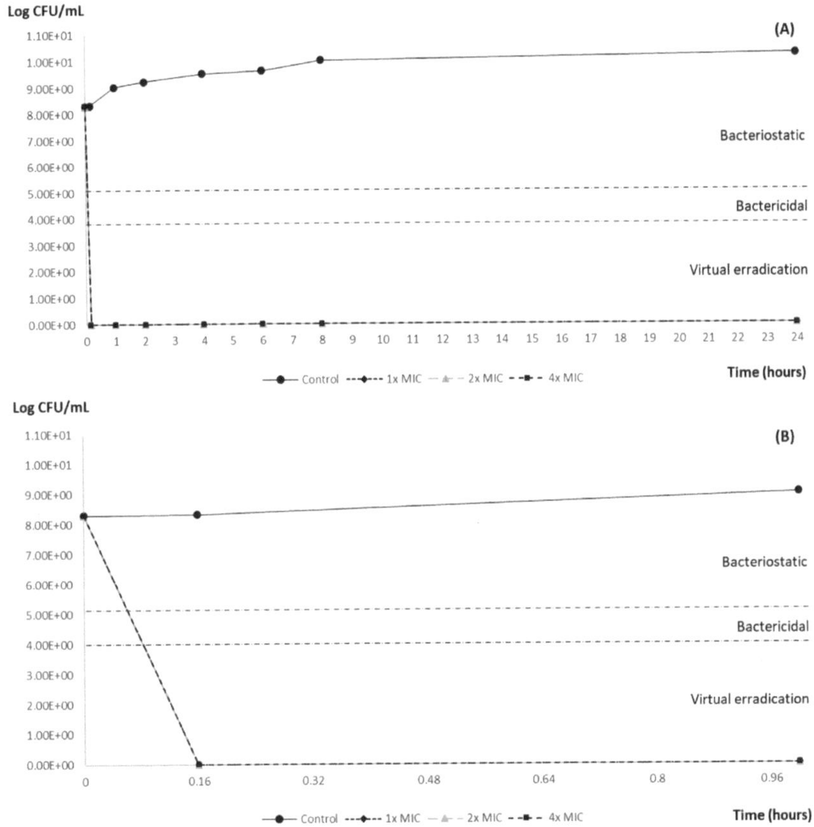

Figure 2. (**A**) Lethality curve of different concentrations of carvacrol against a 10^8 CFU/mL inoculum of *Salmonella* Typhimurium ATCC 14028. (**B**) Enlargement of the results obtained in the first hour of the test. The horizontal dotted lines represent the theoretical cut-off points to evaluate the efficacy of the antimicrobial (reduction with respect to the initial inoculum): bacteriostatic effect (reduction $\geq 2 \log_{10}$), bactericidal (reduction $\geq 3 \log_{10}$) and virtual eradication of bacteria (reduction $\geq 4 \log_{10}$). MIC 600 µg/mL.

3.3. PAE and PA-SME Assays Results

Based on the time–kill curve, PAE and PA-SME were assessed for an initial carvacrol exposure of 10 min. Table 2 shows the PAE and PA-SME (0.25×) effects observed for the standard inoculum (10^6 CFU/mL) and the high-density inoculum (10^8 CFU/mL) of *S.* Typhimurium ATCC 14028. Figures 3 and 4 illustrate the variations in bacterial counts over time for the standard and the high-density inocula, respectively.

As the graphs demonstrate, the PAE was only observed after the exposure of the inocula to supra-inhibitory concentrations of carvacrol. In particular, exposure to a 2× MIC concentration (1.2 mg/mL) was found to inhibit the growth of inoculum 10^6 CFU/mL for 2 h and of inoculum 10^8 CFU/mL for 1 h ($p < 0.05$). Following this period, bacterial growth reactivated exponentially, reaching values close to the positive control in both cases. However, exposure to a 4× MIC concentration (2.4 mg/mL) resulted in a significant increase in the PAE ($p < 0.05$), exceeding 44.5 h with the standard inoculum and 43.5 h with the high-density inoculum.

Table 2. Carvacrol post-antibiotic effect (PAE) and post-antibiotic sub-minimum inhibitory concentration effect (PA-SME) against a standard inoculum and a high-density inoculum of *Salmonella* Typhimurium ATCC 14028 strain.

Inoculum CFU/mL	PAEs (h)			PA-SMEs (h)	
	1× MIC (0.6 mg/mL)	2× MIC (1.2 mg/mL)	4× MIC (2.4 mg/mL)	1× + 0.25 MIC (0.6 + 0.15 mg/mL)	2× + 0.25 MIC (1.2 + 0.15 mg/mL)
10^6	0 ± 0	2 ± 0	>44.5 ± 0	>44.5 ± 0	>44.5 ± 0
10^8	0 ± 0	1 ± 0	>43.5 ± 0	>43.5 ± 0	>43.5 ± 0

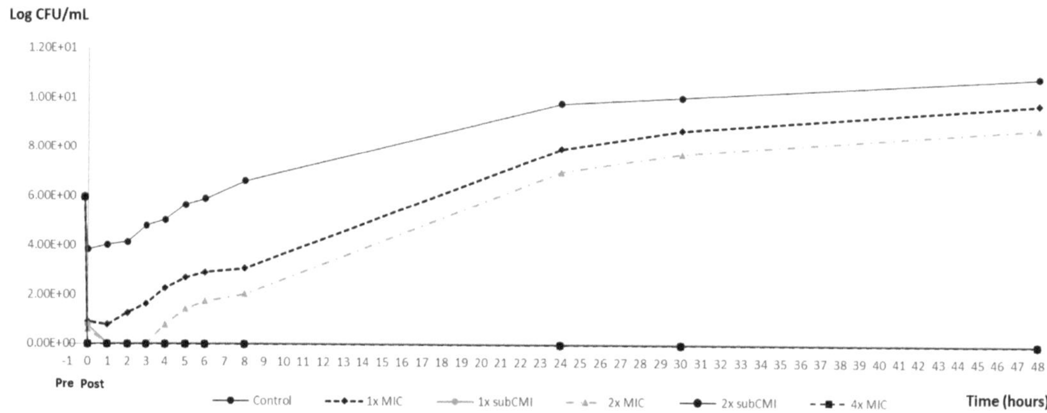

Figure 3. PAE and PA-SME of different carvacrol concentrations for a standard-density inoculum (10^6 CFU/mL) of *S.* Typhimurium ATCC 14028. Pre: The time of beginning exposure to the carvacrol. Post: The time of discontinuing exposure to the carvacrol. MIC = 600 μg/mL; subMIC tested = 150 μg/mL.

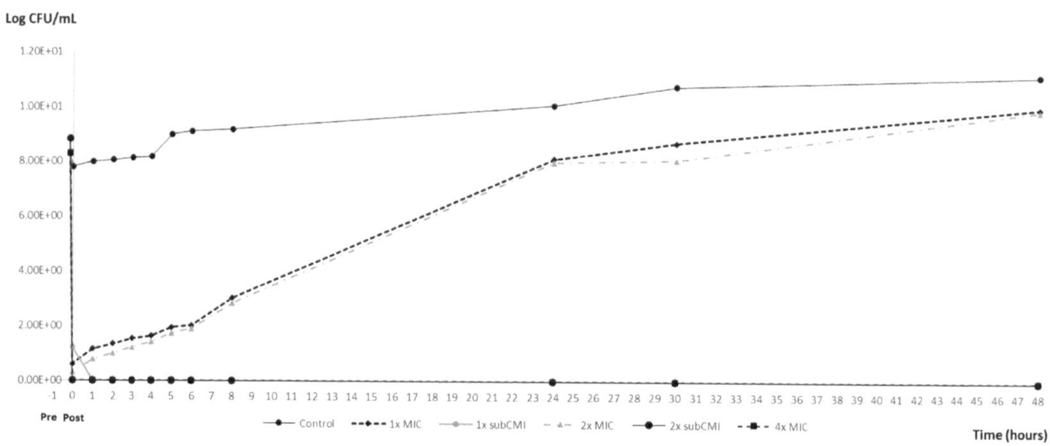

Figure 4. PAE and PA-SME of different carvacrol concentrations for a high-density inoculum (10^8 CFU/mL) of *S.* Typhimurium ATCC 14028. Pre: The time of beginning exposure to the carvacrol. Post: The time of discontinuing exposure to the carvacrol. MIC = 600 μg/mL; subMIC tested = 150 μg/mL.

Further exposure of bacteria that had previously been exposed to antibiotics (1× and 2× MIC) to carvacrol at 0.25× MIC (0.15 mg/mL) resulted in an increased PA-SME over 44.5 h for the standard inoculum and 43.5 h for the high-density inoculum (Figures 2 and 3), which represents a notable divergence from the PAE observed with these same concentrations.

4. Discussion

Pharmacodynamics allows us to evaluate the therapeutic efficacy of antimicrobials based on the plasma concentration they reach and the MIC of each microorganism. There are concentration-dependent antimicrobials (their action is related to the plasma concentration) and time-dependent antimicrobials (their action is related to the time they are present in concentrations over the MIC). In general, concentration-dependent drugs (e.g., aminoglycosides, fluoroquinolones) have a longer PAE, which allows for dose spacing (single daily dose in the case of aminoglycosides) and reduced toxicity. In contrast, time-dependent antimicrobials (e.g., beta-lactams) have a shorter PAE, making continuous administration a more convenient option [22].

Despite the evident utility of the data yielded by PAE and other parameters, and the concurrent necessity to identify alternative antimicrobial agents for disease control, the available literature lacks pertinent information. This study represents the first evaluation of the PAE in natural antimicrobials, with the results demonstrating that carvacrol exhibits concentration-dependent bactericidal activity against *Salmonella* Typhimurium and a notable PAE at supra-inhibitory doses. In terms of bacterial-killing kinetics, the findings differ significantly from those reported by Guimarães et al. [23] using the same product and strain. These authors state that carvacrol exhibits inhibitory activity at a MIC of 0.015 mg/mL (test range 0.002 to 0.25 mg/mL) and no bactericidal activity. In contrast, our assay yielded an estimated MIC of 0.6 mg/mL (assay range 0.15–1.3 mg/mL), accompanied by a 100% reduction in the test inoculum after 10 min of exposure. Our findings align with those previously reported by De Aguiar et al. [24] for *Streptococcus suis*, demonstrating complete eradication of the bacterium after 5 min of exposure to doses of carvacrol at 2× and 4× MIC, 0.3125 and 0.625 mg/mL, respectively.

With regard to the PAE exhibited by carvacrol for the inoculum of 10^6 CFU/mL, which is equivalent to the minimum dose associated with subclinical infection in immunocompetent animals [14], the period was 2 h after a 10 min exposure to 2× MIC, and more than 44.5 h after exposure to 4× MIC.

Both effects are significantly higher than those described by Wain et al. [25] for *Salmonella* Typhi strains after exposure to orbifloxacin (4× MIC/30 min; PAE = 3 h, range 1–4 h) and ceftriaxone (4× MIC/1 h; PAE = 0 h). Other studies showed a PAE of orbifloxacin and enrofloxacin against *E. coli* isolates after 1 h exposure to at 2× MIC of 0.29 h and 0.32 h, respectively [21].

The effect of the antimicrobial concentration on PAE has also been reported in previous studies with respiratory pathogens and veterinary antibiotics, such as amoxicillin, marbofloxacin and enrofloxacin [12]. In this work, exposure for 1 h to 5× MIC antibiotic concentrations induced a PAE for *Actinobacillus pleuropneumoniae* of 8.3 h (6–10 h) with fluoroquinolones and 4 h (2–6 h) with beta-lactams. By increasing the antibiotic concentration to 10× MIC, a significant increase in PAE was detected up to 9.0 h (8–12 h) with marbofloxacin, 8.6 h (8–10 h) with enrofloxacin and 5.7 h (3–6 h) with amoxicillin, respectively. Furthermore, differences in PAE depending on the microorganism were observed. In the case of *Pasteurella multocida*, the PAE of marbofloxacin against *P. multocida* was of 5.1 h (2–6 h), whereas with enrofloxacin was of 3.5 h (2–6 h) and 1.3 h (1–4 h) when amoxicillin was used. At 10× MIC concentrations, the PAE of the fluoroquinolones increased significantly to 5.6 h (3–6 h) and 4.4 h (3–6 h). In the case of amoxicillin, no changes were observed. These results are significantly lower than those obtained in our study with carvacrol at 4× MIC (PAE $\geq$ 44.5 h).

In our study, the effect of inoculum size on the PAE was evaluated, for that, the assays were carried out with an inoculum. In this case, at a concentration of carvacrol 1× and 2×, PAE was of 0 h and 1 h, respectively, in agreement with those obtained for rifampicin

5× MIC (PAE 0.88 h) and ceftriaxone 10× MIC (PAE 0 h) [26]. However, in our work, carvacrol at 4× MIC showed a PAE ≥ 43.5 h.

The chemo-computational toxicity prediction, performed by Akermi et al. [27], with Protox II webserver, showed that carvacrol could be safely and effectively used as drug candidates to tackle bacterial, fungal, and viral infections. On the other hand, in vivo acute oral toxicity studies in rats showed a lethal dose 50 (LD50) of 800 mg/kg in rats and 100 mg/kg in rabbits [20], both below the threshold of 2000 mg/kg established by the Organisation for Economic Cooperation and Development (OECD, 2008). The genotoxic potential of carvacrol is very weak, although its action on DNA cannot be ruled out due to the nuclear fragmentation observed at concentrations of 10 to 160 mg/L [18,19].

The prolongation of PAE by re-exposure of bacteria to concentrations below the MIC has been evidenced. This phenomenon is designated as the post-antibiotic subminimum inhibitory concentration effect (PA-SME), as previously described. Although the action mechanism is not known, early studies by Yan et al. [28] and Odenholt-Tornqvist et al. [9], with beta-lactams and fluoroquinolones, suggested that the PAE is the time needed for the bacteria to recover altered functions (PBP protein synthesis, protein translocation, etc.) and re-enter a logarithmic growth phase. During this time, only small amounts of antibiotic would be needed to limit the growth of these bacteria (PAE-SME). Also, based on the presence in the bacterial populations of more sensitive subpopulations that are inhibited at sub-MIC concentrations, these authors suggested that the PA-SME would represent the time it takes for the most resistant subpopulation to become dominant.

Essential oils and their active ingredients are commonly used in animal feed as additives and probiotics and in the food industry for the development of new active packaging systems [29,30]. The EFSA expert evaluation of the safety of a carvacrol-based feed additive in various animal species (broilers, weaned piglets and dairy cows) carried out in 2017 [31] demonstrated that at the recommended use level (150 mg additive/kg feed), it is safe for broilers and weaned piglets. This conclusion extends to all poultry and swine species reared for meat production. Furthermore, doses of 500 mg of the additive per head per day (equivalent to approximately 25 mg per kg of complete feed) have been demonstrated to be safe for the dairy cow. The residue studies (meat, liver, fatty milk and eggs) demonstrate that consumer exposure to products from animals administered the additive at the recommended use level did not give rise to safety concerns. However, contact with the pure additive may cause skin and mucosal irritation and has the potential for sensitisation in susceptible individuals. Moreover, the utilisation of carvacrol in animal husbandry is not anticipated to engender any adverse effects on the surrounding environment. Although bacterial strains may potentially evolve resistance to this compound over time, there is currently a paucity of documented instances of such resistance to carvacrol, which leads it an efficacious alternative for the management of bacterial populations.

In our study, continuous post-antibiotic exposure of both inocula of *S.* Typhimurium to carvacrol at 0.25× MIC (0.15 mg/mL) resulted in an increase in PAE of the 1× and 2× MIC concentrations up to 43.5 h. This increase was much higher than that found by Harada et al. [21] for *E. coli* strains after re-exposure to orbifloxacin and enrofloxacin at 0.1×, 0.2× and 0.3× MIC concentrations (0.55 and 0.58 h, 1.11 and 0.87 h, and 2.03 and 1.38 h, respectively).

5. Conclusions

Our study demonstrated for the first time that carvacrol could have an important PAE and PA-SME against *Salmonella* spp., which would support its use as an alternative to traditional antimicrobials in the treatment of infected animals and as a disinfectant in agri-food facilities. However, further research is needed with a larger number of strains and to extend the time to measure the effects beyond 48 h, in order to adjust the application interval.

Author Contributions: Conceptualization, B.H.L.; methodology, B.H.L., F.C.d.A., E.B., Á.G.-R., A.R.-S. and P.B.; investigation, B.H.L., E.B. and Á.G.-R.; writing original—draft preparation, B.H.L. and E.B.; writing—review and editing, B.H.L., L.G.-G., I.L., C.T. and Á.G.-R.; supervision, B.H.L. All authors have read and agreed to the published version of the manuscript.

Funding: This work was supported by the Research Project "Characterisation of genetic factors of virulence and antimicrobial resistance in *Salmonella* mST strains isolated from the food chain" (Córdoba University, reference 1253773-R). Galán-Relaño has received a Margarita Salas grant for the training of young doctors from the Ministry of Science, Innovation and Universities (Spanish Government).

Institutional Review Board Statement: Not applicable.

Informed Consent Statement: Not applicable.

Data Availability Statement: Data are contained within the article.

Conflicts of Interest: The authors declare that they have no known competing financial interest or personal relationships that could have appeared to influence the work reported in this research.

References

1. Mith, H.; Duré, R.; Delcenserie, V.; Zhiri, A.; Daube, G.; Clinquart, A. Antimicrobial activities of commercial essential oils and their components against food-borne pathogens and food spoilage bacteria. *Food Sci. Nutr.* **2014**, *2*, 403–416. [CrossRef] [PubMed]
2. EFSA; European Centre for Disease Prevention and Control (ECDC). The European Union One Health 2022 Zoonoses Report. *EFSA J.* **2023**, *21*, e8442. [CrossRef]
3. Solarte, A.L.; Astorga, R.J.; De Aguiar, F.C.; De Frutos, C.; Barrero-Domínguez, B.; Huerta, B. Susceptibility Distribution to Essential Oils of *Salmonella enterica* Strains Involved in Animal and Public Health and Comparison of the Typhimurium and Enteritidis Serotypes. *J. Med. Food* **2018**, *21*, 946–950. [CrossRef] [PubMed]
4. Motta Felício, I.; Limongi de Souza, R.; de Oliveira Melo, C.; Gervázio Lima, K.Y.; Vasconcelos, U.; Olímpio de Moura, R.; Eleamen Oliveira, E. Development and characterization of a carvacrol nanoemulsion and evaluation of its antimicrobial activity against selected food-related pathogens. *Lett. Appl. Microbiol.* **2021**, *72*, 299–306. [CrossRef] [PubMed]
5. Burt, S. Essential oils: Their antibacterial properties and potential applications in foods—A review. *Int. J. Food Microbiol.* **2004**, *94*, 223–253. [CrossRef]
6. Nazzaro, F.; Fratianni, F.; De Martino, L.; Coppola, R.; De Feo, V. Effect of essential oils on pathogenic bacteria. *Pharmaceuticals* **2013**, *6*, 1451–1474. [CrossRef] [PubMed]
7. Sharifi-Rad, M.; Varoni, E.M.; Iriti, M.; Martorell, M.; Setzer, W.N.; del Mar Contreras, M.; Salehi, B.; Soltani-Nejad, A.; Rajabi, S.; Tajbakhsh, M.; et al. Carvacrol and human health: A comprehensive review. *Phyther. Res.* **2018**, *32*, 1675–1687. [CrossRef]
8. Speranza, B.; Bevilacqua, A.; Campaniello, D.; Altieri, C.; Corbo, M.R.; Sinigaglia, M. Minimal Inhibitory Concentrations of Thymol and Carvacrol: Toward a Unified Statistical Approach to Find Common Trends. *Microorganisms* **2023**, *11*, 1774. [CrossRef] [PubMed]
9. Odenholt-Tornqvist, I.; Lowdin, E.; Cars, O. Postantibiotic effects and postantibiotic sub-MIC effects of roxithromycin, clarithromycin, and azithromycin on respiratory tract pathogens. *Antimicrob. Agents Chemother.* **1995**, *39*, 221–226. [CrossRef]
10. Majtán, V.; Majtánová, L. Postantibiotic Effects and Postantibiotic Sub-MIC Effects of Ciprofloxacin, Pefloxacin and Amikacin on the Biological Properties of *Salmonella* Strains. *Folia Microbiol.* **1997**, *42*, 327–332. [CrossRef] [PubMed]
11. Blasco, A.C.; Alfaro, L.A.; Reinoso, J.C.; Mestre, M.J.G.; Rodríguez-Gascón, A. Análisis farmacocinético-farmacodinámico en microbiología: Herramienta para evaluar el tratamiento antimicrobiano. *Enferm. Infecc. Microbiol. Clin.* **2015**, *33*, 48–57. [CrossRef] [PubMed]
12. Nedbalcova, K.; Zouharova, M.; Sperling, D. Post-antibiotic effect of marbofloxacin, enrofloxacin and amoxicillin against selected respiratory pathogens of pigs. *Vet. Med.* **2019**, *64*, 67–77. [CrossRef]
13. Peñalver, P.; Huerta, B.N.; Borge, C.; Astorga, R.; Romero, R.; Perea, A. Antimicrobial activity of five essential oils against origin strains of the *Enterobacteriaceae* family. *APMIS* **2005**, *113*, 1–6. [CrossRef] [PubMed]
14. Gonzalez-Zorn, B.; Goñi, J.G.; Rodríguez, L.J.D. El destino de *Salmonella* en el interior de un ser vivo. *Porci* **2004**, *81*, 23–28.
15. CLSI. *Performance Standards for Antimicrobial Disk and Dilution Susceptibility Tests for Bacteria Isolated from Animals*, 7th ed.; VET01S; CLSI: Wayne, PA, USA, 2024.
16. Baroni, E.; Russi, N.; Rubio, M.; Picco, E.; Formentini, E. Actividad antibacteriana in vitro de ciprofloxacina sobre *Escherichia coli*: Efecto del suero bovino y tamaño del inóculo. *InVet* **2014**, *16*, 7–14.
17. Sidhu, P.K.; Landoni, M.F.; Aliabadi, F.S.; Lees, P. PK-PD integration and modeling of marbofloxacin in sheep. *Res. Vet. Sci.* **2010**, *88*, 134–141. [CrossRef]
18. De Vincenzi, M.; Stammati, A.; De Vincenzi, A.; Silano, M. Constituents of aromatic plants: Carvacrol. *Fitoterapia* **2004**, *75*, 801–804. [CrossRef]

19. Belato, K.K.; de Oliveira, J.R.; de Oliveira, F.S.; de Oliveira, L.D.; Camargo, S.E.A. Cytotoxicity and genotoxicity of thymol verified in murine macrophages (RAW 264.7) after antimicrobial analysis in *Candida albicans, Staphylococcus aureus*, and *Streptococcus mutans*. *J. Funct. Foods* **2018**, *40*, 455–460. [CrossRef]
20. Suntres, Z.E.; Coccimiglio, J.; Alipour, M. The Bioactivity and Toxicological Actions of Carvacrol. *Crit. Rev. Food Sci. Nutr.* **2015**, *55*, 304–318. [CrossRef]
21. Harada, K.; Shimizu, T.; Kataoka, Y.; Takahashi, T. Post-antibiotic effect of orbifloxacin against *Escherichia coli* and *Pseudomonas aeruginosa* isolates from dogs. *Acta Vet. Scand.* **2012**, *54*, 16. [CrossRef]
22. Labarca, J.L. New concepts in pharmacokinetics. Must we thinj again how to use the antibiotic? *Rev. Chil. Infectol.* **2002**, *19*, S33–S37.
23. Guimarães, A.C.; Meireles, L.M.; Lemos, M.F.; Guimarães, M.C.C.; Endringer, D.C.; Fronza, M.; Scherer, R. Antibacterial activity of terpenes and terpenoids present in essential oils. *Molecules* **2019**, *24*, 2471. [CrossRef] [PubMed]
24. de Aguiar, F.C.; Solarte, A.L.; Tarradas, C.; Luque, I.; Maldonado, A.; Galán-Relaño, Á.; Huerta, B. Antimicrobial activity of selected essential oils against *Streptococcus suis* isolated from pigs. *Microbiologyopen* **2018**, *7*, e00613. [CrossRef] [PubMed]
25. Wain, J.; Simpson, J.A.; Thi Diem Nga, L.; Song Diep, T.; Thanh Duy, P.; Baker, S.; Day, N.P.J.; White, N.J.; Parry, C.M. Bactericidal activities and post-antibiotic effects of ofloxacin and ceftriaxone against drug-resistant *Salmonella enterica* serovar Typhi. *J. Antimicrob. Chemother.* **2021**, *76*, 2606–2609. [CrossRef]
26. Soren, O.; Brinch, K.S.; Patel, D.; Liu, Y.; Liu, A.; Coates, A.; Hu, Y. Antimicrobial peptide novicidin synergizes with rifampin, ceftriaxone, and ceftazidime against antibiotic-resistant *Enterobacteriaceae* in vitro. *Antimicrob. Agents Chemother.* **2015**, *59*, 6233–6240. [CrossRef]
27. Akermi, S.; Smaoui, S.; Chaari, M.; Elhadef, K.; Gentile, R.; Hait, M.; Roymahapatra, G.; Mellouli, L. Combined in vitro/in silico approaches, molecular dynamics simulations and safety assessment of the multifunctional properties of thymol and carvacrol: A comparative insight. *Chem. Biodivers.* **2024**, *21*, e202301575. [CrossRef]
28. Yan, S.; Bohach, G.A.; Stevens, D.L. Persistent acylation of high-molecular-weight penicillin-binding proteins by penicillin induces the postantibiotic effect in *Streptococcus pyogenes*. *J. Infect. Dis.* **1994**, *170*, 609–614. [CrossRef]
29. Horváthová, E.; Sramková, M.; Lábaj, J.; Slamenová, D. Study of cytotoxic, genotoxic and DNA-protective effects of selected plant essential oils on human cells cultured *in vitro*. *Neuro. Endocrinol. Lett.* **2006**, *27*, 44–47. [CrossRef]
30. Radha, R.; Latha, R.; Swaminathan, M.S. Chemical composition and bioactivity of essential oilfrom Syzygium travancoricum Gamble. *Flavour Fragr. J.* **2002**, *17*, 478. [CrossRef]
31. EFSA Panel on Additives and Products or Substances used in Animal Feed (FEEDAP); Rychen, G.; Aquilina, G.; Azimonti, G.; Bampidis, V.; Bastos, M.L.; Bories, G.; Cocconcelli, P.S.; Flachowsky, G.; Gropp, J.; et al. Safety and efficacy of an essential oil from *Origanum vulgare* subsp. *hirtum* letsw. var. Vulkan when used as a sensory additive in feed for all animal species. *EFSA J.* **2017**, *15*, e05095. [CrossRef]

animals

MDPI

Article

Integron-Mediated Antimicrobial Resistance and Virulence Factors in *Salmonella* Typhimurium Isolated from Poultry

Elizabeth Kim [1,†], Nora Jean Nealon [1,2,†], Katherine A. Murray [1,3], Cydney Jardine [1], Roberta Magnuson [1] and Sangeeta Rao [1,*]

1 Department of Clinical Sciences, College of Veterinary Medicine and Biomedical Sciences, Colorado State University, Fort Collins, CO 80523, USA; elizabeth.kim@colostate.edu (E.K.); njnealon@gmail.com (N.J.N.); kamurray401@gmail.com (K.A.M.); cydneyjardine@gmail.com (C.J.); magnusonroberta@gmail.com (R.M.)
2 Department of Diagnostic Medicine & Pathobiology, Shreiber School of Veterinary Medicine, Rowan University, Mullica Hill, NJ 08062, USA
3 California Department of Food & Agriculture, Sacramento, CA 95814, USA
* Correspondence: sangeeta.rao@colostate.edu
† These authors contributed equally to this work.

Simple Summary: This study explores antimicrobial-resistant (AMR) *Salmonella* Typhimurium in poultry, emphasizing the impact of integrons on resistance genes and virulence factors. Using whole-genome sequencing, researchers analyzed 26 poultry-derived isolates, finding integrons in three that correlated with greater resistance to beta-lactams, phenicols, and tetracyclines compared to non-integron isolates. Integron-positive isolates also displayed resistance to aminoglycosides and contained high-priority resistance genes, such as *ges*, *imp*, and *oxa*. While AMR genes like *catB* and *tetA* were more common in integron isolates, resistance patterns sometimes appeared in integron-free samples. Integron-bearing isolates had more significant virulence factors like bacteriocin genes, while non-integron isolates showed more fimbrial and pilus genes. These findings suggest that type-I integrons may enhance AMR and virulence in *S.* Typhimurium from poultry, aiding in better AMR surveillance and more effective antimicrobial stewardship in poultry production.

Abstract: This study investigates antimicrobial-resistant (AMR) *Salmonella* Typhimurium in poultry, focusing on how class I integrons contribute to AMR and virulence. Using whole genome sequencing, researchers analyzed 26 *S.* Typhimurium isolates from U.S. poultry, finding that three isolates contained integrons (1000 base pairs each). These integron-positive isolates exhibited significantly higher resistance to beta-lactams, phenicols, and tetracyclines compared to integron-free isolates (p = 0.004, 0.009, and 0.02, respectively) and harbored genes like *ges*, *imp*, and *oxa*, which are linked to extended-spectrum beta-lactamase resistance. Most AMR gene classes (64%) were chromosome-based, with integron-positive isolates showing a broader array of resistance genes, including *catB* and *tetA*. Integron-bearing isolates had higher occurrences of bacteriocin genes and specific AMR genes like aminoglycoside and beta-lactam resistance genes, while integron-free isolates had more fimbrial and pilus genes. The presence of integrons may trend with increased AMR genes and virulence factors, highlighting the role of integron screening in enhancing AMR surveillance and reducing the need for high-priority antimicrobial treatments in poultry. These findings could support better AMR stewardship practices in poultry production, potentially lowering infection risks in humans and livestock.

Keywords: poultry; integrons; virulence factors; *Salmonella* Typhimurium; antimicrobial resistance

1. Introduction

Antimicrobial resistance (AMR) is one of the most recognizable global public health concerns that has an effect on the overall human and animal health [1]. A notable contributor to this escalating concern is the rise in global antibiotic consumption by humans and animals; this is especially the case seen in animal production as the demand for animal protein continues to increase globally [2,3]. Of the widely known AMR pathogens, fluoroquinolone-resistant non-typhoidal *Salmonella* spp. are one of the leading causes of acute gastroenteritis worldwide and are listed as a high-priority threat under WHO global priority AMR pathogens [4]. It was estimated that non-typhoidal *Salmonella enterica* (*S. enterica*) causes approximately 153 million illnesses and 57,000 deaths worldwide [5,6]. In the United States alone, non-typhoidal *S. enterica* causes approximately 1.35 million infections, 26,500 hospitalizations, and 420 deaths annually [7]. *Salmonella enterica* serovar Typhimurium (*S.* Typhimurium) is one of the *S. enterica* serovars most commonly associated with invasive non-typhoidal *Salmonella* disease and has global relevance in relation to livestock and poultry production as it is a zoonotic pathogen known to have a wide range of hosts [8]. One of the hosts that *S.* Typhimurium can be commonly detected is in clinically healthy and diseased poultry, including chickens and turkeys, which are generally raised in large food production systems [9,10]. This poses a risk to food safety as *Salmonella* infection in people and other animals occurs mainly due to the improper handling and consumption of poultry products as well as interactions with fowl that are shedding the pathogen [11].

The prevalence of AMR in *Salmonella* spp. detected in poultry can range from 30% to 40%, with *S.* Typhimurium being a common serotype [12]. With AMR in *Salmonella* being fairly prevalent, this becomes an issue as it not only has an impact on poultry industries and causes large economic losses, but it also results in more severe outcomes of the infection, along with increased hospitalization and deaths [13]. Furthermore, *S.* Typhimurium has acquired new virulence genes that influence host-tropism over the decades, including the acquisition of horizontally transferred genes, which allows for this pathogen to have a broad range of hosts [11]. Virulence factors, such as host adhesion and phagocytosis resistance, can affect the way that *S.* Typhimurium infects a host as well as evade the host immune system, causing a disseminated infection that could escalate and result in severe disease [11]. The combination of the presence of AMR genes in *S.* Typhimurium along with its virulence determinants results in a dual concern in food safety in relation to not only the animals but also people, which include the animal caretakers, meat handlers, and consumers.

The presence of integrons within the *S.* Typhimurium genome is an emerging area of research as it relates to microbial carriage of AMR genes and virulence determinants. An integron is a genetic element that allows mobile gene cassettes to become captured and expressed using site-specific recombination [14]. Among integron types, class I integrons are genetic elements that play a role in integrating gene cassettes from AMR genes into a bacterial genome, contributing to the escalating concern regarding AMR pathogens [15]. A previous study showed how integron-containing *S.* Typhimurium isolates in various host species displayed an increased number and diversity of AMR genes [16]. Although there have been a few studies that showed a relationship between class I integron presence and AMR genes [16,17], there have been limited studies that explored the association of virulence factor genes with class I integrons [18]. Whole Genome Sequencing (WGS) has been used in several studies to identify AMR genes and explore their association with integrons to characterize AMR in different pathogens [19,20]. Using a WGS approach, the primary objective of this study is to examine the association between integron gene presence and carriage of AMR genes and virulence factor genes. Ultimately, the results from this study will be applied to better predict and manage the spread of AMR in *S.* Typhimurium arising from poultry.

2. Materials and Methods

2.1. Study Design and Sample Collection

This study was designed using the methods described in a previous study by Rao et al., 2020 [16]. To collect samples, multiple United States Veterinary Diagnostics Labs were contacted for *S.* Typhimurium data repositories. Of these veterinary diagnostic laboratories, Colorado State University's Veterinary Diagnostic Laboratory and the University of Pennsylvania's Penn Vet Diagnostic Laboratory provided *S.* Typhimurium samples that were isolated from poultry meat birds and collected between 2009 and 2013. A total of 26 samples were donated from these facilities and used in downstream analysis. More details of the sample metadata were unavailable from the source institutions.

2.2. *S. Typhimurium Isolation and Identification*

Each of the 26 poultry isolates was streaked onto trypticase soy agar plates containing 5% sheep blood (Becton, Dickinson and Company, Franklin Lakes, NJ, USA) and incubated overnight at 37 °C. Each isolate was verified to be serogroup B *Salmonella* using traditional slide agglutination (BD Diagnostic Systems®, Becton, Dickinson and Company, Franklin Lakes, NJ, USA). The isolated colonies were inoculated into 1 mL trypticase soy broth (Becton, Dickinson and Company, Franklin Lakes, NJ, USA) and incubated overnight at room temperature. After incubation, each isolated colony was mixed with 10% v/v glycerol prior and stored at −80 °C until downstream testing.

2.3. *Integron and Antimicrobial Gene Cassette Sequencing*

The conserved 5′ and 3′ segments of the class I integron were used as primers to amplify whole integron segments as described in Lucey et al. 2020 [21]. This approach allowed for concurrent sequencing of any antimicrobial resistance and/or virulence genes contained within the integron. The forward primer sequence was 5′-GGC ATC CAA GCA AGC-3′. The reverse integron sequencing was 5′-AAG CAG ACT TGA CCT GAT-3′. Polymerase chain reaction conditions were applied as described in Rao et al., 2008 [16]. Extracted DNA from each isolate was stored at −20 °C until downstream use.

DNA was extracted and purified from excised integron bands using a QIAquick PCR Purification kit (Qiagen®, Hilden, Germany). A NanoDrop One spectrophotometer (Thermo Scientific, Lafayette, CO, USA) measured DNA concentrations and validated sample purity and quality and then prepared for sequencing using ABI BigDye ® Terminator v3.1 sequencing chemistry. Forward and reverse integron sequences were processed using an ABI 3130xL Genetic Analyzer (Applied Biosystems™, Thermo Fisher, Foster City, CA, USA).

2.4. *Sample Processing for DNA Extraction for Whole Genome Sequencing*

Twenty-six isolates were thawed in preparation for DNA extraction for WGS. Each isolate was streaked onto blood agar (Thermo Scientific™) and incubated overnight at 37 °C. To confirm sample purity, isolated colonies from each blood agar plate were streaked onto brain heart infusion broth (Thermo Scientific™), incubated overnight at 37 °C, and subjected to DNA extraction using a Qiagen DNeasy extraction kit following manufacturer protocols (Qiagen, Valencia, CA, USA). The concentration, quality, and purity of each sample were verified on a NanoDrop One spectrophotometer (Thermo Fisher Scientific, Wilmington, DE, USA). Sterile extraction media was used as a negative control. Isolated DNA extracts were stored at −20 °C until further use.

The extracted DNA from the 26 poultry isolates was shipped on dry ice to the Animal Disease Research and Diagnostic Laboratory at South Dakota University (Dr. Joy Scaria) to perform WGS as per the previously described methods [22]. DNA concentrations of each sample were normalized to 0.3 ng/μL, given unique barcode identifier sequences via a Nextera XT DNA library prep kit (Illumina Inc. ©, San Diego, CA, USA), and pooled as equi-volume aliquots. Sequencing was performed on an Illumina MiSeq apparatus (Illumina Inc. ©, San Diego, CA, USA) using a 2 × 250 paired-end approach and V2 chemistry.

2.5. *S. Typhimurium Whole Genome Sequencing Analysis of Chromosomes, Plasmids, and Integrons*

A total of 26 *Salmonella* isolates from poultry hosts were included in WGS analyses. Sequencing analysis was performed in Geneious Prime (Version 2020.0.5). A de novo paired-ends approach was used to assemble each genome. BBDuk (Version 38.90) [23] was used to trim index sequences from each isolate. Next, each trimmed isolate was assembled using the Geneious Prime SPAdes plugin (Version 3.13.0), where the assembler method was set to "method = error correct + assemble" and the assembler mode set to "careful mode" using default conditions for these choices [24]. The SPAdes program was also used for plasmid assembly [24].

Each assembled genome and plasmid were screened for the presence of antimicrobial resistance genes and virulence factors via a basic local alignment search tool (BLAST) approach via the publicly available MEGARes 2.0 (antibiotic resistance genes) [25] and Virulence Factor Database (virulence factor genes) BLAST libraries [26,27]. The BLAST parameters used for each isolate and both libraries were as follows: "Match Mismatch = 1-2", "Gap Cost (Open-Extend) = Linear", "Max E-value: 10", "Max Target Sequences = 100" [25]. Integron sequences for each isolate where integrons were present were imported into the Geneious program and BLAST-aligned with the respective isolate chromosomes and/or plasmids to establish their location.

2.6. Annotation and Organization of AMR and Virulence Factor Genes

For each *S.* Typhimurium isolate whole genome and/or plasmid, BLAST output was filtered using the following criteria: A positive gene match to a library annotation was defined as a ≥85% pairwise identity match (MEGARes 2.0) or ≥95% pairwise identity (VFDB) of the de novo assembled isolate gene across ≥50% of the reference library gene. Annotations not meeting both of these criteria were removed from downstream analysis. Among the remaining genes, for genes with duplicate annotations, the annotation with the highest pairwise identity match over the longest length and lowest E-value was kept as the reference annotation, and the others were discarded from the analysis. In addition, gene entries were removed if they contained the words "hypothetical", "tentative", "possible", or "predicted" when a duplicate annotation with a confirmed identity was also present. AMR genes were categorized by antimicrobial class, and VFDB genes were categorized by virulence factor function for downstream analysis (Supplementary Materials).

2.7. Statistical Analysis

The data on a number of genes were compared between isolates with integrons, and no integrons (integron-free isolates) were compared using a Wilcoxon two-sample test. A Fisher's exact test was used to evaluate the association of the presence of AMR gene and virulence gene classes with integron presence. All statistical analyses were performed using SAS v9.4 (SAS Institute Inc., Cary, NC, USA), and a p-value of <0.05 was used to determine statistical significance.

3. Results

3.1. Integron Presence and Size

Three of the 26 *S.* Typhimurium isolates from poultry contained integrons, all of 1000 base pairs (bp). Among the 26, 100% carried at least nine AMR genes on chromosomes, and 38.5% (n = 10) carried at least one AMR gene on the plasmid. Among the 26 isolates, 96.2% (n = 25) carried at least 13 virulence factor genes on chromosomes and 80.8% (n = 21) virulence genes on plasmids. The antibiogram of the isolates has been published by Rao et al., 2020 [16].

3.2. Integron Presence Impacts Chromosomal and Plasmid Prevalences of AMR Genes in S. Typhimurium

BLAST analysis of *S.* Typhimurium whole genomes was used to establish the AMR genotypes of all isolate chromosomes and/or plasmids. Figures 1 and 2 show the distribution of AMR genes on the chromosomes and plasmids of *S.* Typhimurium isolates, respectively, with and without integrons, where genes are grouped by AMR gene classes (rows), and gene presence versus absence is shown as a column for each isolate. AMR genes were identified on the chromosome of all 26 *S.* Typhimurium isolates (Figure 1) and identified on plasmids of 10 isolates (Figure 2). When examining the chromosomes of *S.* Typhimurium isolates containing integrons (n = 3), AMR drug classes beta-lactams, phenicols, and tetracyclines were significantly higher (100%, 66.7%, and 100%) (p = 0.004, p = 0.009, and p = 0.02, respectively) compared to isolates with no integrons (8.7%, 0%, and 21.7%). However, when comparing the plasmids of isolates with and without integrons, no significant differences in AMR gene prevalences were identified for any gene class, including fluoroquinolone genes (Figure 3), which is in the WHO priority list [5].

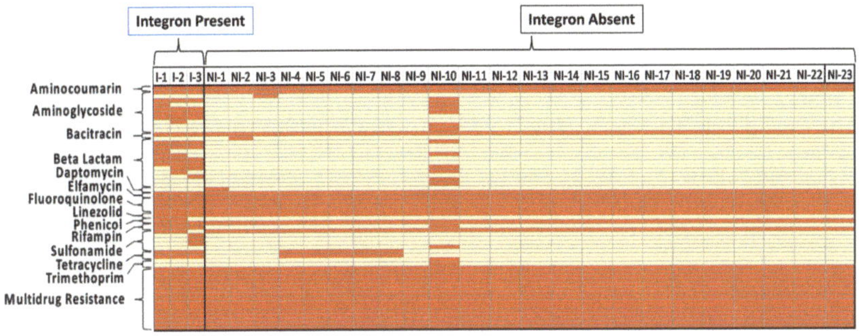

Figure 1. Distribution of antimicrobial resistance genes identified on the chromosome of *S.* Typhimurium isolated from poultry (n = 26) grouped by presence or absence of integrons. Rows are grouped by antimicrobial resistance gene classes. Each column shows the gene profile of an individual *S.* Typhimurium isolates. Each box shows one gene. Red boxes indicate gene presence, and yellow boxes indicate that the gene was not detected in the individual *S.* Typhimurium isolate. Abbreviations: I = Integron Present; NI = Integron Absent.

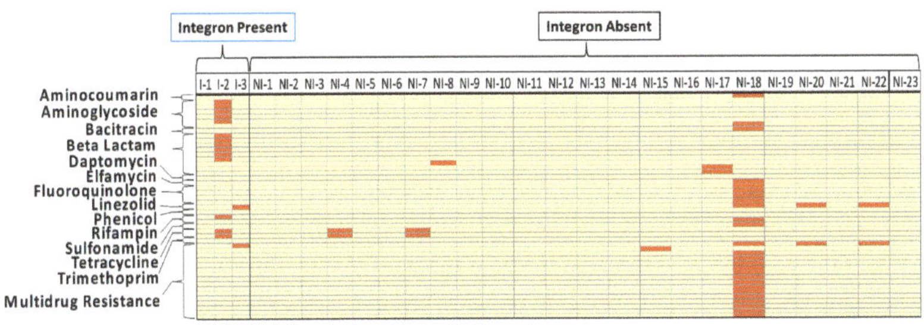

Figure 2. Distribution of antimicrobial resistance genes identified on the plasmid of *S.* Typhimurium isolated from poultry (n = 26) grouped by presence or absence of integrons. Rows are grouped by antimicrobial resistance gene classes. Each column shows the gene profile of an individual *S.* Typhimurium isolates. Each box shows one gene. Red boxes indicate gene presence, and yellow boxes indicate that the gene was not detected in the individual *S.* Typhimurium isolate. Abbreviations: I = Integron Present; NI = Integron Absent.

The distribution of individual AMR chromosomal genes was additionally compared between isolates with and without integrons (Figure 1). Within the aminoglycoside genes harbored on the chromosome, the AMR gene *aac(6′)*, which encodes for an aminoglycoside 6′-N-acetyltransferase enzyme, was found within all 26 isolates regardless of integron presence. The aminoglycoside AMR genes *ant(2″)*, which encodes for an adenine nucleotide translocase enzyme, and *aac(3)*, which encodes for a 3-N-aminoglycoside acetyltransferase enzyme, were found in the chromosome of all three integron-containing isolates and were absent in all but one of the isolates without integrons. The aminoglycoside AMR gene *ant(3″)*, which encodes for an adenine nucleotide translocase enzyme, was also found within all three integron-containing isolates and was absent in all isolates without integrons.

Among beta-lactam genes, the AMR genes *ges* and *imp*, which encode for a Guiana extended-spectrum beta-lactamase enzyme and an imipenemase enzyme, respectively, were found on all three integron-containing isolates and were absent in all isolates without integrons. The beta-lactam AMR gene *oxa*, which encodes for a class D beta-lactamase enzyme, was also found on all three integron-containing isolates but only found in one of the 23 isolates without integrons. Similarly, the beta-lactam AMR gene *carb*, which encodes a class A beta-lactamase enzyme, was found in two of the integron-containing isolates and one of the 23 isolates without integrons. Other notable AMR gene classes found located on the chromosome were phenicol and tetracycline. For phenicol AMR genes, *catB*, which encodes for a type B chloramphenicol acetyltransferase enzyme, was found in two of the integron-containing isolates and absent in all the isolates without integron. As for tetracycline, the AMR genes *tetA* and *tetR*, which encode for tetracycline efflux pumps, were found in all three integron-containing isolates and in five of the isolates without integrons.

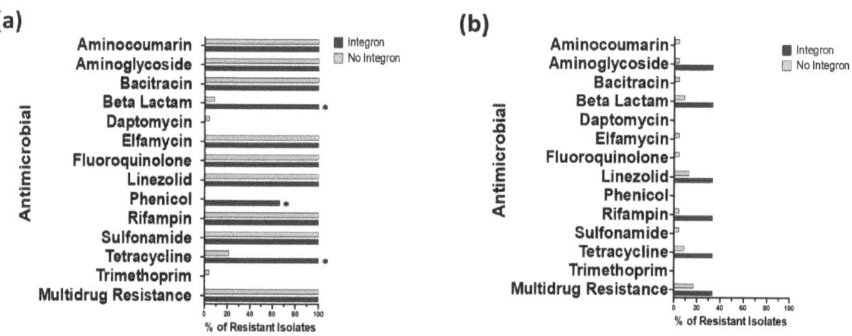

Figure 3. Distribution of antimicrobial resistance gene class prevalences when comparing *S.* Typhimurium isolates with versus without integrons. Blue bars represent isolates with integrons (n = 3), and green bars represent isolates without integrons (n = 23). Panel (**a**) shows gene class distribution prevalences on the chromosome, and (**b**) shows gene class distributions on the plasmid. Prevalences were compared between isolates with and without integrons using Fisher's exact test, and significance was defined as $p < 0.05$. Significance is denoted with an *.

When evaluating the AMR genotypes of *S.* Typhimurium plasmids, two of the three *S.* Typhimurium isolates with integrons contained AMR genes on their plasmid (Figure 2). There was a greater difference in carriage of aminoglycoside and rifampin AMR genes between isolates with integron and with no integron (33.3% and 4.35%, respectively, for both genes), although none of the differences were statistically significant.

3.3. Integron Presence Impacts Chromosomal and Plasmid Prevalences of Virulence Factor Genes in S. Typhimurium

In addition to AMR genes, virulence factor gene type and prevalence were compared on the chromosomes and plasmids of *S.* Typhimurium isolates with and without integrons (Figures 4 and 5). Figure 4 shows the distribution of virulence factor genes on the chromosome (Figure 4a) and plasmid (Figure 4b) of each isolate, where genes are grouped by virulence factor gene classes (rows), and gene presence versus absence is shown as a column for each isolate. Virulence factor genes were more prevalent on the chromosome than on the plasmid for the 26 poultry isolates (Figure 4). When examining virulence factor gene prevalences localized on the chromosome, the isolates with integrons had a higher prevalence of bacteriocin and LPS virulence factor gene classes (66.7% and 33.3%, respectively) than isolates without integrons (13.0% and 4.3%, respectively); however, the difference was not statistically significant ($p = 0.085$ and 0.22, respectively) (Figure 5).

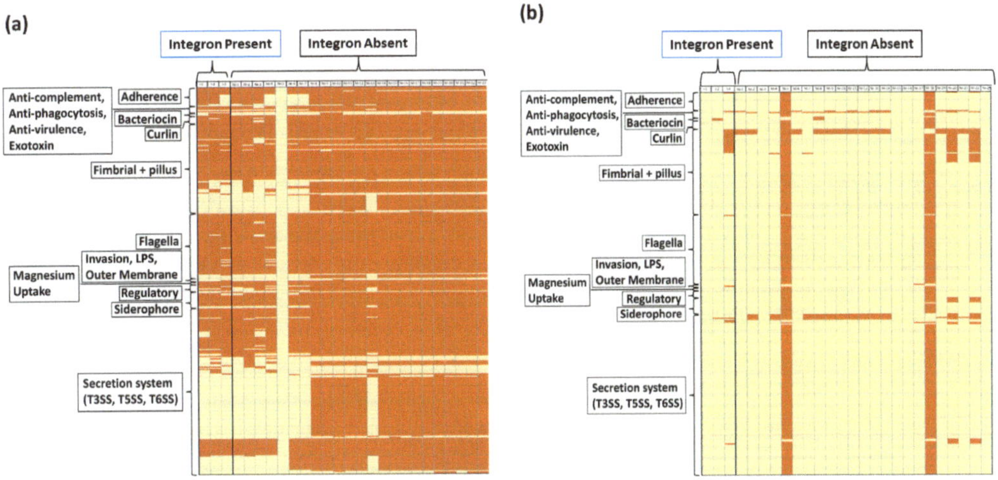

Figure 4. Distribution of virulence factor genes identified on the chromosome (**a**) and plasmid (**b**) of *S.* Typhimurium isolated from poultry (n = 26) grouped by presence or absence of integrons. Rows are grouped by antimicrobial resistance gene classes. Each column shows the gene profile of an individual *S.* Typhimurium isolates. Each box shows one gene. Red boxes indicate gene presence, and white boxes indicate that the gene was not detected in the individual *S.* Typhimurium isolate. Abbreviations: I = Integron Present; NI = Integron Absent.

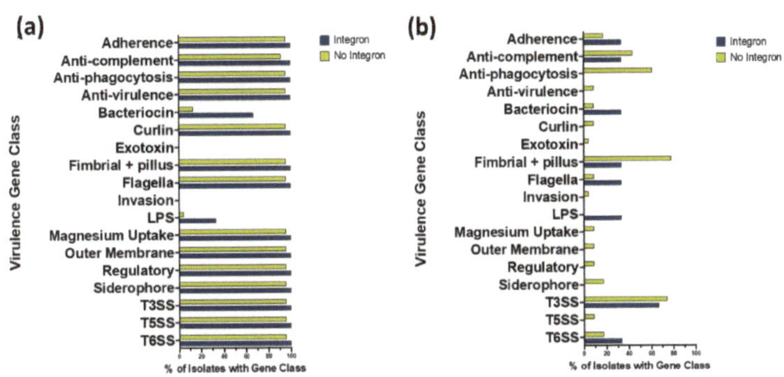

Figure 5. Distribution of virulence factor gene class prevalences when comparing *S.* Typhimurium isolates with versus without integrons. Blue bars represent isolates with integrons (n = 3), and green bars represent isolates without integrons (n = 23). Panel (**a**) shows gene class distribution prevalences on the chromosome, and (**b**) shows gene class distribution prevalences on the plasmid. Abbreviations: LPS = Lipopolysaccharide; T3SS = Type 3 secretion system; T5SS = Type 5 secretion system; T6SS = Type 6 secretion system. Prevalences were compared between isolates with and without integrons using Fisher's exact test, and significance was defined as $p < 0.05$.

Among virulence factor gene classes identified on the chromosome, most were similarly distributed when comparing isolates with versus without integrons (Figure 4). However, the bacteriocin gene class, which encodes for bacteriostatic and bactericidal proteins and peptides, was significantly more prevalent in integron-containing isolates (Mean $\pm$ SE = 1.33 $\pm$ 0.67) versus isolates without integrons (Mean $\pm$ SE = 0.26 $\pm$ 0.14) (p = 0.035). Among bacteriocin genes, the *cib* gene (colicin ib) and *pECS88_0104* gene (colicin Ia) were present in two of the three integron-containing isolates and in only three of the 23 isolates without integrons. In contrast with bacteriocin genes, the chromosomal prevalence of the virulence factor gene classes flagella (Mean $\pm$ SE = 50.67 $\pm$ 0.88 in integron-containing isolates, 51.35 ± 2.38 prevalent in isolates without integrons) and T6SS (Mean ± SE = 13.33 ± 0.33 in integron-containing isolates, Mean ± SE = 23.17 ± 1.63 in isolates without integrons) were significantly lower in integron-containing isolates when compared to isolates without integrons (p = 0.043 and p = 0.046, respectively) (Figure 4).

When examining the prevalences of virulence factor genes on the plasmids, isolates with integrons had a higher prevalence of virulence factor genes under the gene classes bacteriocin, flagella, and lipopolysaccharide (LPS) (33.3%, 33.3%, and 33.3%, respectively) when compared to isolates without integron (8.7%, 8.7%, and 0%, respectively); however, the difference was not statistically significant (p = 0.32, p = 0.32, and p = 0.12, respectively) (Figures 4b and 5). In contrast, the virulence factor gene class anti-phagocytosis was less prevalent in isolates with integrons when compared to isolates without integrons (0%, 60.9%, p = 0.085) (Figure 5).

4. Discussion

S. Typhimurium is among the most frequently identified *Salmonella* serotypes linked to invasive non-typhoidal *Salmonella* infections, significantly affecting both human and animal health [8]. Additionally, AMR in *S.* Typhimurium remains a pressing public health challenge worldwide. Given that poultry commonly hosts *S.* Typhimurium, it is crucial to address AMR within this reservoir to curb new infection cases in both humans and poultry and to reduce economic losses impacting the poultry industry. Studies show that *S.* Typhimurium and other zoonotic pathogens are prevalent in poultry, where high antimicrobial use promotes AMR development, raising health risks and economic burdens by increasing treatment costs and reducing productivity [28,29]. Studies further highlight the productivity losses in poultry production due to AMR, as well as the risk of resistant strains transferring to humans via the food supply chain. This underscores the critical need for enhanced surveillance and biosecurity measures on poultry farms [30,31]. Additionally, class I integrons, frequently associated with AMR in *S.* Typhimurium across human and livestock populations, warrant further research to understand how integrons influence the carriage of AMR and virulence genes in *S.* Typhimurium to address the rising AMR concerns [16]. The CDC (2022) [32] notes that infections with resistant *Salmonella* strains are often more severe and lead to higher hospitalization rates.

Although numerous studies have characterized AMR in *S.* Typhimurium from poultry [33,34], there is limited knowledge on how integron presence specifically influences the types of AMR genes carried by *S.* Typhimurium in this population, both on the chromosome and plasmids. In our study, we observed that two major AMR gene classes, aminoglycosides and beta-lactams, were more frequently localized on the chromosome in isolates containing integrons compared to those without. Prior studies also demonstrate high resistance levels to these antibiotic classes in *Salmonella* spp. isolated from poultry [35–37], suggesting that integrons may be a contributing factor to the persistence of these resistance genes in poultry hosts.

At the gene level, the aminoglycoside resistance gene *aac(6′)* was consistently present on the chromosome in all isolates, regardless of integron presence. Other aminoglycoside resistance genes, such as *ant(2″)*, *aac(3)*, and *ant(3″)*, were found in all three integron-containing isolates but were found in only one of those without integrons. Similarly, for beta-lactam resistance, genes such as *ges*, *imp*, and *oxa* were observed in all integron-positive isolates, while they were infrequent in integron-free isolates. These findings reinforce the potential role of integrons in expanding and stabilizing AMR gene profiles within *S.* Typhimurium in poultry populations.

Two additional notable AMR gene classes, phenicol and tetracycline, were observed on the chromosome in integron-containing isolates compared to those without integrons. Specifically, within the phenicol class, the *catB* gene was present in two of the integron-positive isolates but absent in all integron-free isolates. For tetracycline resistance, *tetA* and *tetR* genes were observed in all three integron-containing isolates, whereas only five integron-free isolates carried these genes. Although the phenicol and tetracycline classes exhibited fewer AMR genes than the aminoglycoside and beta-lactam classes, the consistent association of integrons with specific genes in these categories highlights integrons' potential role in enhancing resistance profiles within S. Typhimurium in poultry.

In addition to the association between AMR genes and integron presence, few studies have examined the relationship between integrons and virulence factor genes in S. Typhimurium isolates from poultry. In this study, we observed that the overall virulence factor profile remained similar across most isolates, regardless of integron presence or isolate location. However, bacteriocin genes were an exception. Within the chromosome, bacteriocin genes were more prevalent in integron-containing isolates (2 of 3) compared to those without integrons (3 of 23). Specifically, the bacteriocin-related virulence genes *cib* (colicin Ib) and *pECS88_0104* (colicin Ia) were identified. Colicins, produced by S. Typhimurium, provide a competitive advantage by killing susceptible bacteria, thus supporting pathogen survival in a host environment [38].

While few studies have explored colicin prevalence in *Salmonella* from poultry [39,40], some research indicates that colicin expression can influence disease severity by enhancing bacterial competitiveness and fitness in the host, particularly under inflammatory conditions. For example, in S. Typhimurium studies, colicin Ib (*ColIb*) expression, intensified by gut inflammation, allowed S. Typhimurium to outcompete sensitive *E. coli* strains, potentially worsening disease in animal models. This intensified competition due to colicin may disrupt the microbial balance, contributing to more severe disease outcomes [41].

A study using a commensal *E. coli* model in mice demonstrated that while colicin production may not be crucial for gut colonization under normal conditions, it provides a competitive edge under environmental stressors like inflammation or resource competition. This suggests that colicins play a role in virulence and competitive survival, particularly in microbiomes disrupted by external pressures [42]. These findings imply that colicin expression, although dependent on specific environmental factors, can influence the severity of bacterial infections by shifting microbial populations and enhancing resistance. Investigating its prevalence and clinical implications among non-typhoidal *Salmonella* isolates, including in infections that have integron-containing isolates, consequently merits further investigation.

To date, no studies have specifically examined how the chromosomal virulence factor genes in S. Typhimurium are associated with class I integrons, which could help in better characterizing the virulence profiles of S. Typhimurium isolates. In our study, we observed no consistent pattern in the localization of AMR or virulence factor genes on plasmids, underscoring the potential importance of chromosomal studies in understanding AMR and virulence determinants in S. Typhimurium and their link to integrons. Future research focusing on chromosomal AMR and virulence genes may clarify the role of integrons in these processes, thereby aiding in the development of predictive tools for pathogenicity in S. Typhimurium.

There are several limitations to consider in this study. First, the total sample size of 26 isolates, spanning only two different US regions and institutions, may not fully represent the prevalence of S. Typhimurium isolates in other US or global regions. In addition, only three out of the 26 poultry isolates contained integrons, all of which were of a single size (1000 bp), limiting the potential for statistical analyses due to the small sample size. Prior research suggests that the low prevalence of integrons may be influenced by varying environmental factors, such as reduced antibiotic exposure [43]. Additionally, recent improvements in antimicrobial stewardship within the poultry industry may also contribute to the lower occurrence of integron-containing isolates, as many major poultry producers have committed to reducing antibiotic use [44,45].

Further, the 26 poultry isolates were collected from two institutions over a span of four years, which may introduce variability related to geographic and temporal differences in microbial populations. Moreover, it remains challenging to assess how integron size influences the carriage of AMR or virulence genes on chromosomal and plasmid locations, as no studies to date have thoroughly examined this association. Further research on integron size and its impact on AMR and virulence gene carriage is needed to clarify these relationships.

5. Conclusions

This study advances our understanding of the complex interplay between AMR genes, virulence factor genes, and integron presence in *S.* Typhimurium isolates from poultry. These insights could aid in addressing AMR concerns by informing the development of early screening tools that better predict *S.* Typhimurium pathogenicity. Such predictive measures would support more effective disease management and potentially guide the creation of targeted antimicrobial therapies to counteract specific virulence mechanisms identified in poultry-associated *S.* Typhimurium.

Supplementary Materials: The following supporting information can be downloaded at: https://www.mdpi.com/article/10.3390/ani14233483/s1.

Author Contributions: Conceptualization, S.R. and N.J.N.; methodology, S.R., N.J.N., R.M. and K.A.M.; software, S.R., N.J.N. and K.A.M.; validation, S.R., N.J.N. and K.A.M.; formal analysis, S.R., N.J.N., K.A.M., E.K. and C.J.; writing—original draft preparation, E.K., N.J.N. and S.R.; writing—review and editing, N.J.N. and S.R.; visualization, E.K., C.J., S.R. and N.J.N.; supervision, S.R. and N.J.N.; project administration, S.R.; funding acquisition, S.R. and N.J.N. All authors have read and agreed to the published version of the manuscript.

Funding: This project was financially supported by United States Department of Agriculture (USDA) Animal Health & Disease funds with project number COLV 2021-03 via the College Research Council (CRC) Grant Program from Colorado State University, and a part of the data analysis was supported by 2018, 2019, and 2022 Veterinary Summer Scholars Program.

Institutional Review Board Statement: Not applicable.

Informed Consent Statement: Not applicable.

Data Availability Statement: Data will be made available on reasonable request.

Acknowledgments: This work would not be possible without the support of Joy Scaria from South Dakota State University and Josh Daniels from Colorado State University. We are also thankful to the Colorado State University and University of Pennsylvania Veterinary Diagnostic Laboratories for providing us with the poultry isolates for this study.

Conflicts of Interest: The authors declare no conflicts of interest.

References

1. Govindaraj Vaithinathan, A.; Vanitha, A. WHO Global Priority Pathogens List on Antibiotic Resistance: An Urgent Need for Action to Integrate One Health Data. *Perspect. Public Health* **2018**, *138*, 87–88. [CrossRef] [PubMed]
2. Klein, E.Y.; Van Boeckel, T.P.; Martinez, E.M.; Pant, S.; Gandra, S.; Levin, S.A.; Goossens, H.; Laxminarayan, R. Global Increase and Geographic Convergence in Antibiotic Consumption between 2000 and 2015. *Proc. Natl. Acad. Sci. USA* **2018**, *115*, E3463–E3470. [CrossRef]
3. Tiseo, K.; Huber, L.; Gilbert, M.; Robinson, T.P.; Van Boeckel, T.P. Global Trends in Antimicrobial Use in Food Animals from 2017 to 2030. *Antibiotics* **2020**, *9*, 918. [CrossRef]
4. Asokan, G.V.; Ramadhan, T.; Ahmed, E.; Sanad, H. WHO Global Priority Pathogens List: A Bibliometric Analysis of Medline-PubMed for Knowledge Mobilization to Infection Prevention and Control Practices in Bahrain. *Oman Med. J.* **2019**, *34*, 184–193. [CrossRef] [PubMed]
5. *WHO Bacterial Priority Pathogens List, 2024: Bacterial Pathogens of Public Health Importance to Guide Research, Development and Strategies to Prevent and Control Antimicrobial Resistance*; World Health Organization: Geneva, Switzerland, 2024.
6. Kirk, M.D.; Pires, S.M.; Black, R.E.; Caipo, M.; Crump, J.A.; Devleesschauwer, B.; Döpfer, D.; Fazil, A.; Fischer-Walker, C.L.; Hald, T.; et al. World Health Organization Estimates of the Global and Regional Disease Burden of 22 Foodborne Bacterial, Protozoal, and Viral Diseases, 2010: A Data Synthesis. *PLoS Med.* **2015**, *12*, e1001921. [CrossRef]
7. Centers for Disease Control and Prevention. *Salmonella—Technical Information on Salmonella*; CDC: Atlanta, GA, USA. Available online: https://www.cdc.gov/Salmonella/general/technical.html (accessed on 29 October 2024).
8. Balasubramanian, R.; Im, J.; Lee, J.-S.; Jeon, H.J.; Mogeni, O.D.; Kim, J.H.; Rakotozandrindrainy, R.; Baker, S.; Marks, F. The Global Burden and Epidemiology of Invasive Non-Typhoidal *Salmonella* Infections. *Hum. Vaccin. Immunother.* **2019**, *15*, 1421–1426. [CrossRef]

9. Dar, M.; Ahmad, S.M.; Bhat, S.; Ahmed, R.; Urwat, U.; Mumtaz, P.T.; Bhat, S.; Dar, T.; Shah, R.; Ganai, N. *Salmonella* Typhimurium in Poultry: A Review. *World's Poult. Sci. J.* **2017**. [CrossRef]
10. Shaji, S.; Selvaraj, R.K.; Shanmugasundaram, R. *Salmonella* Infection in Poultry: A Review on the Pathogen and Control Strategies. *Microorganisms* **2023**, *11*, 2814. [CrossRef] [PubMed]
11. Hoelzer, K.; Moreno Switt, A.I.; Wiedmann, M. Animal Contact as a Source of Human Non-Typhoidal Salmonellosis. *Vet. Res.* **2011**, *42*, 34. [CrossRef]
12. Castro-Vargas, R.E.; Herrera-Sánchez, M.P.; Rodríguez-Hernández, R.; Rondón-Barragán, I.S. Antibiotic Resistance in *Salmonella* spp. Isolated from Poultry: A Global Overview. *Vet. World* **2020**, *13*, 2070–2084. [CrossRef]
13. Krueger, A.L.; Greene, S.A.; Barzilay, E.J.; Henao, O.; Vugia, D.; Hanna, S.; Meyer, S.; Smith, K.; Pecic, G.; Hoefer, D.; et al. Clinical Outcomes of Nalidixic Acid, Ceftriaxone, and Multidrug-Resistant Nontyphoidal *Salmonella* Infections Compared with Pansusceptible Infections in FoodNet Sites, 2006–2008. *Foodborne Pathog. Dis.* **2014**, *11*, 335–341. [CrossRef] [PubMed]
14. Bennett, P.M. Integrons and Gene Cassettes: A Genetic Construction Kit for Bacteria. *J. Antimicrob. Chemother.* **1999**, *43*, 1–4. [CrossRef] [PubMed]
15. Gillings, M.; Boucher, Y.; Labbate, M.; Holmes, A.; Krishnan, S.; Holley, M.; Stokes, H.W. The Evolution of Class 1 Integrons and the Rise of Antibiotic Resistance. *J. Bacteriol.* **2008**, *190*, 5095–5100. [CrossRef] [PubMed]
16. Rao, S.; Linke, L.; Doster, E.; Hyatt, D.; Burgess, B.A.; Magnuson, R.; Pabilonia, K.L.; Morley, P.S. Genomic Diversity of Class I Integrons from Antimicrobial Resistant Strains of *Salmonella* Typhimurium Isolated from Livestock, Poultry and Humans. *PLoS ONE* **2020**, *15*, e0243477. [CrossRef] [PubMed]
17. Rao, S.; Maddox, C.W.; Hoien-Dalen, P.; Lanka, S.; Weigel, R.M. Diagnostic Accuracy of Class 1 Integron PCR Method in Detection of Antibiotic Resistance in *Salmonella* Isolates from Swine Production Systems. *J. Clin. Microbiol.* **2008**, *46*, 916–920. [CrossRef]
18. Cummins, M.L.; Reid, C.J.; Chowdhury, P.R.; Bushell, R.N.; Esbert, N.; Tivendale, K.A.; Noormohammadi, A.H.; Islam, S.; Marenda, M.S.; Browning, G.F.; et al. Whole Genome Sequence Analysis of Australian Avian Pathogenic *Escherichia coli* That Carry the Class 1 Integrase Gene. *Microb. Genom.* **2019**, *5*, e000250. [CrossRef]
19. Massella, E.; Reid, C.J.; Cummins, M.L.; Anantanawat, K.; Zingali, T.; Serraino, A.; Piva, S.; Giacometti, F.; Djordjevic, S.P. Snapshot Study of Whole Genome Sequences of *Escherichia coli* from Healthy Companion Animals, Livestock, Wildlife, Humans, and Food in Italy. *Antibiotics* **2020**, *9*, 782. [CrossRef]
20. Joddha, H.B.; Mathakiya, R.A.; Joshi, K.V.; Khant, R.B.; Golaviya, A.V.; Hinsu, A.T.; Desai, M.R.; Jakhesara, S.J.; Koringa, P.G. Profiling of Antimicrobial Resistance Genes and Integron from *Escherichia coli* Isolates Using Whole Genome Sequencing. *Genes* **2023**, *14*, 1212. [CrossRef]
21. Lucey, B.; Crowley, D.; Moloney, P.; Cryan, B.; Daly, M.; O'Halloran, F.; Threlfall, E.J.; Fanning, S. Integron like Structures in *Campylobacter* spp. of Human and Animal Origin. *Emerg. Infect. Dis.* **2000**, *6*, 50–55. [CrossRef]
22. Thomas, M.; Fenske, G.J.; Antony, L.; Ghimire, S.; Welsh, R.; Ramachandran, A.; Scaria, J. Whole Genome Sequencing-Based Detection of Antimicrobial Resistance and Virulence in Non-Typhoidal *Salmonella enterica* Isolated from Wildlife. *Gut Pathog.* **2017**, *9*, 66. [CrossRef]
23. Bushnell, B. BBTools Software Package. 2014. Available online: http://bbtools.jgi.doe.gov (accessed on 5 September 2023).
24. Bankevich, A.; Nurk, S.; Antipov, D.; Gurevich, A.A.; Dvorkin, M.; Kulikov, A.S.; Lesin, V.M.; Nikolenko, S.I.; Pham, S.; Prjibelski, A.D.; et al. SPAdes: A New Genome Assembly Algorithm and Its Applications to Single-Cell Sequencing. *J. Comput. Biol.* **2012**, *19*, 455–477. [CrossRef] [PubMed]
25. Doster, E.; Rovira, P.; Noyes, N.R.; Burgess, B.A.; Yang, X.; Weinroth, M.D.; Linke, L.; Magnuson, R.; Boucher, C.; Belk, K.E.; et al. A Cautionary Report for Pathogen Identification Using Shotgun Metagenomics: A Comparison to Aerobic Culture and Polymerase Chain Reaction for *Salmonella enterica* Identification. *Front. Microbiol.* **2019**, *10*, 2499. [CrossRef] [PubMed]
26. Liu, B.; Zheng, D.; Zhou, S.; Chen, L.; Yang, J. VFDB 2022: A General Classification Scheme for Bacterial Virulence Factors. *Nucleic Acids Res.* **2022**, *50*, D912–D917. [CrossRef] [PubMed]
27. Chen, L.; Yang, J.; Yu, J.; Yao, Z.; Sun, L.; Shen, Y.; Jin, Q. VFDB: A Reference Database for Bacterial Virulence Factors. *Nucleic Acids Res.* **2005**, *33*, D325–D328. [CrossRef]
28. Nhung, N.T.; Chansiripornchai, N.; Carrique-Mas, J.J. Antimicrobial Resistance in Bacterial Poultry Pathogens: A Review. *Front. Vet. Sci.* **2017**, *4*, 126. [CrossRef] [PubMed]
29. Abreu, R.; Semedo-Lemsaddek, T.; Cunha, E.; Tavares, L.; Oliveira, M. Antimicrobial Drug Resistance in Poultry Production: Current Status and Innovative Strategies for Bacterial Control. *Microorganisms* **2023**, *11*, 953. [CrossRef]
30. Hafez, H.M.; El Adawy, H. Foodborne Diseases of Poultry and Related Problems. *J. Food Nutr. Metab.* **2019**, *1*, 3–5. [CrossRef]
31. Hedman, H.D.; Vasco, K.A.; Zhang, L. A Review of Antimicrobial Resistance in Poultry Farming within Low-Resource Settings. *Animals* **2020**, *10*, 1264. [CrossRef]
32. Centers for Disease Control and Prevention (CDC). *Salmonella* Homepage. Available online: https://www.cdc.gov/Salmonella/index.html (accessed on 27 July 2022).
33. Maasjost, J.; Mühldorfer, K.; Cortez de Jäckel, S.; Hafez, H.M. Antimicrobial Susceptibility Patterns of Enterococcus faecalis and Enterococcus faecium Isolated from Poultry Flocks in Germany. *Avian Dis.* **2015**, *59*, 143–148. [CrossRef]
34. Wang, C.; Wang, X.; Hao, J.; Kong, H.; Zhao, L.; Li, M.; Zou, M.; Liu, G. Serotype Distribution and Antimicrobial Resistance of *Salmonella* Isolates from Poultry Sources in China. *Antibiotics* **2024**, *13*, 959. [CrossRef]

35. Cortés, V.; Sevilla-Navarro, S.; García, C.; Marín, C.; Catalá-Gregori, P. Monitoring Antimicrobial Resistance Trends in *Salmonella* spp. from Poultry in Eastern Spain. *Poult. Sci.* **2022**, *101*, 101832. [CrossRef] [PubMed]
36. Hu, L.; Cao, G.; Brown, E.W.; Allard, M.W.; Ma, L.M.; Khan, A.A.; Zhang, G. Antimicrobial Resistance and Related Gene Analysis of *Salmonella* from Egg and Chicken Sources by Whole-Genome Sequencing. *Poult. Sci.* **2020**, *99*, 7076–7083. [CrossRef]
37. Siddique, A.; Azim, S.; Ali, A.; Andleeb, S.; Ahsan, A.; Imran, M.; Rahman, A. Antimicrobial Resistance Profiling of Biofilm Forming Non-Typhoidal *Salmonella enterica* Isolates from Poultry and Its Associated Food Products from Pakistan. *Antibiotics* **2021**, *10*, 785. [CrossRef] [PubMed]
38. Spriewald, S.; Glaser, J.; Beutler, M.; Koeppel, M.B.; Stecher, B. Reporters for Single-Cell Analysis of Colicin Ib Expression in *Salmonella enterica* Serovar Typhimurium. *PLoS ONE* **2015**, *10*, e0144647. [CrossRef] [PubMed]
39. Hull, D.M.; Harrell, E.; Harden, L.; Thakur, S. Multidrug Resistance and Virulence Genes Carried by Mobile Genomic Elements in *Salmonella enterica* Isolated from Live Food Animals, Processed, and Retail Meat in North Carolina, 2018–2019. *Int. J. Food Microbiol.* **2022**, *378*, 109821. [CrossRef]
40. Olah, P.A.; Sherwood, J.S.; Logue, C.M. Molecular Analysis of *Salmonella* Isolates Recovered from Processed Turkey Carcasses. *J. Food Prot.* **2005**, *68*, 845–849. [CrossRef]
41. Anderson, S.E.; Ghosh, S. Inflammation Fuels Colicin Ib-Dependent Competition of *Salmonella* Serovar Typhimurium and *E.* coli in Enterobacterial Blooms. *PLOS Pathog.* **2022**, *18*, e1010281.
42. Zhu, J.; Goulian, M. Deciphering the Role of Colicins during Colonization of the Mammalian Gut by Commensal *E. coli*. *Microorganisms* **2020**, *8*, 664.
43. Díaz-Mejía, J.J.; Amábile-Cuevas, C.F.; Rosas, I.; Souza, V. An Analysis of the Evolutionary Relationships of Integron Integrases, with Emphasis on the Prevalence of Class 1 Integrons in *Escherichia coli* Isolates from Clinical and Environmental Origins. *Microbiology* **2008**, *154*, 94–102. [CrossRef]
44. NRDC: Natural Resources Defense Council. Antibiotic Consumption in U.S. Pork, Beef, and Turkey Industries Vastly Outstrips Comparable Industries in Europe, and the U.S. Chicken Industry [Issue Brief]. Food Animal Concerns Trust. *NRDC*. 2018. Available online: https://www.nrdc.org/sites/default/files/antibiotic-consumption-us-pork-beef-and-turkey-industries-ib.pdf (accessed on 11 May 2024).
45. Friends of the Earth; Consumers Union; Natural Resources Defense Council; Center for Food Safety; Food Animal Concerns Trust; U.S. Public Interest Research Group Education Fund. Chain Reaction III: How Top Restaurants Rate on Reducing Antibiotics in Their Meat Supply. September 2017. Available online: https://www.nrdc.org/sites/default/files/restaurants-antibiotic-use-es-2017.pdf (accessed on 9 April 2024).

animals

MDPI

Article

miR-215 Modulates Ubiquitination to Impair Inflammasome Activation and Autophagy During *Salmonella* Typhimurium Infection in Porcine Intestinal Cells

Carmen Entrenas-García [1], José M. Suárez-Cárdenas [1,2], Raúl Fernández-Rodríguez [1,2], Rocío Bautista [3,4], M. Gonzalo Claros [3,4,5,6,7], Juan J. Garrido [1,2,*,†] and Sara Zaldívar-López [1,2,*,†]

1 Immunogenomics and Molecular Pathogenesis Group, UIC Zoonosis and Emergent Diseases ENZOEM, Department of Genetics, University of Cordoba, 14014 Cordoba, Spain; b82engac@uco.es (C.E.-G.); b52sucaj@uco.es (J.M.S.-C.); b42feror@uco.es (R.F.-R.)
2 GA-14 Research Group, Maimónides Biomedical Research Institute of Córdoba (IMIBIC), 14004 Cordoba, Spain
3 Plataforma Andaluza de Bioinformática, Supercomputing and Bioinnovation Center (SCBI), Universidad de Málaga, 29590 Malaga, Spain; rociobm@uma.es (R.B.); claros@uma.es (M.G.C.)
4 Institute of Biomedical Research in Malaga (IBIMA), IBIMA-RARE, 29590 Malaga, Spain
5 Institute for Mediterranean and Subtropical Horticulture "La Mayora" (IHSM-UMA-CSIC), 29590 Malaga, Spain
6 Department of Molecular Biology and Biochemistry, Universidad de Málaga, 29010 Malaga, Spain
7 CIBER de Enfermedades Raras (CIBERER) U741, 29071 Malaga, Spain
* Correspondence: ge1gapaj@uco.es (J.J.G.); szaldivar@uco.es (S.Z.-L.)
† These authors share senior authorship.

Simple Summary: In this study, we investigated the role of microRNA-215 (miR-215) during *Salmonella* Typhimurium (*S*. Typhimurium) infection using a combination of in vivo porcine models and in vitro epithelial cell lines. Several dysregulated microRNAs (miRNAs) were identified in the infected porcine ileum, with miR-215 showing significant downregulation. Gain-of-function experiments in porcine (IPEC-J2) and human (HT29) intestinal epithelial cells revealed that miR-215 overexpression inhibits ubiquitination, subsequently downregulating key host immune response pathways such as autophagy and inflammasome activation. Our findings demonstrate that miR-215, downregulated during *Salmonella* Typhimurium infection in porcine ileum, plays a crucial role in regulating the host immune response by targeting proteins involved in ubiquitination, thereby influencing inflammasome activation and autophagy.

Abstract: The host response to *S*. Typhimurium infection can be post-transcriptionally regulated by miRNAs. In this study, we investigated the role of miR-215 using both in vivo porcine infection models and in vitro intestinal epithelial cell lines. Several miRNAs were found to be dysregulated in the porcine ileum during infection with wild-type and SPI2-defective mutant strains of *S*. Typhimurium, with some changes being SPI2-dependent. Notably, miR-215 was significantly downregulated during infection. To explore its functional role, gain-of-function experiments were performed by transfecting porcine intestinal epithelial cells (IPEC-J2) with a miR-215-5p mimic, followed by label-free quantitative (LFQ) proteomic analysis. This analysis identified 157 proteins, of which 35 were downregulated in response to miR-215 overexpression, suggesting they are potential targets of this miRNA. Among these, E2 small ubiquitin-like modifier (SUMO)-conjugating enzyme UBC9 and E3 ubiquitin-ligase HUWE1 were identified as key targets, both of which are upregulated during *S*. Typhimurium infection. The miR-215-mediated downregulation of these proteins resulted in a significant decrease in overall ubiquitination, a process crucial for regulating inflammasome activation and autophagy. Consistently, inflammasome markers caspase 1 (CASP1) and apoptosis-associated speck-like protein containing a CARD (ASC), as well as autophagy markers microtubule-associated protein 1A/1B-light chain 3 (LC3B) and Ras-related protein Rab-11 (RAB11A), showed decreased expression in miR-215 mimic-transfected and infected IPEC-J2 cells. To further validate these findings, human intestinal epithelial cells (HT29) were used as a complementary model, providing additional insights into conserved immune pathways and extending the observations made in the porcine system. Overall, our findings demonstrate that miR-215 plays a significant role in modulating host inflammasome activation and autophagy by targeting proteins involved in ubiquitination during *S.* Typhimurium infection.

Keywords: microRNA; salmonellosis; inflammation; zoonosis; immune response

1. Introduction

Salmonellosis is one of the leading foodborne zoonoses worldwide, ranking second within the European Union (EU) and affecting over 60,000 people every year [1]. With the establishment of control programs in poultry, swine have emerged as an important source of infection, along with turkey and cattle. Pigs remain as carriers after infection, not showing any signs of disease and thus silently spreading bacteria through the food chain, which represents a considerable risk to public health. Given the importance of the pork industry in the economy of many producing and exporting countries, economic losses caused by a widespread disease such as salmonellosis can be very significant. Among all disease-causing serovars, one of the most frequently isolated serovars in pigs in Europe is *Salmonella enterica* serovar Typhimurium (hereinafter *S.* Typhimurium) [1,2].

The interaction between bacterial pathogens and the host immune system is a complex process involving numerous molecular mechanisms. In this context, the regulation of inflammasome activation and autophagy plays a pivotal role in the innate immune response to bacterial infections. MicroRNAs have emerged as key regulators of the infection process, modulating gene expression at the post-transcriptional level [3–5]. Mature miRNAs have the capacity to bind to hundreds of target genes by complementarity with the mRNA, typically in the 3′ untranslated region (UTR). This binding results in the silencing of the target mRNA through mechanisms such as mRNA degradation or inhibition of its translation by ribosomes [6]. It has been demonstrated that infections by bacterial pathogens can be modulated by miRNAs in a very complex manner, altering not only immune pathways, but also cell death, cytoskeletal organization and autophagy [7,8]. It has been demonstrated that miRNAs such as miR-194, miR-155, miR-125, let-7i and miR-21 can regulate host responses to *S.* Typhimurium infection [9,10]. For instance, miR-194 has been shown to modulate inflammatory response by regulating the toll-like receptor 4 (TLR4)-mediated signaling pathway [11]; miR-21 induces macrophage activation [12]; and let-7i-3p is involved in bacterial adhesion and intracellular replication [9]. More recently, it has been shown that alteration of the intestinal mucosal miRNA profile (miRNome) during *S.* Typhimurium infection contributes positively to the spread of infection, mainly as a consequence of the downregulation of miR-15a, miR-15b, miR-16, miR-22, miR-421, miR-744 and let-7i [13].

To gain deeper insight into the regulatory process exerted by miRNAs, it is crucial to understand the molecular pathogenesis of *Salmonella* infection. After traveling through the gastrointestinal tract, *S.* Typhimurium invades the intestinal mucosal epithelium, causing a strong inflammatory reaction [14,15]. *S.* Typhimurium employs effector proteins, encoded by pathogenicity islands 1 and 2 of the bacterial chromosome (SPI1 and SPI2, respectively) [16]. These proteins are secreted by the bacterial type III secretion system and allow entry into epithelial mucosa enterocytes, forming the *Salmonella*-containing vacuole and thereby creating a suitable environment for intracellular permanence and replication [17]. Bacterial entry into cells triggers the innate immune response, characterized by synthesis of pro-inflammatory cytokines, which further promote the recruitment of innate immunity cells (e.g., macrophages, neutrophils) and magnify the inflammatory response. *Salmonella* invasion is recognized by the nucleotide oligomerization domain producing leucine rich repeat-like receptors (NLR), such as the NLR family caspase recruitment (CARD) domain-containing protein 4 (NLRC4) and the NLR family pyrin domain-containing 3 (NLRP3), which activate the inflammasome complex. Inflammasome is composed of an NLR, an adaptor protein (ASC), and an effector protein (CASP1), and leads to the production of interleukin 1β (IL1β) and interleukin 18 (IL18) [18]. Elimination of the pathogen and unwanted proteins is carried out by autophagy, which is a cellular degradation and recycling process that also contributes to the regulation of inflammation.

Ubiquitination, a post-translational modification involving the covalent attachment of ubiquitin molecules to target proteins, is a key regulatory mechanism for both inflammasome activation and autophagy [19]. The conjugation of ubiquitin is mediated by E1 ubiquitin-activating enzymes, E2 ubiquitin-conjugating enzymes, and E3 ubiquitin ligases. Upon infection, ubiquitin-coated proteins or *Salmonella* are recognized by autophagy receptors and sent to autophagosomes for degradation, which will fuse with lysosomes to form the autolysosome [19,20]. Also, the ubiquitination/deubiquitination of inflammasome components such as ASC, NLRP3 or CASP1 regulates the activation of the inflammasome, which helps maintaining inflammatory homeostasis [21]. *Salmonella* has developed mechanisms to counteract host defense by hijacking host ubiquitin pathways, thus evading the immune response [19]. Some *Salmonella* SPI1 and SPI2 effectors have E3 ligase activity (i.e., SopA, SspH1, SspH2 and SlrP) [22–24]. SopA can stimulate host E3 ubiquitin ligases, resulting in their proteasomal degradation during infection, SspH1 inhibits nuclear factor kappa-light-chain-enhancer of activated B cells (NFκB) pathway by manipulating host ubiquitination, and SspH2 enhances interleukin 8 secretion [19]. On the contrary, other virulence factors such as SseL have deubiquitinase (DUB) activity [25].

As previous studies have demonstrated that expression of miRNAs changes during *S.* Typhimurium infection in porcine ileum [11], here, we aimed to evaluate if a *S.* Typhimurium strain lacking SPI2, which is essential for intracellular replication and immune response activation, induces miRNA dysregulation during infection. We identified several miRNAs that are differentially expressed, choosing downregulated miR-215 for further functional analysis in intestinal cell lines. MiR-215 is a highly conserved microRNA implicated in various biological processes, including cell cycle regulation, differentiation, and stress responses [26,27], with emerging evidence suggesting a potential role in modulating immune responses and inflammation [28,29]. However, its precise function in regulating ubiquitination and its effects on inflammasome activation and autophagy during *S.* Typhimurium infection in porcine intestinal cells remain poorly understood. We hypothesize that miR-215 influences the expression of genes involved in the ubiquitination machinery, thereby impacting inflammasome activation and the induction of autophagy in response to *S.* Typhimurium infection.

This study aims to uncover the molecular mechanisms underlying innate immune responses in porcine intestinal cells during *S.* Typhimurium infection, offering potential insights for developing innovative therapeutic strategies to combat porcine salmonellosis and enhance food safety.

2. Materials and Methods

2.1. *Experimental Infection and Tissue Sample Collection*

Twelve recently weaned piglets (commercial hybrids of Landrace × Large White × Pietrain) of approximately 4 weeks of age were used in this study. Experimental procedures were performed at University of León (Spain) following a methodology previously described [14]. Briefly, piglets were divided into the following three groups: four pigs were orally inoculated with 5 mL of saline solution (control group, C); four pigs were orally inoculated with 5 mL of culture broth (brain heart infusion, BHI, Sigma-Aldrich, St. Louis, MO, USA) containing 10^8 colony forming units (CFU)/mL of the *S*. Typhimurium strain SL1344 (ST group); and the remaining four pigs were inoculated with 5 mL of culture broth (BHI, Sigma-Aldrich, St. Louis, MO, USA) containing 10^8 CFU/mL of a SL1344 SPI2-defective mutant (MUT group) kindly provided by Dr. Carmen Aguilar (University Würzburg, Germany). Prior to in vivo infection and following a previously published methodology [30], bacterial invasion was assessed in both strains, confirming comparable invasion capacities. All animals in the ST and MUT groups tested positive for *Salmonella* in fecal samples by microbiological cultures, while all animals from the C group tested negative. At 2 days post-infection (dpi), animals were humanely euthanized, and ileum sections were collected and snap-frozen in liquid nitrogen for further processing.

2.2. *RNA Isolation*

RNA was isolated from ileal mucosa, as previously described [11]. Briefly, intestinal mucosa was scraped from the intestinal luminal surface with a sterile razor blade, and immediately homogenized in mirVana miRNA isolation kit lysis buffer (Ambion Inc., Austin, TX, USA) using a rotor–stator homogenizer. RNA extraction was performed following manufacturer instructions. Eluted RNA was treated with TURBO DNA-free™ Kit (Ambion Inc., Austin, TX, USA) and RNA integrity was assessed in the Agilent Bioanalyzer 2100 (Agilent Technologies, Palo Alto, CA, USA); only samples with RNA integrity numbers (RIN) ≥ 8 were used for sequencing.

2.3. *Small RNA Sequencing*

Library preparation and small RNA sequencing were performed at the Genomics core service at the Institute of Biomedical Research Barcelona (IRB), and data obtained were bioinformatically pre-processed and analyzed at the Supercomputing and Bioinnovation Center of the University of Malaga (UMA-SCBI), following a methodology previously described [11]. Data were graphed with Volcanoser [31], miRNA sequences were retrieved from miRbase and aligned using Clustal Omega [32], and structure was predicted using mfold [33].

2.4. *Cell Culture*

IPEC-J2 cells, a non-transformed epithelial cell line derived from the jejunum of a neonatal piglet, were used as the primary model for our experiments due to their porcine origin, which aligns with our in vivo infection model. These cells were cultured in Dulbecco's modified Eagle's and Ham's F-12 medium (Capricorn Scientific GmbH, Ebsdorfergrund, Germany) supplemented with 5% fetal bovine serum (Gibco; Thermo Fisher Scientific, Inc., Waltham, MA, USA), in a humidified atmosphere with 5% CO_2 at 37 °C. To evaluate the invasion capacities of both *Salmonella* strains (ST and MUT) used in the in vitro experiments, gentamicin protection assays and quantification using *S.* Typhimurium-specific probes and primers were performed as previously reported [30], which confirmed their ability to invade the cells with no statistical differences between them. For specific functional validation experiments requiring protein-level analysis, significant challenges due to the limited availability and poor performance of porcine-specific antibodies were encountered. To overcome this technical limitation and ensure robust protein detection, the study was supplemented with the human epithelial HT29-MTX-E12 cell line (hereinafter HT29). This well-characterized human cell line allows for more reliable protein quantification using widely available human-specific antibodies. HT29 cells were cultured in Dulbecco's modified Eagle's medium high glucose (Biowest) supplemented with 10% fetal bovine serum and 1% non-essential amino acids (both from Gibco; Thermo Fisher Scientific, Inc., Waltham, MA, USA). By using both cell lines, it was possible to leverage the species-specific advantages of IPEC-J2 cells while addressing the technical constraints in protein analysis using HT29 cells.

2.5. Transfection and Infection

microRNA mimic hsa-miR-215-5p (C-300570-05, homologous to ssc-miR-215), and mimic negative control (CN-001000-01-05) were obtained from miRIDIAN (Dharmacon Inc., Lafayette, CO, USA). Mimic miR-215 and negative control were transfected using Lipofectamine™ RNAiMAX (Thermo Scientific Inc., Waltham, MA, USA), following the direct transfection protocol recommended by the manufacturer, at a final miRNA concentration of 50 nM. Briefly, 7.5×10^4 or 2.5×10^5 cells (per well) were seeded, 24 h before transfection, on 24- or 6-well plates, respectively. Following the objectives of the experiment, 24-well plates were used for RNA isolation and 6-well plates for protein isolation. Transfected cells were incubated at 37 °C in a 5% CO_2 humidified atmosphere for 48 h, and then infected with *S.* Typhimurium phage type DT104 [11] (optical density at 600 nm (OD_{600}) = 0.8, multiplicity of infection (MOI) = 100). After 1 h of infection, the medium was replaced with fresh medium containing gentamicin (50 µg/mL) to kill extracellular bacteria and incubated for 2 h and 24 h. At that time, cells were scraped and collected using radio-immunoprecipitation assay (RIPA) buffer. The transfection of miRNA was confirmed using quantitative polymerase chain reaction (qPCR) analysis, as previously described [11].

2.6. Protein Sample Preparation and Liquid Chromatography with Tandem Mass Spectrometry (LC-MS/MS) Proteomics

Protein concentration was measured using the Pierce™ BCA Protein Assay Kits (Thermo Scientific Inc., Waltham, MA, USA), and 100 µg were sent to the Proteomics Facility at Research Support Central Service, University of Cordoba (SCAI-UCO), for sample preparation and analysis. Protein extracts underwent clean-up in one dimensional sodium dodecyl sulfate-polyacrylamide gel electrophoresis with 10% polyacrylamide. After electrophoresis at 100 V, gel was stained with Coomassie Blue and protein bands were cut out and stored in water until digestion. Protein bands were first destained in 200 mM ammonium bicarbonate (AB)/50% acetonitrile, then in 100% acetonitrile, reduced with 20 mM dithiothreitol and incubated in 25 mM AB at 55 °C. After cooling to room temperature, free thiols were alkylated with 40 mM iodoacetamide. Following two AB washes, proteins were digested with trypsin (Promega, Madison, WI, USA) and incubated overnight at 37 °C. Digestion was stopped by adding trifluoroacetic acid to a final concentration of 1%. Digested samples were dried and resuspended in 0.1% formic acid (FA).

Nano liquid chromatography (LC) was performed on a Dionex Ultimate 3000 nano ultra performance LC (Thermo Scientific Inc., Waltham, MA, USA) with a C18 75 µm × 50 Acclaim Pepmap column (Thermo Scientific Inc., Waltham, MA, USA). Peptides were loaded onto a 300 μm × 5 mm Acclaim Pepmap precolumn (Thermo Scientific Inc., Waltham, MA, USA) in 2% acetonitrile/0.05% trifluoroacetic acid, then separated at 40 °C with a gradient of buffer A (water with 0.1% FA) and buffer B (20% acetonitrile with 0.1% FA) over 85 min of chromatography. Peptide cations were ionized and analyzed on a Thermo Orbitrap Fusion (Q-OT-qIT, Thermo Scientific Inc., Waltham, MA, USA), operated in positive mode with survey scans at 120K resolution. Tandem mass spectrometry (MS/MS) involved quadrupole isolation, collision-induced dissociation fragmentation, and rapid scan MS analysis, sampling only precursors with charge states 2–5. The instrument operated in top speed mode with 3 s MS2 cycles, excluding precursors dynamically.

MS2 spectra were analyzed using MaxQuant software v. 1.6.17.0 [34], with the Andromeda search engine set against a Uniprot_proteome_Sus-scrofa_Oct21 database. Peptides were generated through tryptic digestion with up to one missed cleavage, fixed carbamidomethylation of cysteines, and variable oxidation of methionine. The mass tolerance was 10 ppm, and product ions were searched with 0.6 Da tolerance. Peptide matches were filtered to a 1% false discovery rate (FDR). Quantification was conducted using the MaxLFQ label-free method [35], using retention time alignment and match-between-runs protocol. Heatmaps were generated using the Morpheus software (https://software.broadinstitute.org/morpheus/, accessed on 12 July 2024).

2.7. Western Blotting and Label-Free Quantitation

Briefly, 30 μg of protein were loaded per lane, separated on 12% SDS-PAGE gels, and blotted onto polyvinylidene difluoride membranes (Immobilon-P, Merck Millipore Inc., Burlington, MA, USA). Membranes were blocked in 5% non-fat milk for 1 h at room temperature, followed by overnight incubation at 4 °C with the following primary antibodies: Glyceraldehyde 3-phosphate dehydrogenase (GAPDH) (G8795; 1:10,000 dilution; Merck Millipore Inc., Burlington, MA, USA); LC3B (#3868; 1:1,000 dilution; Cell Signaling Technology Inc., Danvers, MA, USA); CASP1 (sc-398715; 1:1,000 dilution; Santa Cruz Biotechnology Inc., Dallas, TX, USA); ASC (sc-514414; 1:1,000 dilution; Santa Cruz Biotechnology Inc., Dallas, TX, USA); ubiquitin (sc-271289; 1:5,000 dilution; Santa Cruz Biotechnology Inc., Dallas, TX, USA); IL1β (sc-12742; 1:1,000 dilution, Santa Cruz Biotechnology Inc., Dallas, TX, USA); RAB11A (sc-166912; 1:1,000 dilution; Santa Cruz Biotechnology Inc., Dallas, TX, USA); and caspase 11 (CASP11) (sc-374615; 1:1,000 dilution; Santa Cruz Biotechnology Inc., Dallas, TX, USA). Membranes were further incubated with the secondary antibody Anti-Mouse immunoglobulin G (whole molecule)–peroxidase antibody produced in rabbit (A9044; 1:10,000; Merck Millipore Inc., Burlington, MA, USA) at room temperature during 1h, and immunoblots were visualized using an Immobilon UltraPlus Western HRP Substrate chemiluminescence kit (Merck Millipore Inc., Burlington, MA, USA) in a Chemidoc MP imaging system. The relative protein expression was analyzed using ImageJ software 1.4 (National Institutes of Health), using GAPDH as a control.

2.8. Reverse Transcription-Quantitative Real-Time PCR (RT-qPCR) and Statistical Analysis

Total RNA from cell cultures was extracted with TRIzol reagent (Thermo Scientific Inc., Waltham, MA, USA) as described in the manufacturer's protocol. For the miRNA expression analysis, 100 ng of total RNA per sample was reverse transcribed to cDNA as previously reported [36]; for mRNA analysis, 500 ng of RNA was reverse transcribed using qScript™ cDNA synthesis kit (QuantaBio LLC., Beverly, MA, USA), following the manufacturer's instructions. PCR reaction mixes and settings were as previously described [11], and primer sets are included in Table S3. Differential gene expression was calculated using the

$2^{-\Delta\Delta Ct}$ method [37] using the GenEx software (GenEx 6, bioMCC, Freising, Germany), and statistical analysis and graphs were performed using GraphPad Prism software (v9.4.1). Comparison between groups was performed using unpaired *t*-test or analysis of variance with Tukey's multiple comparison's test, setting $p = 0.05$ as threshold of significance.

3. Results

3.1. *miRNA Profiling and Characterization of miR-215 During S. Typhimurium Infection in Porcine Ileum*

To investigate the role of miRNAs during *S.* Typhimurium (ST) infection, we performed small RNA sequencing on ileal samples from pigs infected with wild-type ST, a SPI2-defective mutant (MUT), and uninfected controls (C). Our analysis identified a total of 334 known porcine miRNAs across all samples (Table 1 and Table S1). Comparison between infected and control samples revealed differences in miRNA expression patterns, with 43 miRNAs differentially expressed in ST-infected samples and 27 in MUT-infected samples (p value < 0.05). It was observed that some miRNAs behaved similarly in both groups ST and MUT, such as upregulated miR-29b, miR-144-5p/3p, miR-181b-3p, miR-301b-3p, miR-374b-3p and miR-20b, as well as downregulated miR-31-5p, miR-676-3p, miR-574-5p and miR-9.

Table 1. Top 10 differentially up- (red) and down-regulated (green) miRNAs in ileum samples from the two infected study groups (ST and MUT).

	ID	Fold Change	*p*-Value	FDR
ST/C	miR-29b-3p	2.83	1.6×10^{-7}	4.9×10^{-5}
	miR-144-5p	2.65	9.0×10^{-5}	5.7×10^{-3}
	miR-181b-3p	2.50	2.2×10^{-4}	9.9×10^{-3}
	miR-21-3p	2.36	6.9×10^{-6}	1.1×10^{-3}
	miR-301a-3p	2.14	4.6×10^{-4}	1.6×10^{-2}
	miR-144-3p	2.11	2.3×10^{-3}	4.3×10^{-2}
	miR-374b-3p	2.07	5.9×10^{-4}	1.9×10^{-2}
	miR-9-3p	2.07	1.1×10^{-3}	2.8×10^{-2}
	miR-301b-3p	2.05	3.3×10^{-3}	4.7×10^{-2}
	miR-20b-3p	1.99	7.3×10^{-3}	7.5×10^{-2}
	miR-31-3p	−2.00	2.0×10^{-2}	1.3×10^{-1}
	miR-582-5p	−2.18	1.7×10^{-3}	3.8×10^{-2}
	miR-31-5p	−2.20	4.2×10^{-4}	1.6×10^{-2}
	miR-215-5p	−2.34	1.6×10^{-2}	1.2×10^{-1}
	miR-676-3p	−2.53	4.3×10^{-5}	3.4×10^{-3}
	miR-574-5p	−2.54	1.8×10^{-4}	9.4×10^{-3}
	miR-196b-5p	−2.72	1.3×10^{-1}	3.7×10^{-1}
	miR-196a-5p	−2.78	6.2×10^{-2}	2.4×10^{-1}
	miR-215-3p	−2.87	2.3×10^{-3}	4.3×10^{-2}
	miR-192-3p	−2.93	1.5×10^{-5}	1.6×10^{-3}
MUT/C	miR-451a	2.44	3.6×10^{-3}	5.9×10^{-1}
	miR-184	1.97	2.4×10^{-2}	9.4×10^{-1}
	miR-144-5p	1.83	4.0×10^{-2}	9.4×10^{-1}
	miR-374b-3p	1.77	5.6×10^{-3}	5.9×10^{-1}
	miR-20b-3p	1.69	3.8×10^{-2}	9.4×10^{-1}
	miR-144-3p	1.59	8.9×10^{-2}	9.7×10^{-1}
	miR-301b-3p	1.57	6.5×10^{-2}	9.4×10^{-1}
	miR-181b-3p	1.56	5.6×10^{-2}	9.4×10^{-1}
	miR-486-5p	1.54	2.2×10^{-2}	9.4×10^{-1}
	miR-29b-3p	1.52	2.4×10^{-2}	9.4×10^{-1}
	miR-676-3p	−1.91	5.6×10^{-3}	5.9×10^{-1}
	miR-802	−1.88	1.8×10^{-2}	9.4×10^{-1}
	miR-141-3p	−1.70	5.9×10^{-2}	9.4×10^{-1}
	miR-31-5p	−1.61	6.9×10^{-2}	9.4×10^{-1}
	miR-22-3p	−1.58	2.8×10^{-2}	9.4×10^{-1}
	miR-615-3p	−1.57	6.5×10^{-2}	9.4×10^{-1}
	miR-574-5p	−1.56	5.6×10^{-2}	9.4×10^{-1}
	miR-200c-5p	−1.55	9.2×10^{-2}	9.7×10^{-1}
	miR-200c-3p	−1.55	3.1×10^{-2}	9.4×10^{-1}
	miR-10b-3p	−1.49	4.0×10^{-2}	9.4×10^{-1}

Notably, we observed that the modulation of several miRNAs was influenced by the presence of a functional SPI2. Among these, miR-215 showed significant downregulation in ST-infected samples compared to the C group. No significant changes were observed in MUT-infected samples, suggesting that the downregulation of miR-215 during ST infection may be associated with SPI2 function, potentially through direct or indirect mechanisms.

Bioinformatic analysis revealed that miR-215 is highly conserved across species, including humans, pigs, and other mammals. Notably, only miR-215-5p has been reported in pigs, exhibiting complete sequence homology with its human counterpart. However, the sequencing of small RNAs carried out in the present work and subsequent mapping into the human database allowed us to detect that miR-215-3p is also present in swine, although it is expressed in the *S.* Typhimurium-infected ileum to a significantly lesser extent than miR-215-5p (Figure S1, Table S1). Consequently, we used the latter to investigate the function of miR-215 during *S.* Typhimurium infection using a mimic-based gain-of-function approach.

3.2. *miR-215 Modulates Ubiquitination, Inflammasome Activation and Autophagy*

To investigate the regulatory role of miR-215, porcine intestinal epithelial cells (IPEC-J2) were transfected with a miR-215-5p mimic, subsequently infected with *S.* Typhimurium, and analyzed using label-free quantitative (LFQ) proteomics. Four experimental groups were established: mock control cells (transfected with a mimic negative control), mock control cells infected with *S.* Typhimurium, cells transfected with the miR-215 mimic, and cells transfected with the miR-215 mimic and infected with *S.* Typhimurium. Transfection resulted in a 15-fold overexpression of miR-215 compared to negative controls. The primary focus of the analysis was on the regulatory capacities of the miR-215 mimic under condition (a), where miR-215 mimic-transfected cells were compared to mock control cells. This comparison aimed to identify proteins directly downregulated by the mimic, as these are potential post-transcriptional targets of miR-215. Additional comparisons explored the role of miR-215 in the context of bacterial infection: cells transfected with the miR-215 mimic and infected with *S.* Typhimurium were compared to mock control cells infected with *S.* Typhimurium (b) to study the combined effects of miR-215 overexpression and bacterial infection, while mock control cells infected with *S.* Typhimurium were compared to mock control cells (c) to examine proteomic changes induced by *S.* Typhimurium infection alone. In total, LFQ proteomics identified 162 proteins/peptides across these conditions, of which 127 were upregulated and 35 were downregulated following miR-215 mimic transfection. Proteins downregulated in condition (a) were prioritized for further analysis as potential direct targets of miR-215-5p, while the additional conditions provided a deeper understanding of how miR-215-5p modulates host responses during bacterial infection (Figure 1, Table S2).

Proteins that were downregulated in condition (a) and upregulated in condition (c) are likely to be regulated by miR-215-5p. This is because miR-215 was overexpressed in condition (a) due to mimic transfection, leading to the repression of its targets, while in condition (c), miR-215 was endogenously downregulated as a result of *S.* Typhimurium infection, allowing the expression of those same targets to increase. These patterns suggest a direct regulatory relationship between miR-215 and these proteins. Among these proteins, UBE2I (ubiquitin conjugating enzyme E2) and HUWE1 (E3 ubiquitin ligase) play central roles in the ubiquitination pathway. The modulation of these proteins suggests that miR-215 plays a critical role in host–pathogen interactions, particularly in processes such as inflammasome activation and autophagy. To evaluate the overall ubiquitination status and its regulation by miR-215, ubiquitin expression was analyzed in miR-215 mimic-transfected IPEC-J2 intestinal cells 2 h after infection with *S.* Typhimurium, using a human antibody capable of detecting mono- and polyubiquitinated proteins. Quantification of sample lanes, normalized to the internal control GAPDH, revealed a significant decrease ($p < 0.01$) in overall ubiquitination in miR-215 mimic-overexpressing cells infected with *S.* Typhimurium, compared to *S.* Typhimurium-infected IPEC-J2 cells without mimic transfection (Figure 2A). Additionally, HUWE1 and UBE2I gene expression levels were evaluated during infection, showing a decrease in expression when miR-215 mimic was overexpressed (Figure 2B). Although IPEC-J2 cells were chosen as the primary model due to their porcine origin, it was observed that human intestinal epithelial cells (HT29), with more physiologically relevant characteristics in their mucus layer formation, exhibit higher efficiency of *Salmonella* infection and a more robust immune response compared to other cell lines [38]. Given the conserved nature of ubiquitination pathways across species, the use of HT29 cells provides additional insights into the dynamics of ubiquitination and miR-215 regulation during infection. To further investigate ubiquitination over time during the infection process, HT29 cells were included, with an additional time point at

24 h post-infection. In this model, ubiquitination was significantly inhibited by miR-215 mimic overexpression at both 2 h ($p < 0.001$) and 24 h ($p < 0.05$) post-infection, compared to infected mock controls (Figure 2C). These findings highlight the relevance of miR-215 in regulating ubiquitination in the context of *S.* Typhimurium infection and demonstrate the utility of complementary human cell models to validate and expand observations made in the porcine system.

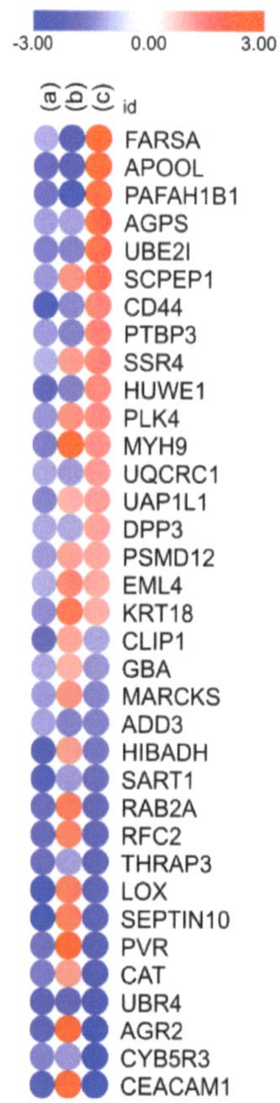

Figure 1. Fold change protein expression on IPEC-J2 cells in the following three experimental situations (columns) studied: (**a**) miR-215 mimic-transfected cells compared to mock control cells; (**b**) cells transfected with the miR-215 mimic and infected with *S.* Typhimurium compared to mock control cells infected with *S.* Typhimurium (infection duration 2 h); and (**c**) mock control cells infected with *S.* Typhimurium compared to mock control cells (infection duration 2 h). Red indicates overexpression and blue indicates downregulation.

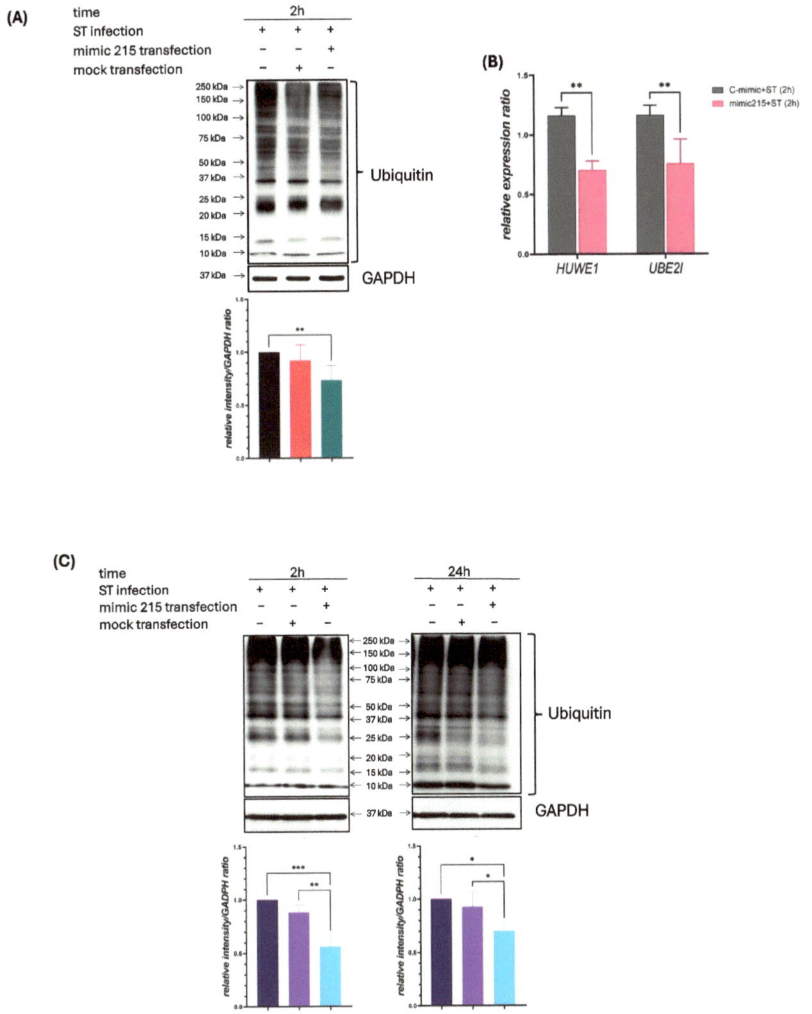

Figure 2. (**A**) Ubiquitin expression in IPEC-J2 cells, IPEC-J2 cells transfected with negative controls (mock) and IPEC-J2 cells transfected with miR-215 mimic, 2 h after infection with ST (ratio of the relative quantification intensity to GAPDH below each lane). (**B**) Relative gene expression (to cyclophilin A) of HUWE1 and UBE2I genes. (**C**) Ubiquitin expression in HT29 cells, HT29 cells transfected with negative controls (mock) and HT29 cells transfected with miR-215 mimic, 2 h and 24 h after infection with ST (ratio of the relative quantification intensity to GAPDH below each lane). * $p < 0.05$; **: $p < 0.01$; ***: $p < 0.001$.

Given the crucial role of ubiquitination in inflammasome activation, inflammasome markers were further analyzed. In IPEC-J2 cells, miR-215 overexpression during infection significantly reduced the *IL1β* and *CASP1* gene expression ($p < 0.01$ and $p < 0.05$, respectively) (Figure 3A). This reduction was confirmed at the protein level in HT29 cells, with a significant decrease in IL1β and CASP1 expression observed 2 h post-infection ($p < 0.05$) and further downregulation at 24 h post-infection ($p < 0.01$) (Figure 3B,C). Additionally, ASC was downregulated at 24 h ($p < 0.01$), while no significant changes were observed in CASP11 or IL1β at this later time point (Figure 3B,C).

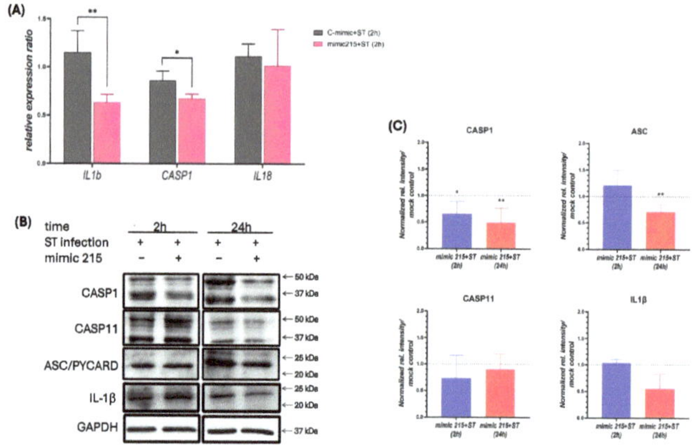

Figure 3. (**A**) Relative gene expression (to cyclophilin A) of inflammasome markers *IL1β*, *CASP1* and *IL18* in IPEC-J2 cells after mock or miR-215 mimic transfection and 2 h infection with *S.* Typhimurium. (**B**) Western blot of CASP1, ASC, CASP11 and IL1b in HT-29 cells transfected either with miR-215 or mock controls, 2h and 24h after infection by *S.* Typhimurium and (**C**) quantification of Western blot by expression ratio (ratio of relative intensity normalized to GAPDH compared to mock controls). Expression in transfected and infected samples were compared to their infected mock controls for each time point. * $p < 0.05$; **: $p < 0.01$.

Ubiquitination also plays a pivotal role in autophagy, as ubiquitinated proteins and pathogens are recognized and directed to autophagosomes for elimination. In IPEC-J2 cells, miR-215 mimic transfection during infection resulted in a significant reduction in genes coding for autophagy-related markers, including LC3B, SQSTM1 (sequestosome-1, p62) and lysosomal marker LAMP1 (Lysosomal-associated membrane protein 1) (Figure 4A). Consistently, in HT29 cells, protein expression of LC3B (specifically the LC3B-I form) and RAB11A also decreased significantly 24 h post-infection in miR-215 mimic-transfected cells (Figure 4B,C).

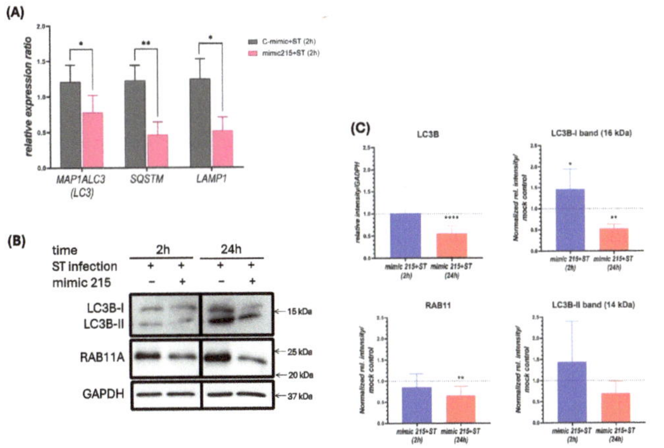

Figure 4. (**A**) Relative gene expression (to cyclophilin A) of autophagy markers *MAP1ALC3* (LC3), *SQSTM1* (p62) and *LAMP1* in IPEC-J2 cells after mock or miR-215 mimic transfection and 2 h infection with *S.* Typhimurium. (**B**) Western blot of LC3B and RAB11A in HT-29 cells transfected either with miR-215 or mock controls, 2 h and 24 h after infection by *S.* Typhimurium. (**C**) Quantification of Western blot by expression ratio (ratio of relative intensity normalized to GAPDH compared to mock controls). Expression in transfected and infected samples were compared to their infected mock controls for each time point. * $p < 0.05$; **: $p < 0.01$; ****: $p < 0.0001$.

4. Discussion

Reducing the prevalence of *S.* Typhimurium in pigs is crucial for improving both animal and public health. However, the complexity of the interaction between *S.* Typhimurium and the host immune system, combined with the pathogen's ability to evade host defenses, makes this a particularly challenging task. Our research on the role of miR-215 in regulating the innate immune response, especially in inflammasome activation and autophagy, provides new insights that could contribute to more effective control strategies.

Infection with *S.* Typhimurium leads to the differential expression of miRNAs, which regulate host response pathways. Some of the miRNAs identified in the present study (miR-215, miR-192, miR-9 and miR-31) have been previously reported to be altered during infection with a different *S.* Typhimurium isolate in pigs [11]. In our observations, we found similar expression patterns of certain miRNAs in both *S.* Typhimurium (ST) and mutant (MUT)-infected ileum, suggesting dysregulation in response to *Salmonella* infection. For instance, we found upregulation of miR-29b, which has been previously associated with positive regulation of *Shigella flexneri* infection, repressing UNC5C in a Rho-guanosine triphosphate hydrolases (GTPase)-dependent manner, thereby enhancing filopodia formation [39]. Similarly, upregulated miR-144 agrees with previous reports correlating its expression with inflammation in peripheral blood mononuclear cells from patients infected with *Mycobacteroides abscessus*, where it increased the expression of cytokines and chemokines such as IL1β, tumor necrosis factor (TNF), chemokine (C-X-C motif) ligand 2 (CXCL2) or interleukin 6 [40]. Also, miR-144-5p has been shown to modulate inflammatory response by targeting the NLR pathway, downregulating the NLRP3 inflammasome pathway [41].

Interestingly, some miRNAs were found differently expressed depending on the presence of *Salmonella* SPI2: miR-451a was overexpressed only in the SPI2 defective strain (MUT), while downregulation of the family of miR-215 (miR-215-5p, miR-215-3p and miR-192-3p) was found only in ST. As SPI2 is essential for intracellular survival and replication, we hypothesize that SPI2 may play a role in miRNA-mediated regulation during *Salmonella* infection. Contrary to our findings, a reduced expression of miR-451a has been previously found in COVID-19 patients, where miR451a was described as a down-regulator of the interleukin 6 receptor (IL6R), contributing to the cytokine storm characteristic of this disease [42]. Mir-451a has also been associated with other infections such as malaria [43].

Regarding miR-215, previous studies have reported a downregulation of this miRNA in porcine ileum during acute *S.* Typhimurium infection (log2FC = −4.1) [11]. Those results agree with the present study, where miRNA sequencing analysis of ST samples have revealed downregulation of all members of the miR-215/192 family, including miR-215-3p (FC = −2.9, $p < 0.001$), miR-215-5p (FC = −2.3, $p < 0.05$), miR-192-3p (FC = −2.9, $p < 0.001$) and miR-192-5p (FC = −1.6, $p < 0.05$). However, these findings contrast with a study on *Salmonella*-infected chicken ceca, in which the upregulation of miR-215-5p was reported [44]. Additionally, miR-215 has been found to be downregulated in skeletal muscle following lipopolysaccharide (LPS) challenge [45], and it has been suggested to regulate resistance to *Escherichia coli* in pigs [46]. MiR-215 is also associated with various biological processes and diseases, including pulmonary fibrosis [47] and several types of cancer [26,27]. Moreover, it has been identified as a regulator of inflammatory pathways, indicating that miR-215 may play a critical role in *Salmonella* infection [28,29,48]. However, the mechanism by which miR-215 is involved in the *Salmonella* infection process is still poorly understood. In related findings, it has been shown that hepatitis B virus can induce autophagy by downregulating miR-192-3p in hepatocellular carcinoma cell lines [49]. However, a study by Deng et al. observed the upregulation of miR-215 in *Mycobacterium tuberculosis* patients, concluding that this miRNA inhibits autophagosome-lysosome fusion in macrophages [29]. Interestingly, while only miR-215-5p had been reported in the porcine miRNA database, our work, by mapping the sequenced miRNAs to the human database, revealed the presence of miR-215-3p in porcine ileum, albeit at much lower expression than miR-215-5p.

To explore the function of miR-215-5p during *S.* Typhimurium infection, we used a mimic-based gain-of-function approach, identifying several potential targets involved in the ubiquitination process. Among these proteins we found UBE2I, also known as UBC9, which is a SUMO-conjugating enzyme, key in the SUMOylation pathway [50]. Depletion of UBC9 has been associated with overexpression of miR-30c and miR-30e during *S.* Typhimurium infection [51], suggesting that multiple miRNAs may regulate this protein. Another target, HUWE1, is an E3 ubiquitin ligase responsible for conjugating ubiquitin molecules to substrates [52], and we hypothesize that miR-215 may influence this process. Our ubiquitination assays indicated that miR-215 decreased ubiquitination both 2 h and 24 h after infection in porcine and human cell lines. Ubiquitination is a post translational modification (PTM) that can mark proteins for degradation or regulate processes such as endocytosis and intracellular trafficking [53]. This PTM can be modulated by the host or the pathogen during infection. *Salmonella* Typhimurium uses ubiquitination to counteract host immune responses [54], leading to selective autophagy, contributing to inflammasome assembly, or playing important roles in interferon- or NFκB-mediated immune signaling. The regulation of this process by miRNAs has been previously described. For instance, miR-339 targets ubiquitin-specific peptidase 8 (USP8), and has been reported to be involved in deubiquitination during tumorigenesis [55]. Similarly, let-7 downregulates E3 ubiquitin ligase TRIM71 in liver cancer [56]. Other miRNAs, including miR-7, miR-9 and miR-181c, are involved in the precise regulation of ubiquitination and deubiquitination processes in neurodegenerative diseases, where they interfere with proteasome-mediated degradation [57].

Ubiquitination plays a crucial role in the host autophagy process. It allows the host to ubiquitinate *Salmonella* for degradation, but *Salmonella* can also evade this process by manipulating or mimicking deubiquitinases, thereby reverting the PTM [54]. During *Salmonella* infection, the host cell attempts to remove unwanted cytosolic materials or bacteria through autophagy, forming autophagosomes that will fuse to lysosomes leading to degradation. In this selective process, ubiquitin binds to autophagy receptors (e.g., sequestosome-1 SQSTM1/p62 protein, optineurin) and microtubule-associated protein 1A/1B-light chain 3 (LC3), recruiting autophagy-related proteins (ATGs) and subsequently forming the autophagosome. During autophagy, the cytosolic form of LC3B (LC3B-I) is lipidated to form LC3B-II, which is associated with the autophagosomal membrane. LC3B-II is involved in the elongation and closure of the membrane, playing an essential role in autophagosome biogenesis. LC3B interacts with autophagy receptors that recognize and bind ubiquitinated cargo, ensuring the selective degradation of damaged organelles, misfolded proteins, and intracellular pathogens. We observed downregulation of LC3B-I but no changes in LC3B-II. We also found that miR-215 overexpression during infection leads to the decreased expression of the small GTPase RAB11, a small GTPase implicated in the maturation of autophagosomes by mediating autophagosomes–lysosome fusion, an essential step for the degradation of autophagic cargo [58]. RAB11 also facilitates the transport of immune receptors and other molecules to the cell surface, enhancing the cell's ability to respond to pathogens [58].

Ubiquitination also regulates the assembly and activation of inflammasomes. Canonical NLRC4 and NLRP3 inflammasomes, as well as the non-canonical caspase-11 pathways, are activated in *S.* Typhimurium infection [18,59]. The inflammasome is a cytosolic multiprotein complex that induces inflammation and pyroptotic cell death in response to both pathogen (PAMPs) and endogenous activators (DAMPs). Recognition of PAMPs or DAMPs leads to formation of the inflammasome complex, which results in activation of caspase-1, followed by cleavage and release of pro-inflammatory cytokines such as IL-1β and IL-18. In this study, we found that miR-215 induces the downregulation of caspase-1 and ASC, suggesting a decrease in inflammasome activation through the canonical pathways, given that caspase-11 expression was not altered. Previous studies have shown that E3 ligases, such as TRIM31 and MARCH7, ubiquitinate NLRP3, targeting it for proteasomal degradation or modulating its activity [60,61]. The ubiquitination of ASC is crucial for the recruitment and activation of pro-caspase-1, leading to the processing and release of IL-1β and IL-18 [19]. Therefore, further studies on the ubiquitination of these proteins will provide additional information on this process.

The importance of this study in controlling *S.* Typhimurium infections in pigs lies in its potential to identify novel targets for intervention. Reducing the prevalence of *S.* Typhimurium in pigs is essential for improving animal health, food safety and public health, particularly as antibiotic resistance becomes a growing concern. By elucidating the role of miR-215 in regulating immune responses, such as autophagy and inflammasome activation, the research provides a deeper understanding of the molecular mechanisms involved in host–pathogen interactions. Additionally, by revealing how the pathogen manipulates host processes like ubiquitination and autophagy, the study offers insights into how *S.* Typhimurium evades immune responses, potentially opening the door for interventions that boost the host ability to clear the pathogen. This research holds promise for more effective control measures, reducing the burden of infection in pigs and ultimately mitigating the risk of zoonotic transmission to humans.

5. Conclusions

Enhanced understanding of how microRNAs regulate critical biological processes, such as ubiquitination, in response to infection paves the way for developing alternative therapies to combat this disease. In summary, this study found that miR-215 is down-regulated in *S.* Typhimurium infection, which may cause increased ubiquitination, autophagy and inflammasome activation during infection.

Supplementary Materials: The following supporting information can be downloaded at: https://www.mdpi.com/article/10.3390/ani15030431/s1. Table S1: Reads and differential expression analysis from small RNA-seq of ST- and MUT-infected ileum. Table S2: Results from LFQ proteomics of miR-215 mimic transfected and ST-infected IPEC-J2 cells. Table S3: Primers used in this study. Figure S1: (A) Sequencing reads from porcine ileum mapped to human miRNA database. (B) alignment of human and porcine mature miR-215-3p sequence. (C) Structures of human and porcine stem-loop sequences, with the only human-porcine sequence mismatch highlighted.

Author Contributions: Conceptualization, S.Z.-L. and J.J.G.; methodology, C.E.-G. and R.F.-R.; bioinformatic analysis, R.B. and M.G.C.; validation, C.E.-G., J.M.S.-C. and R.F.-R.; formal analysis, C.E.-G. and S.Z.-L.; investigation, C.E.-G., S.Z.-L. and J.J.G.; resources, S.Z.-L. and J.J.G.; data curation, S.Z.-L.; writing—original draft preparation, S.Z.-L.; writing—review and editing, C.E.-G., R.B., M.G.C., J.M.S.-C., R.F.-R. and J.J.G.; supervision, J.J.G.; funding acquisition, S.Z.-L. and J.J.G. S.Z.-L. and J.J.G. equally contributed to this work and share senior authorship. All authors have read and agreed to the published version of the manuscript.

Funding: This work was supported by the Spanish Ministry of Economy and Competitiveness (AGL2017-87415-R), the Spanish Ministry of Science, Innovation and Universities (PID2022-142887OB-I00) and Plan Propio de Investigación 2020 from Universidad de Cordoba (Submodalidad 2.6. UC♀IMPULSA).

Institutional Review Board Statement: All animal procedures were performed in accordance with European regulations on the protection of animals used for experimental and other scientific purposes and with the approval of the Ethics and Animal Welfare Committee of the University of León, Spain (Ref. ULE OEBA-ULE-009-2017, registration number ES240890000172).

Informed Consent Statement: Not applicable.

Data Availability Statement: Sequencing data have been deposited at National Institutes of Health Sequencing Reading Archive (NCBI SRA) database under Bioproject accession ID PRJNA1160452. The mass spectrometry proteomics data have been deposited to the ProteomeXchange Consortium via the Proteomics Identification Database (PRIDE) (81) partner repository with the dataset identifier PXD054446.

Acknowledgments: The authors would like to thank Hector Argüello and Ana Carvajal from the Department of Animal Health at the Faculty of Veterinary Medicine from the University of León (Spain) for the help provided in the in vivo experimental infection.

Conflicts of Interest: The authors declare no conflicts of interest.

References

1. European Food Safety Authority (EFSA); European Centre for Disease Prevention and Control (ECDC). The European Union One Health 2022 Zoonoses Report. *EFSA J.* **2023**, *21*, e8442. [CrossRef]
2. Majowicz, S.E.; Musto, J.; Scallan, E.; Angulo, F.J.; Kirk, M.; O'Brien, S.J.; Jones, T.F.; Fazil, A.; Hoekstra, R.M.; International Collaboration on Enteric Disease "Burden of Illness" Studies. The global burden of nontyphoidal *Salmonella* gastroenteritis. *Clin. Infect. Dis.* **2010**, *50*, 882–889. [CrossRef] [PubMed]
3. Hong, J.S.; Kim, N.H.; Choi, C.Y.; Lee, J.S.; Na, D.; Chun, T.; Lee, Y.S. Changes in cellular microRNA expression induced by porcine circovirus type 2-encoded proteins. *Vet. Res.* **2015**, *46*, 39. [CrossRef] [PubMed]

4. Kappari, L.; Dasireddy, J.; Applegate, T.; Selvaraj, R.; Shanmugasundaram, R. MicroRNAs: Exploring their role in farm animal disease and mycotoxin challenges. *Front. Vet. Sci.* **2024**, *11*, 1372961. [CrossRef]
5. Hasankhani, A.; Bakherad, M.; Bahrami, A.; Shahrbabak, H.; Pecho, R.; Shahrbabak, M. Integrated analysis of inflammatory mRNAs, miRNAs, and lncRNAs elucidates the molecular interactome behind bovine mastitis. *Sci. Rep.* **2023**, *13*, 13826. [CrossRef]
6. Carthew, R.; Sontheimer, E. Origins and Mechanisms of miRNAs and siRNAs. *Cell* **2009**, *136*, 642–655. [CrossRef]
7. Aguilar, C.; Mano, M.; Eulalio, A. MicroRNAs at the Host-Bacteria Interface: Host Defense or Bacterial Offense. *Trends Microbiol.* **2019**, *27*, 206–218. [CrossRef]
8. Staedel, C.; Darfeuille, F. MicroRNAs and bacterial infection. *Cell. Microbiol.* **2013**, *15*, 1496–1507. [CrossRef]
9. Aguilar, C.; Cruz, A.R.; Rodrigues Lopes, I.; Maudet, C.; Sunkavalli, U.; Silva, R.J.; Sharan, M.; Lisowski, C.; Zaldivar-Lopez, S.; Garrido, J.J.; et al. Functional screenings reveal different requirements for host microRNAs in *Salmonella* and *Shigella* infection. *Nat. Microbiol.* **2020**, *5*, 192–205. [CrossRef]
10. Ro, Y.; Jo, G.; Jung, S.; Lee, E.; Shin, J.; Lee, J. *Salmonella*-induced miR-155 enhances necroptotic death in macrophage cells via targeting RIP1/3. *Mol. Med. Rep.* **2018**, *18*, 5133–5140. [CrossRef]
11. Herrera-Uribe, J.; Zaldívar-López, S.; Aguilar, C.; Entrenas-García, C.; Bautista, R.; Claros, M.G.; Garrido, J.J. Study of microRNA expression in *Salmonella* Typhimurium-infected porcine ileum reveals miR-194a-5p as an important regulator of the TLR4-mediated inflammatory response. *Vet. Res.* **2022**, *53*, 35. [CrossRef] [PubMed]
12. Herrera Uribe, J.L. Integrated Analysis of miRNA Expression in Response to *Salmonella* Typhimurium Infection in Pigs. Ph.D. Thesis, Universidad de Córdoba, Córdoba, Spain, 2017.
13. Aguilar, C.; Costa, S.; Maudet, C.; Vivek-Ananth, R.; Zaldívar-López, S.; Garrido, J.; Samal, A.; Mano, M.; Eulalio, A. Reprogramming of microRNA expression via E2F1 downregulation promotes *Salmonella* infection both in infected and bystander cells. *Nat. Commun.* **2021**, *12*, 3392. [CrossRef] [PubMed]
14. Herrera-Uribe, J.; Collado-Romero, M.; Zaldivar-Lopez, S.; Arce, C.; Bautista, R.; Carvajal, A.; Cirera, S.; Claros, M.G.; Garrido, J.J. Transcriptional analysis of porcine intestinal mucosa infected with *Salmonella* Typhimurium revealed a massive inflammatory response and disruption of bile acid absorption in ileum. *Vet. Res.* **2016**, *47*, 11. [CrossRef] [PubMed]
15. Ménard, S.; Lacroix-Lamandé, S.; Ehrhardt, K.; Yan, J.; Grassl, G.; Wiedemann, A. Cross-Talk Between the Intestinal Epithelium and *Salmonella* Typhimurium. *Front. Microbiol.* **2022**, *13*, 906238. [CrossRef]
16. Galán, J. *Salmonella* Typhimurium and inflammation: A pathogen-centric affair. *Nat. Rev. Microbiol.* **2021**, *19*, 716–725. [CrossRef]
17. de Souza Santos, M.; Orth, K. Subversion of the cytoskeleton by intracellular bacteria: Lessons from *Listeria*, *Salmonella* and *Vibrio*. *Cell. Microbiol.* **2015**, *17*, 164–173. [CrossRef]
18. Bryant, C. Inflammasome activation by *Salmonella*. *Curr. Opin. Microbiol.* **2021**, *64*, 27–32. [CrossRef]
19. Wang, L.; Yan, J.; Niu, H.; Huang, R.; Wu, S. Autophagy and Ubiquitination in *Salmonella* Infection and the Related Inflammatory Responses. *Front. Cell. Infect. Microbiol.* **2018**, *8*, 78. [CrossRef]
20. Gomes, L.; Dikic, I. Autophagy in antimicrobial immunity. *Mol. Cell* **2014**, *54*, 224–233. [CrossRef]
21. Xu, T.; Guo, Y.; Qi, X. Ubiquitination-Mediated Inflammasome Activation during Bacterial Infection. *Int. J. Mol. Sci.* **2019**, *20*, 2110. [CrossRef]
22. Kamanova, J.; Sun, H.; Lara-Tejero, M.; Galán, J. The *Salmonella* Effector Protein SopA Modulates Innate Immune Responses by Targeting TRIM E3 Ligase Family Members. *PLoS Pathog.* **2016**, *12*, e1005552. [CrossRef] [PubMed]
23. Rohde, J.; Breitkreutz, A.; Chenal, A.; Sansonetti, P.; Parsot, C. Type III secretion effectors of the IpaH family are E3 ubiquitin ligases. *Cell Host Microbe* **2007**, *1*, 77–83. [CrossRef] [PubMed]
24. Cordero-Alba, M.; Ramos-Morales, F. Patterns of expression and translocation of the ubiquitin ligase SlrP in *Salmonella enterica* serovar Typhimurium. *J. Bacteriol.* **2014**, *196*, 3912–3922. [CrossRef] [PubMed]
25. Rytkönen, A.; Poh, J.; Garmendia, J.; Boyle, C.; Thompson, A.; Liu, M.; Freemont, P.; Hinton, J.; Holden, D. SseL, a *Salmonella* deubiquitinase required for macrophage killing and virulence. *Proc. Natl. Acad. Sci. USA* **2007**, *104*, 3502–3507. [CrossRef]
26. Han, J.; Zhang, M.; Nie, C.; Jia, J.; Wang, F.; Yu, J.; Bi, W.; Liu, B.; Sheng, R.; He, G.; et al. miR-215 suppresses papillary thyroid cancer proliferation, migration, and invasion through the AKT/GSK-3β/Snail signaling by targeting ARFGEF1. *Cell Death Dis.* **2019**, *10*, 195. [CrossRef]
27. Liu, S.; Zhang, Y.; Huang, C.; Lin, S. miR-215-5p is an anticancer gene in multiple myeloma by targeting RUNX1 and deactivating the PI3K/AKT/mTOR pathway. *J. Cell. Biochem.* **2020**, *121*, 1475–1490. [CrossRef]
28. Cai, J.; Yang, J.; Liu, Q.; Gong, Y.; Zhang, Y.; Zheng, Y.; Yu, D.; Zhang, Z. Mir-215-5p induces autophagy by targeting PI3K and activating ROS-mediated MAPK pathways in cardiomyocytes of chicken. *J. Inorg. Biochem.* **2019**, *193*, 60–69. [CrossRef]
29. Deng, F.; Xu, P.; Miao, J.; Jin, C.; Tu, H.; Zhang, J. Pulmonary tuberculosis biomarker miR-215-5p inhibits autophagosome-lysosome fusion in macrophages. *Tuberculosis* **2023**, *143*, 102422. [CrossRef]

30. Bellido-Carreras, N.; Argüello, H.; Zaldívar-López, S.; Jiménez-Marín, Á.; Martins, R.P.; Arce, C.; Morera, L.; Carvajal, A.; Garrido, J.J. *Salmonella* Typhimurium Infection Along the Porcine Gastrointestinal Tract and Associated Lymphoid Tissues. *Vet. Pathol.* **2019**, *56*, 681–690. [CrossRef]
31. Goedhart, J.; Luijsterburg, M. VolcaNoseR is a web app for creating, exploring, labeling and sharing volcano plots. *Sci. Rep.* **2020**, *10*, 20560. [CrossRef]
32. Sievers, F.; Higgins, D. Clustal omega. *Curr. Protoc. Bioinform.* **2014**, *48*, 3–13. [CrossRef] [PubMed]
33. Zuker, M. Mfold web server for nucleic acid folding and hybridization prediction. *Nucleic Acids Res.* **2003**, *31*, 3406–3415. [CrossRef] [PubMed]
34. Cox, J.; Mann, M. MaxQuant enables high peptide identification rates, individualized p.p.b.-range mass accuracies and proteome-wide protein quantification. *Nat. Biotechnol.* **2008**, *26*, 1367–1372. [CrossRef] [PubMed]
35. Cox, J.; Hein, M.; Luber, C.; Paron, I.; Nagaraj, N.; Mann, M. Accurate proteome-wide label-free quantification by delayed normalization and maximal peptide ratio extraction, termed MaxLFQ. *Mol. Cell. Proteom.* **2014**, *13*, 2513–2526. [CrossRef]
36. Balcells, I.; Cirera, S.; Busk, P.K. Specific and sensitive quantitative RT-PCR of miRNAs with DNA primers. *BMC Biotechnol.* **2011**, *11*, 70. [CrossRef]
37. Livak, K.J.; Schmittgen, T.D. Analysis of relative gene expression data using real-time quantitative PCR and the $2^{-\Delta\Delta C_T}$ Method. *Methods* **2001**, *25*, 402–408. [CrossRef]
38. Gagnon, M.; Zihler Berner, A.; Chervet, N.; Chassard, C.; Lacroix, C. Comparison of the Caco-2, HT-29 and the mucus-secreting HT29-MTX intestinal cell models to investigate *Salmonella* adhesion and invasion. *J. Microbiol. Methods* **2013**, *94*, 274–279. [CrossRef]
39. Sunkavalli, U.; Aguilar, C.; Silva, R.; Sharan, M.; Cruz, A.; Tawk, C.; Maudet, C.; Mano, M.; Eulalio, A. Analysis of host microRNA function uncovers a role for miR-29b-2-5p in *Shigella* capture by filopodia. *PLoS Pathog.* **2017**, *13*, e1006327. [CrossRef]
40. Kim, H.; Kim, I.; Lee, S.; Kim, Y.; Silwal, P.; Kim, J.; Kim, J.; Seo, W.; Chung, C.; Cho, H.; et al. MiR-144-3p is associated with pathological inflammation in patients infected with *Mycobacteroides abscessus*. *Exp. Mol. Med.* **2021**, *53*, 136–149. [CrossRef]
41. Li, Y.; Su, L.; Guo, B.; Zhou, S.; Li, C.; Xiu, Y. Bacterial induced miR-144-5p modulates intestinal inflammatory response of Japanese flounder by targeting Hsp90α-dependent NLR signaling pathway. *Aquaculture* **2023**, *564*, 739053. [CrossRef]
42. Yang, P.; Zhao, Y.; Li, J.; Liu, C.; Zhu, L.; Zhang, J.; Yu, Y.; Wang, W.; Lei, G.; Yan, J.; et al. Downregulated miR-451a as a feature of the plasma cfRNA landscape reveals regulatory networks of IL-6/IL-6R-associated cytokine storms in COVID-19 patients. *Cell. Mol. Immunol.* **2021**, *18*, 1064–1066. [CrossRef] [PubMed]
43. Oxendine Harp, K.; Bashi, A.; Botchway, F.; Addo-Gyan, D.; Tetteh-Tsifoanya, M.; Lamptey, A.; Djameh, G.; Iqbal, S.; Lekpor, C.; Banerjee, S.; et al. Sickle Cell Hemoglobin Genotypes Affect Malaria Parasite Growth and Correlate with Exosomal miR-451a and let-7i-5p Levels. *Int. J. Mol. Sci.* **2023**, *24*, 7546. [CrossRef] [PubMed]
44. Chen, Q.; Tong, C.; Ma, S.; Zhou, L.; Zhao, L.; Zhao, X. Involvement of MicroRNAs in Probiotics-Induced Reduction of the Cecal Inflammation by *Salmonella* Typhimurium. *Front. Immunol.* **2017**, *8*, 704. [CrossRef] [PubMed]
45. Zhang, J.; Fu, S.; Liu, Y.; Liu, Y.; Wang, W. Analysis of MicroRNA Expression Profiles in Weaned Pig Skeletal Muscle after Lipopolysaccharide Challenge. *Int. J. Mol. Sci.* **2015**, *16*, 22438–22455. [CrossRef] [PubMed]
46. Dai, C.; Wang, F.; Wang, S.; Wu, Z.; Wu, S.; Bao, W. miR-215 Targeting Novel Genes EREG, NIPAL 1 and PTPRU Regulates the Resistance to *E. coli* F18 in Piglets. *Genes* **2020**, *11*, 1053. [CrossRef]
47. Huang, J.; Cao, Y.; Li, X.; Yu, F.; Han, X. E2F1 regulates miR-215-5p to aggravate paraquat-induced pulmonary fibrosis via repressing BMPR2 expression. *Toxicol. Res.* **2022**, *11*, 940–950. [CrossRef]
48. Pareek, S.; Sanchenkova, X.; Sakaguchi, T.; Murakami, M.; Okumura, R.; Kayama, H.; Kawauchi, S.; Motooka, D.; Nakamura, S.; Okuzaki, D.; et al. Epithelial miR-215 negatively modulates Th17-dominant inflammation by inhibiting CXCL12 production in the small intestine. *Genes Cells* **2022**, *27*, 243–253. [CrossRef]
49. Wang, J.; Chen, J.; Liu, Y.; Zeng, X.; Wei, M.; Wu, S.; Xiong, Q.; Song, F.; Yuan, X.; Xiao, Y.; et al. Hepatitis B Virus Induces Autophagy to Promote its Replication by the Axis of miR-192-3p-XIAP Through NF kappa B Signaling. *Hepatology* **2019**, *69*, 974–992. [CrossRef]
50. Desterro, J.; Thomson, J.; Hay, R. Ubch9 conjugates SUMO but not ubiquitin. *FEBS Lett.* **1997**, *417*, 297–300. [CrossRef]
51. Verma, S.; Mohapatra, G.; Ahmad, S.; Rana, S.; Jain, S.; Khalsa, J.; Srikanth, C. *Salmonella* Engages Host MicroRNAs To Modulate SUMOylation: A New Arsenal for Intracellular Survival. *Mol. Cell. Biol.* **2015**, *35*, 2932–2946. [CrossRef]
52. Guo, Y.; Li, L.; Xu, T.; Guo, X.; Wang, C.; Li, Y.; Yang, Y.; Yang, D.; Sun, B.; Zhao, X.; et al. HUWE1 mediates inflammasome activation and promotes host defense against bacterial infection. *J. Clin. Investig.* **2020**, *130*, 6301–6316. [CrossRef] [PubMed]
53. Hochstrasser, M. Origin and function of ubiquitin-like proteins. *Nature* **2009**, *458*, 422–429. [CrossRef] [PubMed]
54. Narayanan, L.; Edelmann, M. Ubiquitination as an efficient molecular strategy employed in *Salmonella* infection. *Front. Immunol.* **2014**, *5*, 558. [CrossRef] [PubMed]

55. Zhou, X.; Gao, F.; Gao, W.; Wang, Q.; Li, X.; Li, X.; Li, W.; Liu, J.; Zhou, H.; Luo, A.; et al. Bismuth Sulfide Nanoflowers Facilitated miR339 Delivery to Overcome Stemness and Radioresistance through Ubiquitin-Specific Peptidase 8 in Esophageal Cancer. *ACS Nano* **2024**, *18*, 19232–19246. [CrossRef]
56. Chang, J.; Lin, T.; Jhang, X.; Chan, S. hnRNP Q/SYNCRIP interacts with LIN28B and modulates the LIN28B/let-7 axis in human hepatoma cells. *PLoS ONE* **2024**, *19*, e0304947. [CrossRef]
57. Azam, H.; Rößling, R.; Geithe, C.; Khan, M.; Dinter, F.; Hanack, K.; Prüß, H.; Husse, B.; Roggenbuck, D.; Schierack, P.; et al. MicroRNA biomarkers as next-generation diagnostic tools for neurodegenerative diseases: A comprehensive review. *Front. Mol. Neurosci.* **2024**, *17*, 1386735. [CrossRef]
58. Guichard, A.; Nizet, V.; Bier, E. RAB11-mediated trafficking in host-pathogen interactions. *Nat. Rev. Microbiol.* **2014**, *12*, 624–634. [CrossRef]
59. Broz, P.; Ruby, T.; Belhocine, K.; Bouley, D.; Kayagaki, N.; Dixit, V.; Monack, D. Caspase-11 increases susceptibility to *Salmonella* infection in the absence of caspase-1. *Nature* **2012**, *490*, 288–291. [CrossRef]
60. Song, H.; Liu, B.; Huai, W.; Yu, Z.; Wang, W.; Zhao, J.; Han, L.; Jiang, G.; Zhang, L.; Gao, C.; et al. The E3 ubiquitin ligase TRIM31 attenuates NLRP3 inflammasome activation by promoting proteasomal degradation of NLRP3. *Nat. Commun.* **2016**, *7*, 13727. [CrossRef]
61. McKee, C.; Coll, R. NLRP3 inflammasome priming: A riddle wrapped in a mystery inside an enigma. *J. Leukoc. Biol.* **2020**, *108*, 937–952. [CrossRef]

animals

MDPI

Article

Phenotypic and Genomic Assessment of Antimicrobial Resistance and Virulence Factors Determinants in *Salmonella* Heidelberg Isolated from Broiler Chickens

Arthur de Almeida Figueira [1], Thomas Salles Dias [1,*], Gisllany Alves Costa [1], Dayse Lima da Costa Abreu [1,2], Luciana dos Santos Medeiros [2] and Virginia Léo de Almeida Pereira [1,2]

[1] Postgraduate Program in Veterinary Medicine (Veterinary Hygiene and Processing Technology of Animal Products), Faculdade de Veterinária, Universidade Federal Fluminense, Niterói 24230-320, RJ, Brazil; arthurfigueira@id.uff.br (A.d.A.F.); gisllanyalves@id.uff.br (G.A.C.); dayseabreu@id.uff.br (D.L.d.C.A.); virginialeo@id.uff.br (V.L.d.A.P.)

[2] Department of Preventive Veterinary Medicine, Faculdade de Veterinária, Universidade Federal Fluminense, Niteroi 24230-320, RJ, Brazil; luciana_medeiros@id.uff.br

* Correspondence: thomassalles@id.uff.br

Simple Summary: In our study, we investigated antimicrobial resistance in *Salmonella* Heidelberg, a bacterium commonly found in poultry and poultry food products. We analyzed 14 strains isolated from chickens in Brazil between 2013 and 2019, testing their phenotypic resistance to various antimicrobials and performing genomic analysis to identify the genes responsible for this resistance. Our results revealed a high frequency of resistance to commonly used antibiotics, such as cephalosporins, tetracyclines, and sulfonamides. All samples carried specific genetic mutations and resistance genes that could enhance their resistance and potentially increase their virulence. Notably, we identified genes typically associated with *Yersinia*, a pathogen that has been linked to more severe infections, suggesting that these *Salmonella* strains may pose a greater risk than previously expected. These findings underscore the need for continued surveillance to prevent foodborne illness and ensure the effectiveness of infection treatments.

Abstract: *Salmonella* Heidelberg is frequently found in poultry and poultry products and is associated with antimicrobial resistance strains and infections and mortality in humans. Whole-genome sequencing is used to monitor and understand epidemiological factors related to antimicrobial resistance. This study aimed to characterize the phenotypic resistance and sequence the whole genome of *Salmonella* Heidelberg strains isolated from poultry products in Brazil. Fourteen *Salmonella* Heidelberg strains isolated from whole broiler chicken carcasses and portions in Brazil between 2013 and 2019 were used in this study. Genus confirmation was performed by polymerase chain reaction. The disk diffusion test was conducted to assess the phenotypical antimicrobial susceptibility of the strains. Whole-genome sequencing was carried out to investigate the presence of antimicrobial resistance genes, plasmids, multilocus sequence typing, and virulence-associated genes. A high frequency of phenotypic resistance to cephalosporins, tetracyclines, and sulfonamides was detected. All strains had mutations in *gyrA* and *parC* and contained the genes *tet(A)*, *fosA7*, and *sul*. The presence of genes originating from *Yersinia* pathogenicity islands was also detected. This study identified a high frequency of antimicrobial resistance in *Salmonella* Heidelberg strains from broilers slaughtered in different regions of Brazil, all belonging to the same sequence type (ST15) and associated with multiple resistance and virulence genes. The presence of the *Yersinia* high-pathogenicity island was detected, indicating potential virulence. These findings highlight the importance of continuously monitoring antimicrobial resistance to control and prevent foodborne infections and maintain the efficacy of treatments for human salmonellosis.

Keywords: antimicrobial resistance; broiler chicken; foodborne infections; *Salmonella*; whole-genome sequencing

1. Introduction

Salmonella enterica can colonize animals and humans and is one of the most common foodborne microorganisms worldwide. The consumption of raw, undercooked, or contaminated chicken meat is considered to be one of the main sources of human infection [1]. There are more than 2400 known serotypes of *Salmonella*, and Heidelberg serovar is frequently associated with poultry and poultry products [2]. In Brazil, the world's leading exporter of poultry meat, four serovars are controlled by the official surveillance program: Enteritidis, Typhimurium, Gallinarum and Pullorum [3]. Despite the lack of specific legislation focused on the *Heidelberg* serovar, in 2016, as a result of Normative Instruction 16, control and monitoring of *Salmonella* spp. were established in commercial poultry establishments for broiler chickens and turkeys, as well as in slaughterhouses for chickens, hens, turkeys, and breeders registered with the Federal Inspection Service [4]. Research on strains from the Brazilian poultry chain has detected a high prevalence of *Salmonella* Heidelberg, making it one of the serovars most commonly associated with poultry products [5–7]. The Heidelberg serovar has shown a significant increase in occurrence in both poultry and humans, and is often associated with severe infections. It can carry various virulence factors, including those found in the *Yersinia* high-pathogenicity island (HPI), which are associated with increased virulence and enhanced biofilm formation capacity, potentially facilitating its persistence in the environment [8]. Additionally, concern over antimicrobial resistance is particularly relevant, as multidrug-resistant strains have been found to be linked to outbreaks of salmonellosis in humans [9–12].

Antimicrobial resistance is a global threat to human and animal health and represents one of the greatest challenges of this generation. Antimicrobial resistance in *Salmonella* spp. requires continuous study, given the significance of this bacterial genus among foodborne pathogens [13]. Currently, the Centers for Disease Control and Prevention (CDC) and World Health Organization (WHO) consider non-typhoidal *Salmonella* strains that are resistant to fluoroquinolones and cephalosporins to be urgent threats to public health [14,15]. In Brazil, several studies have reported the emergence of *Salmonella* Heidelberg strains from poultry that have proven resistant to multiple drugs, including cephalosporins, fluoroquinolones, tetracyclines, and sulfonamides [5,16,17].

In response to the growing concerns about antimicrobial resistance in *Salmonella* Heidelberg strains that originate from poultry, advanced molecular tools have become critical when it comes to monitoring and understanding the spread of resistance. The use of whole-genome sequencing represents a significant advancement in microbial monitoring, providing definitive information on the relatedness of bacteria from different sources and their transfer of specific genes, including antimicrobial resistance genes [13]. Genomic surveillance of antimicrobial resistance is one of the strategies outlined by the WHO to mitigate the burden of infections caused by resistant microorganisms and guide the optimal use of existing antibiotics [18]. Therefore, we aimed to characterize the phenotypic antimicrobial resistance and perform whole-genome sequencing of *Salmonella* Heidelberg strains isolated from poultry products in Brazil in order to identify antimicrobial resistance genes and genes associated with virulence. This study noted a high frequency of antimicrobial resistance in *Salmonella* Heidelberg strains isolated from broiler chickens in Brazil, all belonging to sequence type ST15 and harboring multiple resistance and virulence genes, including those from *Yersinia* pathogenicity islands.

2. Materials and Methods

2.1. Bacterial Strains

Fourteen strains of *Salmonella* Heidelberg originating from broiler carcasses and portions from the Central-West, Southeast, and South regions of Brazil were analyzed. These strains were stored at the Poultry Health Laboratory, Department of Veterinary Collective

Health and Public Health, Faculty of Veterinary Medicine, Fluminense Federal University, between 2013 and 2019, and previously serotyped in the National Reference Center, Institute Oswaldo Cruz, Rio de Janeiro, Brazil (Figure 1).

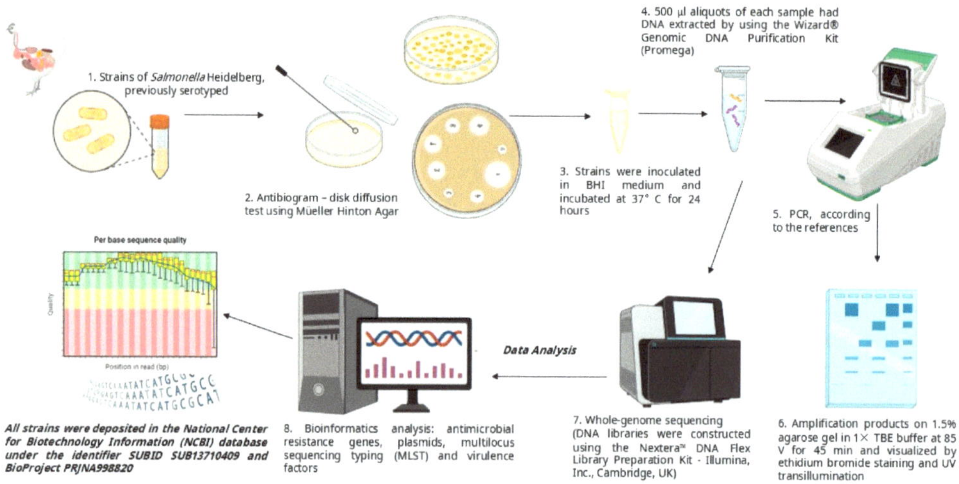

Figure 1. Representative scheme of methodology and experimental design of this study.

2.2. Antibiogram

The disk diffusion test was performed for 19 antimicrobials using plates containing Müeller Hinton Agar (Merck®, Darmstadt, Germany), with the cut-off points defined by the Clinical and Laboratory Standards Institute [19,20]. The antimicrobial categories tested were aminoglycosides (Gentamicin-10 µg), carbapenems (Ertapenem-10 µg, Imipenem-10 µg, Meropenem-10 µg), 1st- and 2nd-generation cephalosporins (Cephalexin-30 µg, Cefalotin-30 µg, Cefoxitin-30 µg), 3rd-generation and 4th-generation cephalosporins (Cefotaxime-30 µg, Ceftazidime-30 µg, Ceftiofur-30 µg, Cefepime-30 µg), penicillins (Ampicillin-10 µg), penicillins + β-lactamase inhibitors (Amoxicillin + clavulanate-20/10 µg), monobactams (Aztreonam-30 µg), fluoroquinolones (Ciprofloxacin-5 µg, Enrofloxacin-5 µg), phenicols (Chloramphenicol-30 µg), tetracyclines (Tetracycline-30 µg), and sulfonamide (300 µg). The *Escherichia coli* strain ATCC 25922 was used as a control for the test.

2.3. DNA Extraction

The strains were inoculated in BHI medium and incubated at 37 °C for 24 h without agitation. After growth, DNA was extracted from 500 µL aliquots of each sample using the Wizard® Genomic DNA Purification Kit (Promega, São Paulo, Brazil), according to the manufacturer's recommendations. The DNA samples were evaluated for purity using a BioDrop Touch® spectrophotometer (Biochrom Ltd., Cambridge, UK).

2.4. Polymerase Chain Reaction

To confirm the isolates at the genus level, we used the ST11 (AGC CAA CCA TTG CTA AAT TGG CGC A) and ST15 (GGT AGA AAT TCC CAG CGG GTA CTG) primer set, as described by Aabo et al. 1993 [21]. This primer set is highly specific to *Salmonella* species and produces an amplified fragment of 429 bp. For PCR analysis, the samples were required to have a minimum concentration of 20 ng/µL, with a 260/280 absorbance ratio ≅ 1.8 and a 260/230 ratio ≅ 2.0. The PCR reactions had a final volume of 25 µL containing 1X PCR buffer, 1.5 mM $MgCl_2$, 5 µL of DNA, 0.2 µM of each primer and, 1U of Taq polymerase (Promega, São Paulo, Brazil), and the amplification was performed on a thermocycler model *T-100* (Bio-Rad Laboratories, Inc., Waltham, MA, USA). Samples were denatured at 94 °C for 2 min, and 35 cycles of amplification were performed at 95 °C for 30 s, 60 °C for 30 s, and 72 °C for 30 s. The reaction was completed by a final 10 min extension at 72 °C. *Salmonella* Enteritidis ATCC 13076 was used as a positive control and ultra-pure water was used as a negative control. Aliquots of amplification products were separated on 1.5% agarose gel in 1× TBE buffer at 85 V for 45 min and visualized by ethidium bromide staining and UV transillumination.

2.5. Whole-Genome Sequencing

The DNA libraries were constructed using the Nextera™ DNA Flex Library Preparation *Kit* (Illumina, Inc., Cambridge, UK). DNA fragments were sequenced using 150 bp paired-end libraries on the Illumina Miseq sequencer. The software FastQC (https://www.bioinformatics.babraham.ac.uk/projects/fastqc/, accessed on 1 June 2024) was used for quality control of the raw sequences for subsequent bioinformatics analyses. Genome assembly and annotation were performed by the Bacterial and Viral Bioinformatics Resource Center (https://www.bv-brc.org/, accessed on 24 June 2024). Bioinformatics analyses were carried out using online tools and databases available at the Center for Genomic Epidemiology (CGE) (www.genomicepidemiology.org) to investigate the presence of antimicrobial resistance genes (ResFinder 3.2), plasmids (PlasmidFinder 2.0), and for multilocus sequence typing (MLST 2.0). A search for virulence factors related to the *Yersinia* high-pathogenicity island was conducted using VFanalyzer in relation to the Virulence Factors Database (VFDB). All analyses were performed using standard parameters available on the platforms. The strains were deposited at the National Center for Biotechnology Information (NCBI) database under the identifier SUBID SUB13710409 and BioProject PRJNA998820 (Figure 1).

3. Results

Fourteen *Salmonella* Heidelberg strains, which were collected from various regions in Brazil between 2013 and 2019, were subjected to phenotypic antimicrobial resistance profiling and whole-genome sequencing. Table 1 details the percentages of antimicrobial resistance and Table 2 presents the sequence types, resistome, and phenotypic resistance profiles of the isolates. In the present study, a high frequency of phenotypic resistance to amoxicillin with clavulanic acid, ampicillin, cephalexin, cefalotin, ceftiofur, cefoxitin, ceftazidime, cefotaxime, tetracycline, and sulfonamide was detected in the *Salmonella* Heidelberg strains analyzed. None of the strains were resistant to the carbapenems tested. In relation to virulence, the genes *ybtT*, *irp2*, *psn/fyuA*, *ybtP*, *ybtQ*, *ybtU*, and *ybtX*, which were all associated with the HPI, were detected in 8 out of the 14 strains analyzed (57.14%).

Table 1. Percentage (%) of antimicrobial susceptibility of *Salmonella* Heidelberg strains isolated from broiler chicken carcasses in the disk diffusion test according to Clinical & Laboratory Standards Institute. R—resistant; I—intermediate resistance; S—susceptible.

Antimicrobial	R	I	S
Amoxicillin with clavulanic acid	71.43	7.14	21.43
Ampicillin	85.71	0.00	14.29
Cephalexin	78.57	0.00	21.43
Cefalotin	78.57	0.00	21.43
Cefoxitin	71.43	0.00	28.57
Ceftazidime	64.29	7.14	28.57
Cefotaxime	78.57	0.00	21.43
Ceftiofur *	57.14	21.43	21.43
Cefepime	7.14	28.57	64.29
Imipenem	0.00	0.00	100.00
Meropenem	0.00	0.00	100.00
Ertapenem	0.00	7.14	92.86
Aztreonam	57.14	7.14	35.71
Gentamicin	7.14	0.00	92.86
Ciprofloxacin	7.14	42.86	50.00
Enrofloxacin *	7.14	7.14	85.71
Chloramphenicol	21.43	0.00	78.57
Tetracycline	92.86	0.00	7.14
Sulfonamide	85.71	0.00	14.29

* Criteria from CLSI VET 2020.

Table 2. Sequence type, resistome, and phenotypic resistance profile of *Salmonella* Heidelberg strains isolated from broiler chicken carcasses.

Strain	Region	Year	Source	ST	Resistance Genotype	Plasmids	Phenotypic Resistance Profile	Intermediate Resistance	HPI
Heidelberg	South	2013	Carcass	15	aac(6′)-Iaa, blaCMY-2, tet(A), sul2, gyrA:p.S83F, parC:p.T57S, fosA7	ColpVC, IncC, IncX1	AMC, AMP, CFE, CFL, CFO, CAZ, CTX, CTF, ATM, TET, SUL	CPM, CIP	Yes
Heidelberg	South	2013	Carcass	15	aac(6′)-Iaa, blaCMY-2, tet(A), sul2, gyrA:p.S83F, parC:p.T57S, fosA7	ColpVC, IncC, IncX1	AMC, AMP, CFE, CFL, CFO, CAZ, CTX, ATM, TET, SUL	CTF, CIP, ENO	No
Heidelberg	South	2013	Carcass	15	aac(6′)-Iaa, blaCMY-2, tet(A), sul2, gyrA:p.S83F, parC:p.T57S, fosA7	ColpVC, IncC, IncX1, IncI1-I(Alpha)	AMC, AMP, CFE, CFL, CFO, CAZ, CTX, CTF, TET, SUL	CPM, ATM, CIP	No
Heidelberg	Midwest	2016	Carcass	15	aac(6′)-Iaa, blaCMY-2, tet(A), sul2, gyrA:p.S83F, parC:p.T57S, fosA7	ColpVC, IncC, IncX1	AMP, CLO, TET	-	No
Heidelberg	South	2017	Carcass	15	aac(6′)-Iaa, gyrA (p.S83F), parC:p.T57S, sul2, tet(A), fosA7,	ColpVC, IncC, IncX1, IncI1-I(Alpha)	AMC, AMP, CFE, CFL, CFO, CAZ, CTX, CTF, ATM, ENO, TET, SUL	CPM, CIP, ENO	Yes
Heidelberg	South	2017	Carcass	15	aac(6′)-Iaa, gyrA (p.S83F), parC:p.T57S, sul2, tet(A), fosA7,	ColpVC, IncC, IncX1	AMC, AMP, CFE, CFL, CFO, CTX, CTF, ATM, TET, SUL	CAZ	Yes
Heidelberg	Southeast	2017	Carcass	15	aac(6′)-Iaa, gyrA (p.S83F), parC:p.T57S, sul2, tet(A), fosA7,	ColpVC, IncC	AMC, AMP, CFE, CFL, CFO, CAZ, CTX, CTF, ATM, TET, SUL	-	No
Heidelberg	Southeast	2017	Carcass	15	aac(6′)-Iaa, gyrA (p.S83F), parC:p.T57S sul2, tet(A), fosA7	ColpVC, IncC, IncX1, IncI1-I(Alpha)	AMC, AMP, CFE, CFL, CFO, CAZ, CTX, CTF, ATM, TET, SUL	-	No
Heidelberg	Southeast	2017	Carcass	15	aac(6′)-Iaa, gyrA (p.S83F), parC:p.T57S sul2, tet(A), fosA7	ColpVC, IncC	AMC, AMP, CFE, CFL, CFO, CAZ, CTX, CTF, ATM, TET, SUL	CPM, ERT, CIP	Yes
Heidelberg	Southeast	2017	Carcass	15	aac(6′)-Iaa, blaCMY-2, tet(A), sul2, gyrA:p.S83F, parC:p.T57S, fosA7	ColpVC, IncC, IncX1, IncI1-I(Alpha)	AMC, AMP, CFE, CFL, CFO, CAZ, CTX, CTF, ATM, TET, SUL	-	Yes
Heidelberg	Southeast	2017	Carcass	15	aac(6′)-Iaa, gyrA (p.S83F), parC:p.T57S sul2, tet(A), fosA7	ColpVC, IncC	AMC, AMP, CFE, CFL, CFO, CAZ, CTX, ATM, TET, SUL	CTF, CIP, ENO	Yes
Heidelberg	South	2019	Carcass	15	aac(6′)-Iaa, blaCMY-2, tet(A), sul2, gyrA:p.S83F, parC:p.T57S, fosA7	ColpVC, IncC, IncX1, IncI1-I(Alpha)	SUL	-	No

Table 2. *Cont.*

Strain	Region	Year	Source	ST	Resistance Genotype	Plasmids	Phenotypic Resistance Profile	Intermediate Resistance	HPI
Heidelberg	Southeast	2019	Retail	15	aac(6′)-Iaa, blaCMY-2, tet(A), sul2, gyrA:p.S83F, parC:p.T57S, fosA7	ColpVC, IncC, IncX1, IncI1-I(Alpha)	AMP, CFE, CFL, CTX, CTF, CPM, ATM, GEN, CLO, SUL	AMC, CIP	Yes
Heidelberg	South	2019	Carcass	15	aac(6′)-Iaa, blaCMY-2, tet(A), sul2, gyrA:p.S83F, parC:p.T57S, fosA7	ColpVC, IncC, IncX1, IncI1-I(Alpha)	CIP, CLO	-	Yes

ST: Sequence type. AMC = amoxicillin + clavulanic acid, AMP = ampicillin, CFE = cephalexin, CFL = cefalotin, CFO = cefoxitin, CAZ = ceftazidime, CTX = cefotaxime, CTF = ceftiofur, CPM = cefepime, IPM = imipenem, MPM = meropenem, ERT = ertapenem; ATM = aztreonam, CIP = ciprofloxacin, ENO = enrofloxacin, GEN = gentamicin, CLO = chloramphenicol, TET = tetracycline, SUL = sulfonamide, HPI = *Yersinia* high-pathogenicity island.

4. Discussion

A high frequency of resistance to clinically important antimicrobials was observed in *Salmonella* Heidelberg strains isolated from cuts and carcasses of slaughtered chickens from different regions of Brazil. In recent years, *Salmonella* Heidelberg has emerged as the predominant serovar associated with the poultry chain in Brazil, exhibiting resistance to multiple classes of antimicrobials [17,22–24], and it has been identified as a common serovar associated with human salmonellosis in North and South America [9,25].

Similar to other studies, we detected high clonality among the *Salmonella* Heidelberg strains [6,7,26]. Studies that used the Pulsed-Field Gel Electrophoresis technique [6,7] reported difficulty in differentiating this serovar due to the identical banding patterns in strains from different origins. In the present study, all Heidelberg strains were classified as ST15, consistent with findings from other genomic studies involving strains isolated in Brazil [26,27]. Kipper et al. (2021) [26] reported that this single lineage of *Salmonella* Heidelberg serovar is widespread in the national poultry chain and is associated with multiple resistance and virulence genes. This may have occurred because serovars such as Enteritidis and Typhimurium are prioritized within the National Avian Health Program, established in 1994 [28]; therefore, the control of these serovars may have created an ecological niche that has allowed *Salmonella* Heidelberg to spread throughout the poultry chain. Possessing various genes related to antimicrobial resistance and survival under stressful conditions, common disinfection measures used in farm and slaughterhouse facilities may not be effective in eliminating this serovar. In a study by Voss-Rech and collaborators [29], *Salmonella* Heidelberg survived in recycled poultry litter, indicating that burning residual feathers and the bedding shaking procedure were not sufficient to interrupt the cycle of residual contamination.

The plasmids replicons detected in the genomes included IncX1, IncI, and IncC2, all of which have been previously described in other isolates of *Salmonella* Heidelberg [26,27,30]. These plasmids have incorporated various resistance genes, facilitating the spread of antimicrobial resistance in diverse environments [26,31]. Horizontal transfer of plasmids between different bacteria can lead to rapid dissemination of resistance, as co-selection due to the use of a single antimicrobial may select for resistance to multiple agents, making eradication of resistant strains more challenging [32]. Therefore, the presence of multiple plasmids with different replicons in *Salmonella* Heidelberg increases the complexity of controlling this microorganism [33,34].

Salmonellosis does not typically require antibiotic treatment, but in patients with immunosuppression, persistent diarrhea, or invasive infection, antibiotic therapy is necessary. Common first-line oral antibiotics for salmonellosis include fluoroquinolones and cephalosporins. Our findings agree with other studies [22,35] in which the main genetic element associated with cephalosporin resistance was the *bla*-CMY-2 gene. This gene is linked to resistance against penicillin, third-generation cephalosporins, β-lactamase inhibitors associated with β-lactams, and cephamycins. It has been identified as the primary determinant of resistance to ceftiofur [35], a cephalosporin used prophylactically in day-old chicks in some hatcheries around the world to mitigate early gastrointestinal infections caused by representatives of Enterobacteriaceae [36]. Touzain et al. (2018) [37] identified Avian Pathogenic *Escherichia coli* strains in diseased birds that carried the *bla*-CMY-2 gene on plasmids. This finding poses a serious risk to both animal welfare and public health, as the presence of *bla*-CMY-2 in poultry and poultry products could facilitate the transfer of these antibiotic-resistance genes to humans, potentially leading to ineffective treatments.

Dias et al., 2022 [16] reported a high frequency of non-susceptibility to ciprofloxacin in *Salmonella* Heidelberg using the minimum inhibitory concentration technique, with a low frequency of mobile elements associated with quinolone resistance. In the present study, all *Salmonella* Heidelberg strains possessed mutations in *gyr*A(S83F) and *par*C (T57S) associated with fluoroquinolone resistance, and no plasmid genes were detected in the analyzed strains. Single-nucleotide polymorphisms in the quinolone resistance-determining regions the *gyrA* and *parC* genes, which encode DNA gyrase and topoisomerase IV, respectively, can result in conformational changes in these enzymes. These changes prevent quinolones from binding to the DNA–enzyme complex, thereby maintaining the enzymes' normal functions [38]. Despite all strains showing the main mutations correlated with phenotypic resistance to fluoroquinolones, 50% (7/14) of the strains exhibited phenotypical resistance to ciprofloxacin or enrofloxacin.

The genes *aac(6′)-Iaa*, *tet(A)*, *sul2*, and *fosA7* were present in all analyzed genomes, representing the main detected genes. The *aac(6′)-Iaa* gene, which is chromosomal and present in most *Salmonella* strains, encodes an enzyme that modifies aminoglycosides by acetylating drugs like tobramycin, kanamycin, and amikacin, but not gentamicin. According to the 2023 CLSI guidelines [19], even if these isolates appear susceptible to these antibiotics in vitro, they should not be reported as susceptible, as the strains may exhibit resistance in vivo. The *tet(A)* gene is linked to tetracycline resistance, *sul2* to sulfonamide resistance, and *fosA7* to fosfomycin resistance. These broad-spectrum antimicrobials have been widely used in animal production for many years due to their effectiveness and low cost. This prolonged use may have favored the selection of strains carrying these resistance genes within the production chain, facilitating their spread to food.

In relation to virulence, *Yersinia* HPI was detected in eight strains. This island was first reported in *Salmonella* Heidelberg from Brazil in 2021 [8]. The role of HPI is to facilitate the synthesis and transport of iron-chelating molecules in an environment that lacks this vital element. This process involves a set of genes organized in specific clusters aimed primarily at achieving iron binding in the surrounding environment. Once HPI binds to iron, it forms a molecular complex recognized by specific sites on the surface of the host cell [39]. *Yersinia* HPI in *Salmonella* can enhance virulence, posing a significant public health concern [8], may affect the regulatory protein of the HPI, and may have broader regulatory effects, influencing the expression of virulence factors or other genes beyond those associated with the island, enhancing the organism's ability to initiate infection. Although *Salmonella* possesses several iron-acquisition systems, yersiniabactin's high affinity for Fe^{3+} and its potential to suppress the host immune response may provide a significant advantage. The HPI could improve the organism environmental fitness and persistence, further aiding its survival [40].

A high degree of clonality was observed among *Salmonella* Heidelberg isolates from various regions and years in Brazil in this study, consistent with previous reports [6,26]. Kipper et al., 2021 [26] identified a multidrug-resistant *Salmonella* Heidelberg lineage circulating within Brazilian commercial poultry flocks, offering critical insights into the recent introduction of this lineage in 2004 and its subsequent population expansion. The rapid dissemination during the initial years is likely attributable to introduction via the poultry production pyramid, particularly through breeder flocks and hatcheries. In the following years, both horizontal and vertical transmission likely played a role in *Salmonella* Heidelberg contamination of broilers during pre-harvest and further along the production chain [26].

5. Conclusions

This study identified a high frequency of phenotypic resistance to medically important antimicrobials in *Salmonella* Heidelberg strains isolated from slaughtered chickens in different regions of Brazil. Genomic analyses revealed high clonality among the strains, all belonging to the same clone (ST15), which was associated with multiple resistance and virulence genes. The presence of the *Yersinia* HPI was also detected, indicating the potential virulence of the studied strains. Given these findings, it is crucial to implement targeted public health interventions. Future research should focus on developing and evaluating effective strategies to control antimicrobial resistance in poultry. Additionally, improving hygiene practices and biosecurity measures in poultry farms, along with stricter regulations in the broiler industry, could significantly reduce the risk of these resistant strains spreading. Enhanced surveillance and monitoring programs are also essential to detect and mitigate resistance trends, ensuring effective treatment options for human salmonellosis and reducing the risk of foodborne infection.

Author Contributions: Conceptualization, V.L.d.A.P., A.d.A.F. and T.S.D.; methodology, A.d.A.F., D.L.d.C.A., L.d.S.M. and T.S.D.; formal analysis, A.d.A.F. and G.A.C.; investigation, A.d.A.F. and G.A.C.; resources V.L.d.A.P.; writing—original A.d.A.F. and T.S.D.; writing—review and editing, T.S.D., A.d.A.F. and L.d.S.M.; supervision, D.L.d.C.A. and L.d.S.M.; project administration, V.L.d.A.P.; funding acquisition, V.L.d.A.P. All authors have read and agreed to the published version of the manuscript.

Funding: This work was funded by the Carlos Chagas Filho Foundation for Research Support of the State of Rio de Janeiro (E-26/010.002133/2019 and E-26/202.465/2022), as well as by the Coordination for the Improvement of Higher Education Personnel (CAPES).

Institutional Review Board Statement: Not applicable.

Informed Consent Statement: Not applicable.

Data Availability Statement: The data presented in this study are available upon reasonable request from the authors.

Conflicts of Interest: The authors declare no conflicts of interest.

References

1. European Food Safety Authority (EFSA). The European Union One Health 2020 Zoonoses Report. *EFSA J.* **2021**, *19*, e06971. [CrossRef]
2. Shah, D.H.; Paul, N.C.; Sischo, W.C.; Crespo, R.; Guard, J. Population Dynamics and Antimicrobial Resistance of the Most Prevalent Poultry-Associated *Salmonella* Serotypes. *Poult. Sci.* **2017**, *96*, 687–702. [CrossRef] [PubMed]
3. Baptista, D.; Borsoi, A.; Reischak, D.; Nascimento, A.; Montesino, L.; Camillo, S.; Abreu, D.; Pereira, V. *Salmonella* Serovars Isolated from Poultry Breeding Flocks under the Brazilian Official Control Programme between 2016 and 2018. *Braz. J. Poult. Sci.* **2023**, *25*, eRBCA-2022. [CrossRef]
4. Brasil, Ministério da Agricultura e Pecurária. Instrução Normativa, No 20 de 21 de Outubro de 2016, Brazil. 2016. Available online: https://www.legisweb.com.br/legislacao/?id=330200 (accessed on 21 August 2024).
5. Rau, R.B.; Ribeiro, A.R.; dos Santos, A.; Barth, A.L. Antimicrobial Resistance of *Salmonella* from Poultry Meat in Brazil: Results of a Nationwide Survey. *Epidemiol. Infect.* **2021**, *149*, e228. [CrossRef]
6. Santos, A.; Machado, S.; Dias, T.; Rodrigues, D.; Pereira, V. High Genetic Similarity Among *Salmonella* Heidelberg Isolated from Poultry Farms, Wild Animals, Beef, Poultry and Pork Meat, and Humans in Brazil. *Braz. J. Poult. Sci.* **2023**, *25*, eRBCA-2022. [CrossRef]
7. Voss-Rech, D.; Kramer, B.; Silva, V.S.; Rebelatto, R.; Abreu, P.G.; Coldebella, A.; Vaz, C.S.L. Longitudinal Study Reveals Persistent Environmental *Salmonella* Heidelberg in Brazilian Broiler Farms. *Vet. Microbiol.* **2019**, *233*, 118–123. [CrossRef] [PubMed]
8. Santos, A.M.P.; Ferrari, R.G.; Panzenhagen, P.; Rodrigues, G.L.; Conte-Junior, C.A. Virulence Genes Identification and Characterization Revealed the Presence of the Yersinia High Pathogenicity Island (HPI) in *Salmonella* from Brazil. *Gene* **2021**, *787*, 145646. [CrossRef]
9. Santos, A.F.d.M.; Amparo, L.F.V.; Machado, S.C.A.; Dias, T.S.; Berto, L.H.; Abreu, D.L.d.C.; Aquino, M.H.C.d.; Rodrigues, D.d.P.; Pereira, V.L.d.A. *Salmonella* Serovars Associated with Human Salmonellosis in Brazil (2011–2020). *Res. Soc. Dev.* **2022**, *11*, e28011830533. [CrossRef]

10. Gieraltowski, L.; Higa, J.; Peralta, V.; Green, A.; Schwensohn, C.; Rosen, H.; Libby, T.; Kissler, B.; Marsden-Haug, N.; Booth, H.; et al. National Outbreak of Multidrug Resistant *Salmonella* Heidelberg Infections Linked to a Single Poultry Company. *PLoS ONE* **2016**, *11*, e0162369. [CrossRef]
11. Folster, J.P.; Pecic, G.; Rickert, R.; Taylor, J.; Zhao, S.; Fedorka-Cray, P.J.; Whichard, J.; McDermott, P. Characterization of Multidrug-Resistant *Salmonella* Enterica Serovar Heidelberg from a Ground Turkey-Associated Outbreak in the United States in 2011. *Antimicrob. Agents Chemother.* **2012**, *56*, 3465–3466. [CrossRef]
12. Crump, J.A.; Medalla, F.M.; Joyce, K.W.; Krueger, A.L.; Hoekstra, R.M.; Whichard, J.M.; Barzilay, E.J. Antimicrobial Resistance among Invasive Nontyphoidal *Salmonella* Enterica Isolates in the United States: National Antimicrobial Resistance Monitoring System, 1996 to 2007. *Antimicrob. Agents Chemother.* **2011**, *55*, 1148–1154. [CrossRef] [PubMed]
13. FDA. Research to Advance Antimicrobial Resistance Monitoring. Available online: https://www.fda.gov/animal-veterinary/national-antimicrobial-resistance-monitoring-system/research-advance-antimicrobial-resistance-monitoring (accessed on 16 July 2024).
14. CDC. *Antibiotic Resistance Threats in the United States, 2019*; CDC: Atlanta, Georgia, 2019.
15. World Health Organization. *2019 Antibacterial Agents in Clinical Development an Analysis of the Antibacterial Development Pipeline*; WHO: Geneva, Switzerland, 2019; ISBN 9789240000193.
16. Dias, T.S.; Figueira, A.A.; Costa, G.A.; Machado, S.C.A.; da Cunha, N.C.; Abreu, D.L.C.; Pereira, V.L.A.; de Aquino, M.H.C. High Frequency of Non-Susceptibility to Ciprofloxacin in Nontyphoid *Salmonella* Recovered from Brazilian Broiler Chicken. *Br. Poult. Sci.* **2022**, *64*, 137–141. [CrossRef] [PubMed]
17. Rodrigues, I.; Silva, R.; Menezes, J.; Machado, S.; Rodrigues, D.; Pomba, C.; Abreu, D.; Nascimento, E.; Aquino, M.; Pereira, V. High Prevalence of Multidrug-Resistant Nontyphoidal *Salmonella* Recovered from Broiler Chickens and Chicken Carcasses in Brazil. *Braz. J. Poult. Sci.* **2020**, *22*, eRBCA-2019. [CrossRef]
18. World Health Organization. *Global Genomic Surveillance Strategy 2022–2032*; WHO: Geneva, Switzerland, 2022; ISBN 9789240046979.
19. Clinical and Laboratory Standards Institute (CLSI). *M100 Performance Standards for Antimicrobial Susceptibility Testing*, 33rd ed.; CLSI: Wayne, PA, USA, 2023.
20. Clinical and Laboratory Standards Institute (CLSI). *Performance Standards for Antimicrobial Disk and Dilution Susceptibility Tests for Bacteria Isolated from Animals*; CLSI: Wayne, PA, USA, 2020.
21. Aabo, S.; Rasmussen, O.F.; Roseen, L.; Sørensen, P.D.; Olsen, J.E. *Salmonella* Identification by the Polymerase Chain Reaction. *Mol. Cell. Probes* **1993**, *7*, 171–178. [CrossRef] [PubMed]
22. Campos, J.; Mourão, J.; Silveira, L.; Saraiva, M.; Correia, C.B.; Maçãs, A.P.; Peixe, L.; Antunes, P. Imported Poultry Meat as a Source of Extended-Spectrum Cephalosporin-Resistant CMY-2-Producing *Salmonella* Heidelberg and *Salmonella* Minnesota in the European Union, 2014–2015. *Int. J. Antimicrob. Agents* **2018**, *51*, 151–154. [CrossRef] [PubMed]
23. Collineau, L.; Chapman, B.; Bao, X.; Sivapathasundaram, B.; Carson, C.A.; Fazil, A.; Reid-Smith, R.J.; Smith, B.A. A Farm-to-Fork Quantitative Risk Assessment Model for *Salmonella* Heidelberg Resistant to Third-Generation Cephalosporins in Broiler Chickens in Canada. *Int. J. Food Microbiol.* **2020**, *330*, 108559. [CrossRef] [PubMed]
24. Melo, R.T.; Galvão, N.N.; Guidotti-Takeuchi, M.; Peres, P.A.B.M.; Fonseca, B.B.; Profeta, R.; Azevedo, V.A.C.; Monteiro, G.P.; Brenig, B.; Rossi, D.A. Molecular Characterization and Survive Abilities of *Salmonella* Heidelberg Strains of Poultry Origin in Brazil. *Front. Microbiol.* **2021**, *12*, 674147. [CrossRef] [PubMed]
25. Etter, A.J.; West, A.M.; Burnett, J.L.; Wu, S.T.; Veenhuizen, D.R.; Ogas, R.A.; Oliver, H.F. *Salmonella* Enterica Subsp. Enterica Serovar Heidelberg Food Isolates Associated with a Salmonellosis Outbreak Have Enhanced Stress Tolerance Capabilities. *Appl. Environ. Microbiol.* **2019**, *85*, e01065-19. [CrossRef] [PubMed]
26. Kipper, D.; Orsi, R.H.; Carroll, L.M.; Mascitti, A.K.; Streck, A.F.; Fonseca, A.S.K.; Ikuta, N.; Tondo, E.C.; Wiedmann, M.; Lunge, V.R. Recent Evolution and Genomic Profile of *Salmonella* Enterica Serovar Heidelberg Isolates from Poultry Flocks in Brazil. *Appl. Environ. Microbiol.* **2021**, *87*, e01036-21. [CrossRef]
27. Saidenberg, A.B.S.; Franco, L.S.; Reple, J.N.; Hounmanou, Y.M.G.; Casas, M.R.T.; Cardoso, B.; Esposito, F.; Lincopan, N.; Dalsgaard, A.; Stegger, M.; et al. *Salmonella* Heidelberg and *Salmonella* Minnesota in Brazilian Broilers: Genomic Characterization of Third-Generation Cephalosporin and Fluoroquinolone-Resistant Strains. *Environ. Microbiol. Rep.* **2023**, *15*, 119–128. [CrossRef]
28. Brasil, Ministério da Agricultura e Pecurária. Portaria Do Ministério Da Agricultura, Pecuária e Abastecimento—193, de 19/9/1994. 1994. Available online: https://www.defesa.agricultura.sp.gov.br/legislacoes/portaria-mapa-193-de-19-09-1994,369.html (accessed on 21 August 2024).
29. Voss-Rech, D.; Trevisol, I.M.; Brentano, L.; Silva, V.S.; Rebelatto, R.; Jaenisch, F.R.F.; Okino, C.H.; Mores, M.A.Z.; Coldebella, A.; Botton, S.d.A.; et al. Impact of Treatments for Recycled Broiler Litter on the Viability and Infectivity of Microorganisms. *Vet. Microbiol.* **2017**, *203*, 308–314. [CrossRef] [PubMed]
30. van den Berg, R.R.; Dissel, S.; Rapallini, M.L.B.A.; van der Weijden, C.C.; Wit, B.; Heymans, R. Characterization and Whole-genome sequencing of Closely Related Multidrug-Resistant *Salmonella* Enterica Serovar Heidelberg Isolates from Imported Poultry Meat in the Netherlands. *PLoS ONE* **2019**, *14*, e0219795. [CrossRef]
31. Lyu, N.; Feng, Y.; Pan, Y.; Huang, H.; Liu, Y.; Xue, C.; Zhu, B.; Hu, Y. Genomic Characterization of *Salmonella* Enterica Isolates From Retail Meat in Beijing, China. *Front. Microbiol.* **2021**, *12*, 636332. [CrossRef]

32. Orlek, A.; Anjum, M.F.; Mather, A.E.; Stoesser, N.; Walker, A.S. Factors Associated with Plasmid Antibiotic Resistance Gene Carriage Revealed Using Large-Scale Multivariable Analysis. *Sci. Rep.* **2023**, *13*, 2500. [CrossRef] [PubMed]
33. Han, J.; Lynne, A.M.; David, D.E.; Tang, H.; Xu, J.; Nayak, R.; Kaldhone, P.; Logue, C.M.; Foley, S.L. DNA Sequence Analysis of Plasmids from Multidrug Resistant *Salmonella* Enterica Serotype Heidelberg Isolates. *PLoS ONE* **2012**, *7*, e51160. [CrossRef] [PubMed]
34. Oladeinde, A.; Cook, K.; Orlek, A.; Zock, G.; Herrington, K.; Cox, N.; Plumblee Lawrence, J.; Hall, C. Hotspot Mutations and ColE1 Plasmids Contribute to the Fitness of *Salmonella* Heidelberg in Poultry Litter. *PLoS ONE* **2018**, *13*, e0202286. [CrossRef]
35. Souza, A.I.S.; Saraiva, M.M.S.; Casas, M.R.T.; Oliveira, G.M.; Cardozo, M.V.; Benevides, V.P.; Barbosa, F.O.; Freitas Neto, O.C.; Almeida, A.M.; Junior, A.B. High Occurrence of β-Lactamase-Producing *Salmonella* Heidelberg from Poultry Origin. *PLoS ONE* **2020**, *15*, e0230676. [CrossRef]
36. Lemos, M.P.L.; Saraiva, M.M.S.; Leite, E.L.; Silva, N.M.V.; Vasconcelos, P.C.; Giachetto, P.F.; Freitas Neto, O.C.; Givisiez, P.E.N.; Gebreyes, W.A.; Oliveira, C.J.B. The Posthatch Prophylactic Use of Ceftiofur Affects the Cecal Microbiota Similar to the Dietary Sanguinarine Supplementation in Broilers. *Poult. Sci.* **2020**, *99*, 6013–6021. [CrossRef]
37. Touzain, F.; Le Devendec, L.; de Boisséson, C.; Baron, S.; Jouy, E.; Perrin-Guyomard, A.; Blanchard, Y.; Kempf, I. Characterization of Plasmids Harboring BlaCTX-M and BlaCMY Genes in *E. Coli* from French Broilers. *PLoS ONE* **2018**, *13*, e0188768. [CrossRef]
38. Mahmud, Z.; Shabnam, S.A.; Mishu, I.D.; Johura, F.-T.; Mannan, S.B.; Sadique, A.; Islam, L.N.; Alam, M. Virotyping, Genotyping, and Molecular Characterization of Multidrug Resistant Escherichia Coli Isolated from Diarrheal Patients of Bangladesh. *Gene Rep.* **2021**, *23*, 101182. [CrossRef]
39. Kramer, J.; Özkaya, Ö.; Kümmerli, R. Bacterial Siderophores in Community and Host Interactions. *Nat. Rev. Microbiol.* **2020**, *18*, 152–163. [CrossRef] [PubMed]
40. Petermann, S.R.; Sherwood, J.S.; Logue, C.M. The Yersinia High Pathogenicity Island Is Present in *Salmonella* Enterica Subspecies I Isolated from Turkeys. *Microb. Pathog.* **2008**, *45*, 110–114. [CrossRef] [PubMed]

MDPI AG
Grosspeteranlage 5
4052 Basel
Switzerland
Tel.: +41 61 683 77 34

Animals Editorial Office
E-mail: animals@mdpi.com
www.mdpi.com/journal/animals

www.ingramcontent.com/pod-product-compliance
Lightning Source LLC
Chambersburg PA
CBHW041352210926
51850CB00013B/3056